Sustainable Environmental Management

Sustainable Environmental Management

(Dr. Jayashree Deshpande Festschrift Volume)

— Editors —

Dr. L.V. Gangawane

Ex Professor and Head
President, Indian Phytopathological Society, New Delhi
Soil Microbiology and Pesticides Laboratory
Department of Botany, Dr. Babasaheb Ambedkar Marathwada University
Aurangabad – 431004

and

Dr. V.C. Khilare

P.G. Department of Botany
Vasantrao Naik Mahavidyalya
Aurangabad–431003

2014

Daya Publishing House®

A Division of

Astral International Pvt. Ltd.

New Delhi – 110 002

First Impression, 2007
ISBN 9789351243793

Publisher's note:
Every possible effort has been made to ensure that the information contained in this book is accurate at the time of going to press, and the publisher and author cannot accept responsibility for any errors or omissions, however caused. No responsibility for loss or damage occasioned to any person acting, or refraining from action, as a result of the material in this publication can be accepted by the editor, the publisher or the author. The Publisher is not associated with any product or vendor mentioned in the book. The contents of this work are intended to further general scientific research, understanding and discussion only. Readers should consult with a specialist where appropriate.
Every effort has been made to trace the owners of copyright material used in this book, if any. The author and the publisher will be grateful for any omission brought to their notice for acknowledgement in the future editions of the book.

Published by : **Daya Publishing House**®
A Division of
Astral International Pvt. Ltd.
– ISO 9001:2008 Certified Company –
4760-61/23, Ansari Road, Darya Ganj
New Delhi-110 002
Ph. 011-43549197, 23278134
E-mail: info@astralint.com
Website: www.astralint.com

Laser Typesetting : **Classic Computer Services**, Delhi - 110 035

Printed at : **Thomson Press India Limited**

PRINTED IN INDIA

डा. चारू दत्त मायी
अध्यक्ष
Dr. C.D. Mayee
CHAIRMAN

कृषि वैज्ञानिक चयन मंडल
भारतीय कृषि अनुसंधान परिषद
कृषि अनुसंधान भवन-1, पूसा,
नई दिल्ली - 110 012

AGRICULTURAL SCIENTIST RECRUITMENT BOARD
INDIAN COUNCIL OF AGRICULTURAL RESEARCH
KRISHI ANUSANDHAN BHAVAN-1, PUSA,
NEW DELHI – 110 012
Telephone: (O) 25843295, 25846540
Fax: 91-11-25846540 Telegram: AGRECBOARD
E-mail: cdmayee@icar.org.in

Foreword

It has been my good luck to know Dr. Jayashree Deshpande for the past 25 years. When I say luck, I mean it truly because being an ardent teacher in a traditional science college of a small place like Aurangabad. Dr. Jayashree not only inspired colleagues but a large number of students to love science and art of Mycology, Plant Pathology: and Environment by remaining a core Botanist at heart. Who has time in the current fast life to go beyond the four walls of class room; dedicate oneself fully to the cause and take along a big bunch of students on a right path. Dr. Jayashree did exactly that and hence she deserve all the special applause on the occasion of her superannuation. I consider it to be fortune to know her. If Dr. Brahmhanand, her husband found music in the art of paintings and carvings of the famous Ajanta and Ellora Caves, Dr. Jayashree found the rhythm in the science of these paintings for making them an everlasting attractions for mankind. Dr. Jayashree contributed substantially to the emerging science of environment by practicing the methods of prevention of environmental pollutions. So diversified her interests have been that you found her active in scientific seminars on draught management, desertification, tree planting, nature lover group, bird watching, antipollution, plant pathology, microbiology, pesticides, forest management and what not, all directly related to environment.

It is, therefore, most befitting of friends and students to bring out a festschrift volume "*Sustainable Environmental Management*" on the occasion of her formal retirement. I say formal only because informally she cannot be detached from the science. She is bound to inspire youngsters in years to come. The volume is going to be very useful to researchers in the field and I am sure that it is the best tribute to a teacher researcher and an environmentalist.

(C.D. Mayee)
Chairman,
Agricultural Scientist Recruitment Board

Prof. C. Manoharachary
MSc, Ph.D, D.Sc, FN.ASc., F.B.S, F.PS.I., F.A.P.A.S.
Professor Emeritus (CSIR)
Former Head, Chairman, Principal.
Co-ordinator (UGC SAP)
Dean, Development & UGC Affairs

DEPARTMENT OF BOTANY
Osmania University,
Hyderabad – 500 007, A.P. India
Off: +91-40-2768 2244
Mobile: 9391164243
E-mail : cmchary@rediffmail.com

Foreword

I am happy to note that a festschrift volume will be brought out in honour of Dr. Jayashree Deshpande who is superannuating in January 2007. This volume consists of 31 scientific articles contributed by learned scholars from different parts of the country. This volume deals with diversified aspects of plant sciences such as pesticide usage, ethanol production, environmental sciences, aspects dealing with Mycology and Plant Pathology, Microbiology, Biotechnology, Oceanography and other related aspects. I am sure that readers will find it quite interesting and subject oriented.

I am happy to note a very good response for this academic venture and a speaks lot about her academic stature, concern for teaching, co-cooperativeness and popularity among academicians. I know Dr. Jayashree Deshpande for the last 20–25 years and I found her hardworking, innovative and knowledgeable. At first instance she has started as plant-pathologist and later the developments around the Globe made her to shift to Environmental Sciences. She has established the Department of Environmental Science and served as Chairman for the last 20 years at SBES college of Science, Aurangabad. As a teacher and researcher she has served the needs of students and scholars through her dedicated and committed teaching. She believed that a teacher is a learner, hence participated in many training courses, refresher courses and national seminars, conferences/symposia documenting her strong aptitude and attitude, towards teaching and research besides keeping her self abreast with advancement of the subject.

She is also the life member and member of different scientific societies. Her rich experience in academics and administration has made the Environmental Science a focussed subject as she has served on different bodies like Board of studies, Editorial boards, COSIP programme, Affiliation committees etc. She has to her credit 30 research papers, five books published and guided 3 students for Ph.D. She has served as president of Maharashtra Environment Society, Councillor-IPS, President-Maharashtra Botanical Society, President-West zone IPS and also received Best Teacher award, Gold

Res: # 2–2–1144/23/1, New Nallakunta, Hyderabad – 500 044, India, Phone: +91-40-27562723

medal from NESA, Savitribai Joshi puraskar and many others. Her best part of life being serving society and needy people through extension and co-curricular activities and in particular in bringing awareness about environmental degradation and environmental protection. Her contributions are enormous and her approach is very much humane. Her efforts in making teaching and research, a noble profession are laudable. On the occasion of her superannuation I convey my hearty greetings, congratulations and felicitations. I wish her healthy, prosperous and progressive life in future.

5/7/2006

Prof. C. MANOHARACHARY
Emeritus Scientist (CSIR)
Project on "Biodiversity, Conservation
... ... important Plants"
Department Of Botany
Osmania University Hyderabad-7.
Prof. C. Manoharachary

Preface

This Festschrift volume *'Sustainable Environmental Management'* is a humble tribute for Dr. Jayashree Deshpande for her outstanding contributions in the field of Plant Pathology, Microbiology and Environmental Science.

The articles in the volume cover a wide array of current research topics dealing with both fundamental and applied researches in the field of Environmental Science. The articles are highly authoritative and thought provoking by the scientist in their respective field. The articles on biodiversity, ecolabelling of organic farming, management of Ajanta cave paintings, use of biocontrol agents for management of diseases, transgenic development of rhizobacteria, techniques in the purification of industrial effluents, bioenergy production etc. are important contributions. We hope that this volume will be highly useful to scientist, students, research workers and policy makers in the Environmental Science and Plant Pathological aspects.

We are highly grateful to all the eminent scientists for taking troubles in writing the articles and Prof. Bramhanand Deshpande for constant encouragement to bring out this volume. We are especially grateful to Daya Publishing House, New Delhi for printing volume in stipulated time. We feel happy and honour in presenting this Festschrift Volume to Dr. Jayashree Deshpande as a token of love and affection at this juncture of her supperannuation. We also prey the Lord for her comfortable life.

Dr. L.V. Gangawane

Dr. V.C. Khilare

Contents

Dr. Jayashree Deshpande

A Biographical Sketch of Dr. Jayashree Deshpande

Dr. Jayashree Deshpande was borne on 24th January 1947 in Gulbarga, Karnataka State. After completion of her graduation in 1965 from Marathwada University, she obtained her M.Sc. in 1967 and Ph.D. in 1974 from Marathwada University in the subject of Botany and also in 2004 in Environment Science.

In July 1967 Dr. Deshpande joined the S.B.E.S. Science College, Aurangabad. In 1987 she was elevated to the position of Vice-Principal. Thereafter in 1990 she became as a Director of P.G. Department in Environment Science and in 1992 as a Reader. She has 39 years teaching and research experience in plant pathology, air and water pollution, ecology, soil microbiology, and sewage technology and distillery effluents. During her long research career spanning she guided many M.Sc. and Ph.D. students on very important aspects of plant pathology and environment science. She has successfully completed the Research Projects of Maharashtra Pollution Control Board and UGC on environmental pollution and allied aspects. In addition Dr. Deshpande has published nearly 30 research papers in various scientific journals of national and international repute. Dr. Deshpande delivered several lectures on different aspects of environment science for public awareness also a regular writer of newspapers showing her kindness towards environment. She attended several national and international seminars and conferences. In pursuit of her research interest, she traveled and visited to Philadelphia in USA. She has completed six training courses of NEERI, Nagpur. Dr. Deshpande has distinction of a being a member of a number of learned societies such as Indian Phytopathological Society, Indian Association of Environmental Management, Society of Mycology and Plant Pathology, she was a Member of International Master Trainer (US based) for 'Globe' and Indian Environmental Society. She was also member of Adhoc BOS in Environment Science, Dr. Babasaheb Ambedkar Marathwada University, Aurangabad. She has organized nine national seminars and conferences. During her work she was President of Maharashtra Environment Society, Councillor and then President of Indian Phytopathological Society (WZ), New Delhi and President of Marathwada Botanical Society.

During her research career, Dr. Deshpande has inspired a number of students. All the students have attained high positions in the colleges, State Pollution Control Board, Institutes and even in Private Sector. Dr. Deshpande is a very unassuming, cheerful, ever-courteous and generous person. The students hold her in respectful reverence and we take opportunity to express our respectful regards to the multifaceted colourful personality. We hope she would continue to work and inspire students, teachers and scientists for the upliftment of the subject of Botany and Environment Science.

Some Important Research Publications

1. Deshpande Jayashree, Tukshetty, M. and R. M. Pai (1968). Floral anatomy of *Xyris indica. Current Sci.* 27 (23): 679–680.
2. Deshpande Jayashree and R. M. Pai (1968). On an interesting *Ophioglassum* from Marathwada. *M. U. Sci. J.* IX, Pp. 25–35.
3. Gangawane, L.V. and Jayashree Deshpande (1977). A report on *Colletotrichum gloeosporoides* Penz. on virus infected Papaya fruits. *Nat. sci. J.M.U.* 16 (9): 59.
4. Gangawane, L.V. and Jayashree Deshpande (1978). A report on sunscald of bell pepper in India. *Proc. Ind. Sci. Cong.* Pp. 54.
5. Gangawane, L.V. and Jayashree Deshpande (1981). Effect of nematicide Dasnit on growth, heterocyst differentiation and N-fixation on *Mastigocladus blaminosus* and *Aulosira fertilissima. Pestology*, 5 (1): 31 –32.
6. Deshpande Jayashree, Leela Kulkarni and L.V. Gangawane (1982). Effect of cellulolytic and proteolytic substrates on growth and sporulation of groundnut fungi. *Biology* 3 (1) 23 –25.
7. Gangawane, L.V. and Jayashree Deshpande (1984). Effect of 2,4-D on non-traget microfungi in the rhizosphere of groundnut. *Ind. Bot. Reptr*. 3 (10): 104–105.
8. Deshpande Jayashree and L.V. Gangawane (1995). On the possibility of airborne fungi in the biodeterioration of Buddhist paintings in Ajanta caves. In: *Proc. 3rd International Conference on biodeterioration of cultural property* (ICBCP–3 Committee Publ.) Bangkok, Pp.455-457.
9. Gangawane, L.V. and Jayashree Deshpande (1996). Fungicidal effect on soil and rhizosphere mycoflora of crops. In: *Perspectives in Biological Sciences*. (ed. V. Rai, M.L. Naik and C. Mahanoharachary), Pt. R.S. University Publ. Raipur, Pp 287–289.
10. Gangawane, L.V. and Jayashree Deshpande (1996). Integrated management of *Aspergillus flavus* causing aflaroot in groundnut. In: *Nat. Sem. IDM for sustainable Agril.* Pp. 14.
11. Gangawane, L.V. and Jayashree Deshpande (1996). Studies on microflora associated with mangrove plants at Bombay coast. In: *Proc. Of Conservation of Mangel Ecosystem. UAE Univ.*
12. Gangawane, L.V. and Jayashree Deshpande (1996). Fungicidal effects on soil and rhizosphere mycoflora of crop plants management of soil pollution. *Persp. Biol. Sci.* 247–251.
13. Deshpande Jayashree and L.V. Gangawane (1997). Microorganisms collected during solar eclipse in India. *Aerobiologia.* 13: 289-294.
14. Khilare, V.C. Jayashree Deshpande and L.V. Gangawane (1997). Synergistic effects on the thiophanate methyl resistance in green mold of sweet orange. In: *Golden Jubilee Confr.* IPS, New Delhi, Pp. 321.
15. Gangawane, L.V., N.S. Suryawanshi and Jayashree Deshpande (1998). Use of homeopathic medicines and ornamental cactus extracts in the management of streptomycin tolerant

Xanthomonas campestris pv. citri citrus canker. In: *Plant Pathogenic Bacteria* (ed. A. Mahadevan), CAS in Botany, Chennai Publ. Pp.460-463.

16. Reddy B.R.C., L.V. Gangawane and Jayashree Deshpande (1998). Variation in sensitivity of *Aspergillus flvus* to carbendazim and its management through agrochemicals. *Dr. BAM Univ. Sci. Jr.* Vol. XXIX Pp. 39–42.
17. Gangawane, L.V. and Jayashree Deshpande (1999). Herbicidal effects on soil fungi. In: *Modern Approaches and Innovations in soil Management* (D.J. Bagyraj, *et al.* Eds.) Rastogi publ. Meerut. Pp. 185-188.
18. Deshpande Jayashree (1999). Toxic effect of H_2S on phyllosphere mycoflora of certain ornamental plants in industrial area. In. *'Frontiers in Botany'* Proc. of State confr. on 'Modern Trends in Teaching and Research in Botany' Pp. 35-39.
19. Deshpande Jayashree (2000). Some thoughts on advancement of seed technology for next millennium. *In: Proc. of 2nd Marathwada Botanical Society Confr.* 'Seed Technology 2000' (Ed.Chavan, Wahegaonkar, Sangai, Khilare), Pp. 3-8.
20. Deshpande Jayashree and D.J.Wanmare (2001). Toxic effects of SO_2 on phyllosphere mycoflora of ornamental plants. *Bull. Env. Sci.* Vol. XIX Pp. 11–13.
21. Deshpande Jayashree (2002). Bioconservation of garbage into soil enricher (Biofertilizer)–A case study. In: *Proc. Of Nat. Confr. Polln. Prvevntion & Control in India*. (Ed. Ind. Asso. Enviro. Management, Nagpur). Pp.195–197.
22. Deshpande Jayashree (2002). Tolerence of pesticides by *Rhizobium* nodulating groundnut *in vitro.* In; *Chemical Management of Plant Pathogens in Western India*. (Ed. L.V. Gangawane) pp. 72-74.
23. Wanmare, D.J. and Jayashree Deshpande (2005). Effect of gaseous pollutants on spore germination of phyllosphere mycoflora. *Bioinfolet* 2 (3): 192–198.

Books Published

1. *Pesticides of Crop Plants in India* (1968). Sahayog publishers, Aurangabad.
2. *Environmental Education* (2000). Kailas Publication, Aurangabad.
3. *Water Quality Analysis* (2000). Kailas Publication, Aurangabad.
4. *Plant Disease Management*. (2001).Kailas Publication, Aurangabad.
5. *Indian Tigers and Tiger Projects in India* (2002). Saket Publishers, Auranagbad.
6. *Integrated Disease Management of Crop Plants*. (2005).Saket Publishers, Auranagabad.

Sustainable Environmental Management *Pages* **1–6**
Edited by: **Dr. L.V. Gangawane & Dr. V.C. Khilare**
Published by: **DAYA PUBLISHING HOUSE**

Chapter 1

Eco-friendly Alternatives to Pesticides in Management of Soil and Tuber Borne Diseases of Potato in Organic Farming

R.K. Arora[1] *and Jyotsana Sharma*[2]

[1]*Principal Scientist,* [2]*Senior Scientist*
(Plant Protection), Central Potato Research Station,
PB No. 1, PO Model Town, Jalandhar – 144 003, Punjab

ABSTRACT

Management of soil and tuber borne diseases of potato without use of pesticides in organic farming is a significant challenge that requires identification of suitable alternative eco-friendly control measures. Experiments on control of these diseases using environmentally friendly methods such as soil solarization, organic amendments and biological control were carried out at Central Potato Research Station, Jalandhar, so as to find safe alternatives to the use of hazardous pesticides. Soil solarization carried out through transparent polyethylene mulching for about four weeks during hot summer period reduced black scurf disease of potato to nearly one half and russet scab to one third as compared to the unsolarized field plots. Addition of organic materials such as *Cannabis sativa, Toona ciliata* and *Eucalyptus* sp. @140q/ha together with soil solarization further reduced black scurf disease incidence to 3.0, 6.0 and 6.7 per cent as against 11.0 per cent observed in the un- amended solarized plots. Incorporation of sawdust or *Cannabis sativa* in solarized plot reduced incidence of skin deformations to a level of 26.0 and 22.0 per cent as against47.5 per cent observed in the control plots. Treatment of seed tubers with a bio-control agent *Trichoderma viride* reduced black scurf disease incidence and disease index in the progeny tubers by 37.7 and 43.6 per cent over the untreated control. Results of these experiments suggested that soil solarization, organic amendments and biological control can provide safe alternatives to the use of pesticides in organic farming.

Keywords: *Soil solarization, Organic amendments, Bioformulation, Potato, Solanum tuberosum L., Tuber diseases, Black scurf, Russet scab, Skin deformations.*

Introduction

Modern agriculture largely depends on the use of pesticides for the control of plant diseases and pests, which are hazardous to animal and human life. The best way to protect our crops from pests without pesticides is to rely on the ecological methods of farming. Organic agriculture has become very popular in recent years with an increased demand for 'health products' developed without use of pesticides. Production of potato (*Solanum tuberosum* L), organically without use of pesticides is a significant challenge, which requires identification of suitable alternative eco-friendly plant protection measures.

Soil solarization is an eco-friendly technique and has been used in organic farming in some countries (Grinstein and Ausher 1991, Bell and Laemmlen 1991). Use of organic amendments and biological control agents could be other environmental friendly methods which can be integrated in organic farming. Soil solarization, organic amendments and a bio-formulation of *T. viride* were therefore evaluated for control of soil and tuber borne diseases of potato in experiments carried out at Central Potato Research Station (CPRS), Jalandhar and the findings are reported in this communication.

Materials and Methods

Three replicated field trials were conducted during the crop seasons 2001-02 and 2002-03.

Soil Solarization

Clear transparent polyethylene sheets of approximately 45 µm in thickness were used for covering the soil to be solarized. The soil before mulching with polyethylene was well prepared, leveled and irrigated between 25 to 30 percent water holding capacity and then covered with the sheets for 4 weeks from mid May to mid June. Two large strips each of size 55 m x 3 m. were thus solarized and the alternate strips of similar dimensions were kept as unsolarized control. Tubers of potato variety of cv. Kufri Badshah were planted in these plots after soloarization. At harvest ten plots each of size 4.2m x 3m were selected at random each from the solarized and the unsolarized strips for comparison of the data.

Organic Amendment

In second experiment above ground green parts of 4 common plants *viz., Cannabis sativa, Toona ciliata, Azadirachta indica and Eucaluptus sp.* and sawdust were incorporated in both solarized and unsolarized soil in the month of May prior to soil solarization. *Azadirachta indica* was incorporated @ 70 q/ha and all others @ 140 q/ha. Tubers of variety Kufri Badshah were planted in plots of size 3 m x 2.4 m, each treatment was replicated thrice.

Biocontrol

In third experiment a bio-formulation of *Trichoderma viride* strain A7 prepared at CPRS, Jalandhar, was evaluated for the control of tuber borne diseases. Tubers of cv. Kufri Chandramukhi, heavily infested with black scurf were treated with the bioformulation containing $2x10^{7}$ cfu of *T. viride*/g of the formulation. The bioformulation was used @ 3.5g/kg potato seed. The treated seed tubers were planted in field within 24 hours of the treatment.

In all the three experiments tubers were planted at a spacing of 60 cm x 20 cm and all recommended cultural practices were followed in routine. Observations on incidence and severity (disease index) of tuber diseases like, black scurf, russet scab and skin deformations (abrasions, cracks, lenticel proliferations, russeting, skin spots etc.), were recorded in the progeny tubers at harvest following standard methods.

Results and Discussion

Effect of Soil Solarization on Tuber Diseases

Soil solarization significantly reduced black scurf and russet scab diseases of potato (Table 1.1). The reduction in black scurf disease incidence was 43.5 per cent and disease index 46.1 per cent, and russet scab 67.7 and 77.9 per cent respectively over the unsolarized control plots (Figure 1.1) which confirms our earlier observations (Arora *et al.*, 2000).

Table 1.1: Effect of Soil Solarization on Black Scurf and Russet Scab of Potato

Treatments	*Black Scurf*		*Russet Scab*	
	DI (%)	*DX*	*DI (%)*	*DX*
Control	11.5	0.13	92.3	2.17
Soilsolarization	6.5	0.07	29.8	0.48
CD (*P*=0.05)	3.8	0.03	6.7	0.29

DI: Disease Incidence; DX: Disease Index.

Soil solarization is a non-chemical and non-hazardous method and can successfully be employed in organic farming to control soil borne tuber diseases of potato. Reduction of Soil several pathogenic fungi in soil, including *Rhizoctonia solani* by soil solarization is known (C.P.R.I., 1997-98; Arya 2002).

Effect of Organic Amendments and Soil Solarization on Tuber Diseases

Efficiency of solarized soil improved further when soil was amended with *Cannabis sativa, Toona ciliata, and Eucaluptus sp.* before solarization. *Cannabis sativa* was most effective followed by *Toona ciliata, and Eucaluptus sp.* (Figure 1.2). Patel *et al.*, reported that soil application of various organic materials control root knot nematodes and weeds. Similar effect has been observed on black scurf in

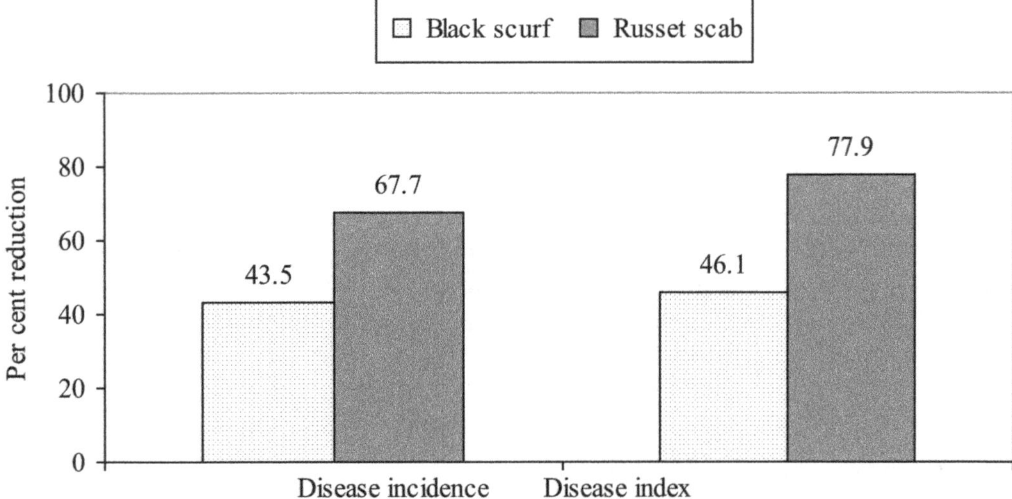

Figure 1.1: Per cent Reduction in Black Scurf and Russet Scab with Soil Solarization

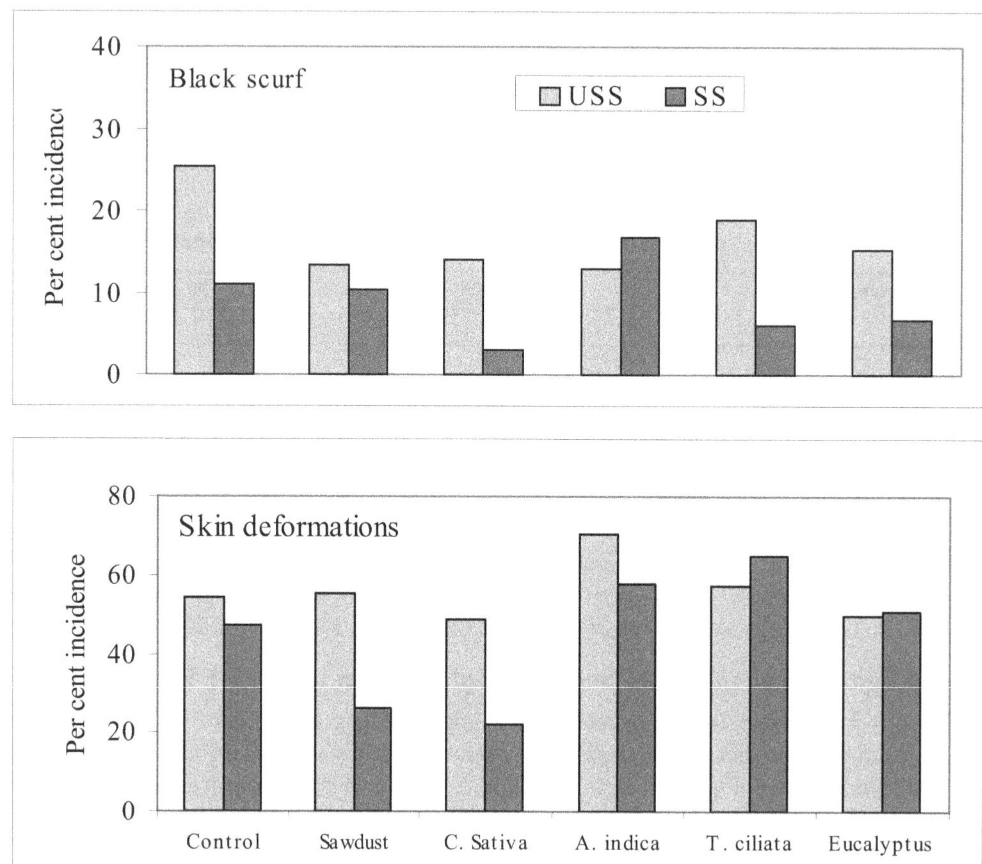

Figure 1.2: Incidence of Black Scurf (BS) and Skin Deformations (SD) in Solarized (SS) and Unsolarized Soil (USS) Amended with Organic Materials

this study. Organic materials during the course of their anaerobic decomposition possibly release some substances that reduce soil borne pathogens.

Organic amendments without soil solarization did not control skin deformations. However, skin appearance improved substantially when incorporation of organic material was followed by soil solarization. This reduced skin deformations by 59.63 per cent with *Cannabis sativa* and 52.3 per cent in case of saw dust. Other organic amendments, however, were not effective (Figure 1.2). Changes in nutrient status of the soil and microflora population due to organic application and solarization may be responsible for improvement in skin appearance, which needs further investigation.

Effect of Bioformulation on Black Scurf

Tuber treatment with a bioformulation containing *T. viride* could reduce black scurf incidence by 37.7 per cent and disease index by 43.6 per cent over the untreated control (Figure 1.3).

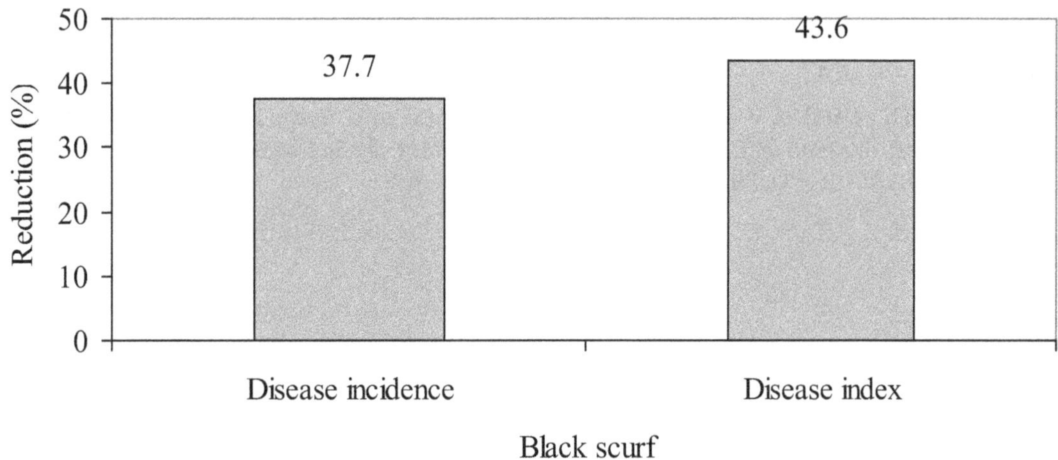

Figure 1.3: Reduction in Black Scurf with *T. viride* Bioformulation Over Control

The control of black scurf disease by *T. viride* was highly significant (Table 1.2). This confirmed the observations recorded earlier (Arora and Somani, 2001). Control of tuber diseases using biocontrol agents is thus viewed as a safe option to the use of pesticides and can successfully be employed in organic farming.

Table 1.2: Control of Black Scurf of Potato with *Trichoderma viride* Bioformulation

Treatments	*Disease Incidence (%)*	*Disease Index*
Control	86.7	2.2
T. viride	54	1. 19
CD (P=0.05)	20.7	0.5

Results of the experiments carried out at this station clearly indicate that soil solarization alone or in combination with organic amendments, and seed treatment with *T. viride* can control soil and tuber borne diseases of potato effectively and thus have a potential to replace the use of pesticides. These methods can safely be employed in organic farming of potato.

References

Arora, R.K. and Somani, A.K., 2001. *J. Indian Potato Assoc.*, 28: 88–89.

Arora, R.K., Trehan, S.P., Sharma, Jyotsana and Khanna, R.N., 2000. In: *Proceedings Indian Phytopathological Soviety-Golden Jubliee. Inter. Confr. on Integrated Plant Dis. Mangmnt. for Sustainable Agril.* IARI, New Delhi, Nov. 11–15, 1997, pp. 1190–1191.

Arya, Arun, 2002. *Ann. Rev. Pl. Path.* Vol. 1. Indian Soc. of Mycol. and Pl. Path., pp. 213–230.

Bell, Carl E. and Laemmlen Franklin, F., 1991. In: *Soil Solarization,* (Eds.) Katan, J. and DeVay, James E.). CRC Press, London, pp. 245–255.

C.P.R.I., 1997–98. *Annual Scientific Report*. Central Potato Research Institute, Shimla, pp. 66.

Grinstein, A. and Ausher, R., 1991. In: *Soil Solarization*, (Eds. Katan, J. and DeVay, James E.). CRC Press, London, pp. 193–204.

Patel, S.K., Patel, D.J., Patel, H.V. and Patil, R.K., 2000. In: *Proceedings Indian Phytopathological Soviety-Golden Jubliee. International conference on Integrated Plant Disease Management for Sustainable Agriculture*. IARI, New Delhi, Nov. 11–15, 1997, pp. 806–809.

Sustainable Environmental Management *Pages 7–13*
Edited by: **Dr. L.V. Gangawane & Dr. V.C. Khilare**
Published by: **DAYA PUBLISHING HOUSE**

Chapter 2
Cellulosic Ethanol: Fuel for the Future?

M.M.V. Baig and S.S. Wadje

Laboratory of Plant and Microbial Biology, Department of Botany, Yeshwant Mahavidyalaya, Nanded – 431 602, MS, India

ABSTRACT

The microbes that degrade cellulose include bacteria and fungi. Microbial degradation is either parasitic or saprophytic. Saprophytic degradation can be specific to certain cell wall components or can involve degradation of all cell wall components. There are many cellulose-degrading fungi such as *Trichoderma, Fusarium, Penicillium, Ceriporiopsis, Humicola, Aspergillus,* and *Phanerochaete.* Cellulolytic bacteria include *Thermonospora, Cellulomonas* and *Clostridium.* Cellulose molecules consist of long chains of glucose molecules as do starch molecules, but have a different structural configuration. Basically fermentation of Ethanol is a biological process in which organic material is converted by microorganisms to simpler compounds, such as sugars. These compounds are then fermented by microorganisms to produce ethanol and CO_2. Ethanol can be made synthetically from petroleum or by microbial conversion of biomass materials through fermentation.

Keywords: *Cellulose, Hemicellulose, Lignin, Ethanol, Fermentation, Recombinant DNA*

Introduction

The increase in population has increased the demands of energy. The conventional and non-renewable sources of energy like fossil fuel are too meager to fulfill these demands now and in future as well. This situation compels to search for alternative sources of energy of renewable type. In recent times there is growing interest for developing such an alternative sources of energy (Arsitidou and Penttila, 2000; John, 2004). This can be met by suitable modification–conversion of the available renewable sources of energy. This can be through the use of biomass–plant and plant based product produced by photosynthesis with biological rather than geological time by making fossil fuel independent. Biomass an umbrella term used to describe all the animal and plant matter on the planet, has gained inevitable importance owing to its rise as alternative to fossil fuel. But biomass can't completely replace the huge volume of petroleum and other fuel that are being used now but it can provide fuel and chemicals. The immediate application will be conversion to fermentable sugars and subsequently to ethanol.

In agriculture-based economy like India, the availability of agricultural wastes/residues, industrial wastes like sugarcane bagasse, forest waste and other lignocellulosic waste are abundant and need to be utilized effectively. The reduction, reuse and recycle of these wastes can save millions of petrodollar, which in turn can be used for other development work in the country. This applicability of the biomass energy can contribute to sustainable development.

The industries involving use of agricultural and forest bio-products have been the backbone of many successful economies of the world. This can be exemplified from the fact that the world's total production of paper and pulp is mainly shared by USA (28 per cent); Canada (14 per cent); China (10 per cent); Sweden (6 per cent) and Finland (6 per cent) (Anonymous, 2000). This biomass energy not only play a decisive role in maintaining environmental health but also is capable enough to generate new opportunities for employment in rural areas as well, thus contributing to social aspect of sustainability. This also provides an permanent solution to agricultural waste management and disposal.

Research in plant biotechnology in general and cellulose in particular is beginning to investigate the effective use. Though, worlds most of the fuel ethanol is product of fermented maize glucose, sucrose or starch, but a country like India with a strong and traditional agricultural backbone can imply the current available technology for production of ethanol for fuel.

This can be possible owing to the development of technology using non food plant sources for large scale production. At the present juncture, this may be a far-fetched idea but can be realized in near future. Thus the available agricultural waste and residue like sugarcane tops, rice straw, wheat straw, cotton and pigeon pea stalks, forest wastes etc and waste from agro-industrial units like sugarcane bagasse, rice bran, groundnut hulls, paper mill discards etc. can be converted into sugars and further to fuel ethanol (Mane, 2000). Though, a lot of improvement in technology development is under process still there are challenges to be met.

The Substrate

In plants, the cell wall is synthesized outside the cell membrane by the material secreted by the cell cytoplasm. Initially formed cell wall is called the primary cell wall and later during course of cell differentiation, expansion of primary cell wall leads to modification with new components forming the secondary cell wall. Thus the organization and subsequently differentiation of the cell wall takes places forming simple and compound tissues and further formation of plants parts.

The secondary cell wall is composed of cellulose, hemicellulose and lignin. The primary xylem cell wall is composed of pectin (47 per cent), cellulose (22 per cent), xylan (11 per cent), xyloglucan (6 per cent), glucomannan (1 per cent) and proteins (10 per cent). The secondary xylem cell has an increased amount of cellulose (43-48 per cent), xylan (18-28 per cent) and glucomannan (5 per cent) following differentiation. The pectin and the xyloglucan content more or less remains the same while lignin content increases up to19-21 per cent, whereas it is completely absent in primary xylem cell walls (Mellerowicz *et al.*, 2001). Besides carbohydrates and lignin the cell wall contains major classes of cell wall proteins such as hydroxyproline-rich glycoproteins, arabinogalactan proteins, glycine-rich proteins, and proline-rich proteins (Ye *et al.*, 1991).

Cellulosic Feedstocks

The worldwide production of plant biomass is lignocellulose which comprises about 200×109 tons per year, where about 8–20×109 tons of the primary biomass remains potentially accessible. However, the effective utilization of the lignocellulosic feedstock is not always practical because of its seasonal availability, scattered stations, and the high costs of transportation and storage of such large amounts of organic material (Polman 1994).

Like sugar materials, starchy materials are also in the human food chain and are thus expensive. Fortunately, a third alternative exists–cellulosic materials. Examples of cellulosic materials are paper, cardboard, wood, and other fibrous plant material. Cellulosic resources are in general very widespread and abundant. For example, forests comprise about 80 per cent of the world's biomass. Being abundant and outside the human food chain makes cellulosic materials relatively inexpensive feedstocks for ethanol production.

Cellulosic materials are comprised of lignin, hemicellulose, and cellulose and are thus sometimes called lignocellulosic materials. One of the primary functions of lignin is to provide structural support for the plant. Thus, in general, trees have higher lignin contents than grasses. Unfortunately, lignin which contains no sugars, encloses the cellulose and hemicellulose molecules, making them difficult to reach.

Cellulose molecules consist of long chains of glucose molecules as do starch molecules, but have a different structural configuration. These structural characteristics plus the encapsulation by lignin makes cellulosic materials more difficult to hydrolyze than starchy materials.

Hemicellulose is also comprised of long chains of sugar molecules; but contains, in addition to glucose (a 6-carbon or hexose sugar), pentoses (5-carbon sugars). To complicate matters, the exact sugar composition of hemicellulose can vary depending on the type of plant. Since 5-carbon sugars comprise a high percentage of the available sugars, the ability to recover and ferment them into ethanol is important for the efficiency and economics of the process. Recently, special microorganisms have been genetically engineered which can ferment 5-carbon sugars into ethanol with relatively high efficiency.

Other researchers have developed microorganisms with the ability to efficiently ferment at least part of the sugars present. Bacteria have drawn special attention from researchers because of their speed of fermentation. In general, bacteria can ferment in minutes as compared to hours by yeast.

Microorganisms for Ethanol Fermentation

Basically fermentation of Ethanol is a biological process in which organic material is converted by microorganisms to simpler compounds, such as sugars. These compounds are then fermented by microorganisms to produce ethanol and CO_2. During the whole process of ethanol fermentation, there are mainly two parts for microorganisms. One is to produce the enzyme to catalyze chemical reactions that hydrolyze the complicated substrates into simpler compounds and the other is the microorganisms which convert fermentable substrates into ethanol.

Several reports and reviews have been published on production of ethanol fermentation by microorganisms, and several bacteria, yeasts, and fungi have been reportedly used for the production of ethanol. Microbes that are capable of yielding ethanol as the major product includes several strains of fungi including *Saccharomyces cerevisiae* (Vallet *et al.*, 1996; da cruz *et al.*, 2003, Kiran *et al.*, 2003; Yu and Zhang, 2004), *Kluyveromyces fragilis* (Vallet *et al.*, 1996), *Candida utilis* (Vallet *et al.*, 1996), *Mucor* sp.

(Ingram *et al.*, 1998) *Fusarium* sp. (Ingram *et al.*, 1998) and bacterial strains like *Clostridium sporogenes* and other clostridial species, *Zymomonas mobilis*, *Spirochaeta* sps., *Erwinia amylovora*, *Leuconostoc mesenteroides*, (Miyamoto, 1997), *Escherichia coli* (Dien *et al*, 2003; Matthews *et al.*, 2005), *Klebsiella* sps. (Ingram *et al.*, 1998).

Microbial Degradation of Cellulose

Microbial degradation of cellulose is an important biological process helping nature to complete the carbon cycle by channeling the carbon stored in carbohydrates back to carbon and hydrogen. The microbes that degrade cellulose include bacteria and fungi. Microbial degradation is either parasitic or saprophytic. Saprophytic degradation can be specific to certain cell wall components or can involve degradation of all cell wall components. There are many cellulose-degrading fungi such as *Trichoderma*, *Fusarium*, *Penicillium*, *Ceriporiopsis*, *Humicola*, *Aspergillus*, and *Phanerochaete*. Cellulolytic bacteria include *Thermonospora*, *Cellulomonas* and *Clostridium*.

Fungi are heterotrophic eukaryotes and one of their major roles in the ecosystem is bioconversion, including cellulose degradation. The saprophytic cellulose degrading fungi are either unicellular or filamentous during the vegetative part of their life cycle. Filamentous fungi are predominant cellulose degraders and their filamentous morphology helps them to creep over and grow into the cellulose structures. In general cellulose–degrading fungi secrete the enzymes that degrade complex cell wall polymers and consumes the resultant smaller polymers, sugars and other products.

There are many fungal strains that are used industrially for various applications. *Trichoderma reesei* is one such extensively used forerunner. Many enzymes from *T. reesei* are used for industrial applications such as pulp and paper, textile, food processing, laundry etc. The white rot fungus *Ceriporiopsis subvermispora*, though not well studied, is an effective delignifier and is used for biopulping (Blanchette, 1991; Akhtar *et al.*, 1992; Wall *et al.*, 1993). In recent times, the white rote fungus *P. chrysosporium*, with its unique ability to degrade lignin and many organic pollutants, has emerged as a potential candidate for future application in hazardous waste remediation and other industrial applications such as paper and textiles.

Hydrolysis of Cellulose and Hemicellulose

T. reesei is a well-studied cellulolytic fungus which degrades cellulose by synergetic action of endoglucanases (endo-1, 4-ß-glucanase), cellobiohydrolases (exo-1, 4- ß-glucanase) and ß-glucosidases (Teeri *et al.*, 1987; Tomme *et al.*, 1988). Endocellulases degrade less crystalline cellulose, exocellulases act from the ends of cellulose chains progressively and ß-glucosidases act on disaccharides and oligosaccharides forming glucose. Similar cellulolytic mechanisms are found in *T. ligonorum* (Baig, 2005; Baig *et al.*, 2003), *T. harzianum* (Baig and Awasthi, 2005; Baig and Muley, 2005), *P. chrysosporium* (Eriksson, 1993).

Enzymatic Hydrolysis

Another basic method of hydrolysis is enzymatic hydrolysis. Enzymes are naturally occurring plant proteins that cause certain chemical reactions to occur. However, for enzymes to work, they must obtain access to the molecules to be hydrolyzed. For enzymatic processes to be effective, some kind of pretreatment process is thus needed to break the crystalline structure of the lignocellulose and remove the lignin to expose the cellulose and hemicellulose molecules. Depending on the biomass material, either physical or chemical pretreatment methods may be used. Physical methods may use high temperature and pressure, milling, radiation, or freezing–all of which require high-energy consumption. The chemical method uses a solvent to break apart and dissolve the crystalline structure.

Similarly chemical methods involve the pretreatment of lignocellulosics with a dilute acid or alkali, then the pretreatmed slurry is neutralized to remove materials that would be toxic to the microorganisms used in the process. Due to the tough crystalline structure, the enzymes currently available require several days to achieve good results. Currently, the cost of enzymes is also too high and research is continuing to bring down the cost of enzymes.

However, if less expensive enzymes can be developed enzymatic processes hold several advantages: (1) their efficiency is quite high and their byproduct production can be controlled; (2) their mild process conditions do not require expensive materials of construction; and (3) their process energy requirements are relatively low.

In the recent processes both enzymes and the fermentation microorganisms are added at the same time to the slurry, and sugar conversion and fermentation occur simultaneously in a process called simultaneous saccharification and co-fermentation (SSCF). Simultaneous saccharification and fermentation (SSF) gives higher ethanol yields and requires lower amounts of enzyme because end-product inhibition from cellobiose and glucose formed during enzymatic hydrolysis is relieved by the yeast fermentation (Banat *et al.*, 1998; McMillan *et al.*, 1999). However, it is not easy to me*et al*l the requirements of industry due to their low rates of cellulose hydrolysis, which is the stage limiting the rate of alcohol production. Another problem arises from the fact that most microorganisms used for converting cellulosic feedstock cannot utilize xylose, a hemicellulose hydrolysis product. Moreover, SSF requires that enzyme and culture conditions be compatible with respect to pH and temperature (Lin and Tanaka, 2006).

Ethanol from Cellulose

The use of ethanol as an alternative motor fuel has been steadily increasing around the world for a number of reasons. Domestic production and use of ethanol for fuel can decrease dependence on foreign oil, reduce trade deficits, create jobs in rural areas, reduce air pollution, and reduce global climate change carbon dioxide buildup. Ethanol, unlike gasoline, is an oxygenated fuel that contains 35 per cent oxygen, which reduces particulate and NOx emissions from combustion.

Ethanol can be made synthetically from petroleum or by microbial conversion of biomass materials through fermentation. In 1995, about 93 per cent of the ethanol in the world was produced by the fermentation method and about 7 per cent by the synthetic method. The fermentation method generally uses three steps: (1) the formation of a solution of fermentable sugars, (2) the fermentation of these sugars to ethanol, and (3) the separation and purification of the ethanol, usually by distillation.

Fermented ethanol has been commercially produced many countries like US and Brazil from cellulosic biomass feedstocks using acid hydrolysis whenver there was fuel shortages,. In recent times some countries with higher ethanol and fuel prices are producing ethanol from cellulosic feedstocks. However, the advancement recent time led to cost-effective technologies for producing cellulosic ethanol started to emerge. There are three basic types of such processes–acid hydrolysis, enzymatic hydrolysis, and thermo-chemical having variations in each process. The most common is acid hydrolysis while any acid can be used but it is easier and cheap to employ sulfuric acid.

The Future

Though a lots of efforts are being taken to optimize the process development for production of ethanol from biomass, the present state of art need to be bought to technological and commercially viable proposition. It would require intensive inputs in terms of the technology improvement. This can be achieved through the application of molecular biology, recombinant DNA technology and systems biology approaches bringing out the efficient metabolically engineered organisms and study system. This includes a better understanding of physiology and molecular biology of cellulose decomposition, pretreatment of cellulosic raw material, cost efficient system for cellulolytic enzymes production and subsequently downstream fermentative production leading to alcohol and other metabolite formation.

In school often children are taught the Grimm Brother's fairy tale, where Rumpelstiltskin spins straw into gold. In near future, we will thank to advances in biotechnology where the researchers will transform straw and other plant wastes into "green" gold–cellulosic ethanol. This ethanol will be essentially identical to ethanol produced from any other source. Cellulosic ethanol exhibits a net energy content with a low net level of greenhouse gases. The advancement of the developed technologies are not only improving yields but are also capable of bringing down the cost of production, thus bringing us closer to the day when cellulosic ethanol will replace expensive, imported "black gold" with a sustainable, indigenously and domestically produced bio–energy.

References

Akhtar, M., Attridge, M.C., Blanchette, R.A., Myers, G.C., Wall, M.B., Sykes, M.S., Koning, J.W., Burgess, R.R., Wegner, T.H., and Kirk, T.K., 1992. *Abstract Pap. Am. Chem. S.*, 204, 12-Cell.

Anonymous, 2000. *Global Forest Resources Assessment*, FAO, UN.

Aristidou, A. and Penttila, M., 2000. 11:187–198.

Baig, M.M.V., 2005. *J. Sci. Ind. Res.*, 64: 57–60.

Baig, M.M.V. and Awasthi, R.S., 2005. *Poll. Res.*, 24: 205–208.

Baig, M.M.V. and Muley, S.M., 2005. *Asian J. Microbiol. Biotechnol. Environ. Sci.*, 4: 169–172.

Baig, M.M.V., Baig, M.L.B., Baig, M.I.A. and Majeda Yasmeen, 2004. *African J. Biotechnol.*, 3: 447–450.

Baig, M.M.V., Mane, V.P., Baig, M.I.A., Shinde, L.P. and More, D.R., 2003. *J. Environ. Biol.*, 24(2):173–176.

Banat, I.M., Nigam, P., Singh, D., Marchant, P., McHale, A.P., 1998. *World J. Microbiol Biotechnol.*, 14: 809–821.

Blanchette, R.A., G.F.L., M. Attridge and M. Akhtar, 1991. *US Patent* 5,055,159.

da Cruz, S.H., Batistote, M. and Ernandes, J.R., 2003. *J. Inst. Brew.*, 109(4): 349–355.

Dien, B.S., Cotta, M.A. and Jeffries, T.W., 2003. 63: 258 –266.

Eriksson, K.E.L., 1993. *J. Biotechnol.*, 30: 149–158.

Ingram, L.O., Gomez, P.F., La, X., Moniruzzaman, M., Wood, B.E., Yomano, L.P. and York, S.W., 1998. *Biotechnol Bioeng.*, 58(2,3): 204–214.

John, T.., 2004. Biofuels for transport. http://www.task39.org/.

Kiran, S., Sikander, A. and Ikram-ul-Haq, 2003. *J. Biol. Sci.*, 3: 984–988.

Lin, Y. and Tanaka, S., 2006. *Appl. Microbiol. Biotechnol.*, 69: 627–642.

Mane, V.P., 2000 Microbiology of agricultural wastes. *Ph.D. Thesis*, Swami Ramanand Teerth Marathwada University, Nanded

Matthew, H., Ashley, O., Brian, K., Alisa, E. and Benjamin, J.S., 2005 Wine making 101. Available at http://www.arches.uga.edu/¡«matthaas/strains.htm.

McMillan, J.D., Newman, M.M., Templeton, D.W. and Mohagheghi, A., 1999. *Appl. Biochem. Biotechnol.*, 77/79: 649–655.

Mellerowicz, E.J., Baucher, M., Sundberg, B. and Boerjan, W., 2001. *Plant Mol. Biol.*, 47: 239–274.

Miyamoto, K., 1997. http://www.fao.org/docrep/w7241e/w7241e00.htm #Contents.

Polman, K., 1994. Review and analysis of renewable feedstocks for the production of commodity chemicals. *Appl. Biochem. Biotechnol.*, 45: 709–722.

Teeri, T.T., Lehtovaara, P., Kauppinen, S., Salovuori, I. and Knowles, J., 1987. *Gene*, 51: 43–52.

Tomme, P., Van Tilbeurgh, H., Pettersson, G., Van Damme, J., Vandekerckhove, J., Knowles, J., Teeri, T., and Claeyssens, M., 1988. *Eur. J. Biochem.*, 170: 575–581.

Vallet, C., Said, R., Rabiller, C. and Martin, M.L., 1996. *Bioorg. Chem.*, 24: 319–330.

Wall, M.B., Cameron, D.C. and Lightfoot, E.N., 1993. *Biotechnol. Adv.* 11: 645–662.

Ye, Z.H., Song, Y.R., Marcus, A. and Varner, J.E., 1991. *Plant J.*, 1: 175–183.

Yu, Z.S. and Zhang, H.X., 2004. *Bioresour. Technol.*, 93: 199–204.

Sustainable Environmental Management *Pages 14–18+*
Edited by: **Dr. L.V. Gangawane & Dr. V.C. Khilare**
Published by: **DAYA PUBLISHING HOUSE**

Chapter 3

Soft Rot of Garlic Caused by *Geotrichum candium* L. Pathogenesis, Biochemical Changes and Control

B. Bhadraiah and P. Rama Rao

Department of Botany, Osmania University, Hyderabad – 500 007, A.P., India

ABSTRACT

Effect of temperature and relative humidity on the soft rot of garlic incited by *Geotrichum candidum*, the biochemical changes during disease development and its fungicidal control were investigated. Increase in temperature enhanced the percentage rot of garlic bulbs with maximum rot at 27° C and minimum at 10°C. Relative humidity was directly proportional to the amount of rot with maximum rot at 100 per cent humidity. Due to infection, an initial increase of alanine, arginine, glutamic acid threonine was recorded while sugars and organic acids depleted. Among the fungicides tested for the control of soft rot, calixin was most effective followed by Dithane M 45.

Keywords: *Geotrichum, Allium sativum, Soft rot, Fungicidal control.*

Introduction

Garlic is an important crop grown in India. It is a compound bulb used as a condiment and also its medicinal values are immense. Considerable losses occur due to rot by several post-harvest pathogens (Manohara Chary and Rama Rao, 1989; Gupta and Singh, 2001). A new post harvest disease of garlic has been described for the first time (Usharani, 1982). Temperature and humidity are two main environmental factors that influence disease advancement and decay of perishable product (Sharma, 1986; Borkar and Patil, 1995). The effect of temperature and relative humidity on the disease development, the biochemical changes due to the disease and fungicide control of the pathogen are presented in this article.

Material and Methods

The rate of disease development and extent of rot under different temperature, humidity levels was studied by inoculating the healthy bulbs with *G. candidum* after surface sterilization with 0.1 per cent $HgCl_2$ and incubated at 10°, 15° and 27°C for 20 days. Samples were collected at 5 day interval to test the extent of rot from all the eight replicates. The visible infected portion was scraped off and the difference in weight was noted. Percentage rot was calculated by the formula W–W' X 100/W, where W = Weight of the bulb before inoculation of W' = Weight of the bulb after the removal of the infected part.

The effect of relative humidity on disease development was studied for 15 days at room temperature. Percentage loss of weight was calculated on 15th day from all the eight replicates. Humidity levels (30 per cent, 50 per cent, 70 per cent, 90 per cent, 100 per cent) were maintenanced in desiccators using H_2SO_4 and water at different dilutions (Buxton and Mellanby, 1934).

Biochemical parameters were estimated from 80 per cent ethanol extracts of host tissue prepared by macerating the tissue in ground glass homogenizer and extracted with hot ethanol for 3-4 times. The extract was filtered and evaporated to dryness and re-dissolved in ethanol (1 ml/1g). The extract thus obtained was analysed chromatographically for amino acids, sugars and organic acids (Kapoor and Tandon, 1969;Bhargava and Arun Arya, 1989).

Results and Discussion

Soft rot caused by *G. candidum* increased with the increase in the temperature. Nearly 50 per cent of the bulbs rotted at 27°C (Table 3.1). There was increase in the rot with the increase in incubation period. Progressive increase in disease development was observed with increasing relative humilities (Table 3.2).

Table 5.1: Percentage of Rot of Garlic Infected with *G. candidum* at Different Temperatures

Days of Incubation	*Percentage of Rot*		
	Temperature		
	10°C	*15 + 1°C*	*27 + 2°C*
5	-	5.8	15.2
10	2.5	12.6	25.4
15	6.5	18.0	36.4
20	8.0	21.5	46.6

Table 5.2: Percentage of Rot Garlic Infected with *G. candidum* at Different Relative Humidities

Pathogen	*Percentage of Rot*				
	Relative Humidity 0%				
	20	*50*	*70*	*90*	*100*
G. candidum	4.5	6.2	18.2	28.4	37.6

Geotrichum candidum infection resulted in the increase of amino acids *viz.*, alanine, arginine, glutamic acid and threonine while cystine and lysine decreased after 10 days of incubation. After 20 days of incubation amino acids were at minimum (Table 3.3). Simple sugars like glucose and fructose depleted due to infection while maltose disappeared after 20 days. Organic acid levels were also depleted in the diseased bulbs. An initial increase of succinic acid was evident after 5 days incubation while it gradually decreased with increasing incubation period (Table 3.4).

Table 3.3: Free Amino Acids in Healthy and *Geotrichum candidum* Infected Garlic at Different Days of Incubation

Amino Acid	*5th day*		*10th day*		*15th day*		*20th day*	
	Healthy	*Diseased*	*Healthy*	*Diseased*	*Healthy*	*Diseased*	*Healthy*	*Diseased*
Alanine	++	+++	++	++++	++	++	++	–
Arginine	++	++++	+++	+++++	+++	+++	+++	++
Asparagine	+	++	+	++	++	++	++	+
Asparatic acid	++++	+++	++++	++	++++	+	++++	–
Cystine	++	+	++	+	++	–	+	–
Glutamic acid	++++	+++++	++++	+++++	++++	+++	++++	++
Histidine	++	+++	++	–	++	–	+	–
Lysine	+	+	+	+	+	–	+	–
Methionine	+	++	++	+++	++	+++	++	+
Threonine	+++	+++++	+++	+++++	+++	++++	+++	+++
Tyrosine	++	+++	++	++++	++	++	++	++

–: Absent; +: Trace; ++: Less; +++: Much; ++++, +++++: Abundant.

Table 3.4: Sugars and Organic Acids in Healthy and *Geotrichum candidum* Infected Garlic at Different Days of Incubation

Sugar	*5th day*		*10th day*		*15th day*		*20th day*	
	Healthy	*Diseased*	*Healthy*	*Diseased*	*Healthy*	*Diseased*	*Healthy*	*Diseased*
Fructose	+++	+++	++++	+++	++++	++	++++	+
Glucose	+++	+	+++	++++	++++	+	+++	+
Maltose	++	+	++	++	++	+	++	–
Sucrose	+++	++	+++	++	+++	++	+++	–
R_6 0.607	+++	+++	+++	++++	+++	+++	+++	++
Organic acid								
Citric acid	++	++	++	+	++	+	++	+
Malic acid	++++	++++	++++	+++	++++	+++	++++	–
Succinic acid	+++	+++++	+++	++	+++	+	+++	–
Tartaric acid	+	+	+	–	+	–	+	–

–: Absent; +: Trace; ++: Less; +++: Much; ++++, +++++: Abundant.

Calixin was the most effective fungicide at and above 100 µg level (Table 3.5). Dithane M 45 inhibited the growth to the greater extent at 300 ppm and no growth was noticed at 500 ppm. *In vivo* calixin and Dithane M 45 curtailed the rot caused by *G. candidum* at 500 pmm and at higher concentrations while Bavinstin and Griseofulvin were effective at 1000 ppm only.

Table 3.5: Assay of Various Fungicides Against *Geotrichum candidum*

Fungicides	*Colony Diameter (mm)*				
	50 ppm	*100 ppm*	*200 ppm*	*300 ppm*	*500 ppm*
Dithane M-45	18.8	12.3	10.4	6.4	–
Bavistin	57.4	41.6	21.0	18.2	12.3
Calixin	16.0	4.3	3.0	–	–
Control	60.5	–	–	–	–

Generally high temperature and high humidity favour the growth of microorganisms resulting in the loss of fruits and vegetables (Bhargava *et al.*, 1965; Harvey, 1978). High temperature (35°C) and high relative humidity (100 per cent) favoured quick rotting of garlic cloves in storage (Rath and Mohanty, 1978). Infection of bulb leads to the decrease in various constituents like amino acids, sugars and organic acid. Kapoor and Tandon (1969) observed additional presence of amino acids like valine, tyrosi, threonine and glutamine in infected fruits. Otazu and Secor (1981) attributed direct relation between soft rot susceptibility of potatoes and the content of reducing sugars. The present observations too indicate and increase in amino acids and depletion of sugars and organic acids due to degradation of various compounds in the infected tissue resulting from the host-pathogen interaction. Several workers obtained positive results against different fungal pathogens employing various fungicides (Thakur and Chenulu, 1974; Raychaudhuri and Dharamvir, 1977; Tandon and Kapoor, 1977; Usharani, 1982;Sharma, 1986;Borkar and Patil,1995). The present data reveals the importance of temperature and relative humidity in storage and fungicides like Calixin and Dithane M 45 in control of the rot.

Acknowledgements

The authors are grateful to UGC for financial assistance.

References

Bhargava, S.N., Ghosh, A.K., Srivastava, M.P., Singh, R.H. and Tandon, R.N., 1965. *Proc. Nat. Acad. Sci. India*, 35: 393–398.

Bhargava, S.N. and Arun Arya, 1989. Postharvest fruit rots and biochemical changes. In: *Recent Advances in Plant Pathology*, (Ed) Akhtar Husain *et al.*, pp. 87–96.

Borkar, S.G. and Patil, B.S., 1995. *Indian Phyto. Path.*, 25: 288–289.

Buxton, P.A. and Mellanby, K., 1934. *Bull. Ent. Res.*, 25: 171–175.

Gupta, R.P. and Singh, D.K., 2001. Diseases of onion and garlic and their management. In: *Diseases of Fruit and Vegetables and their Management*, (Ed) T.S. Thind. Kalyani Publishers, pp. 306–324.

Harvey, J.M., 1978. *Ann. Rev. Phytopathol.*, 16: 321–341.

Kapoor, I.J. and Tandon, R.N., 1969. *Indian Phytopath.*, 22: 408–410.

Manoharachary, C. and Rama Rao, P., 1989. Survey and Patho-physiological studies of fruit rot diseases. In: *Perspective in Plant Pathology,* pp. 141–154. Today and Tomorrow's Printers and Publishers, New Delhi.

Nene, Y.L., 1971. *Fungicides in Plant Disease Control*. Oxford and IBH Publ. Co., New Delhi.

Otazu, V. and Secor, G.A., 1981. *Phytopathology*, 71: 290–295.

Rath, G.C. and Mohanty, G.N., 1978. *Indian Phytopath*., 31: 256–257.

Raychaudri, S.P. and Dharamvir, 1977. *Front. Pl. Sci*., 3: 213–221.

Sharma, S.K., 1986. *Indian Phytopath*., 39: 78–79.

Tandon, R.N. and Kapoor, I.J., 1977. *Front. Pl. Sci*., 3: 223–242.

Thakur, D.P. and Chenulu, V.V., 1974. *Indian Phytopath*., 27: 375.

Usharani, P., 1982. Pre and Postharvest pathology of some vegetables from Hyderabad District, Ph.D. Thesis, Osmania University.

Sustainable Environmental Management *Pages* ***19–21***
Edited by: **Dr. L.V. Gangawane & Dr. V.C. Khilare**
Published by: **DAYA PUBLISHING HOUSE**

Chapter 4

Role of Air-borne Fungi in the Biodeterioration of Buddhist Paintings in Ajanta Caves

Brahmanand Deshpande and L.V. Gangawane

Soil Microbiology and Pesticides Laboratory, Department of Botany, Dr. Babasaheb Ambedkar Marathwada University, Aurangabad – 431 004, MS, India

ABSTRACT

The classical paintings in Ajanta Caves, based on the life of Lord Buddha are being deteriorated day by day due to microbial degradation. Plates containing Martin's rose bengal agar medium, protein pectin and CMC were exposed for two minutes at the entrance of cave Nos. 1, 2, 10 and 16. Simultaneously, plates were also exposed in the horseshoe valley of the 27 fungal species in the premises, 13 were recovered from the cave No. 10, 1, and 18. The dominant species recorded were *Aspergillus niger, A. fumigatus, A. ustus, Cladosporium oxysporum, Fusarium oxysporum* and *Phoma herbarum.* All these species are known as biodegradant by producing proteolytic, cellulolytic and pectolytic enzymes and suggest that they are responsible for spoiling the valuable paintings. None of these species could grow on plates containing 0.1 per cent bavistin and 2 per cent sulphur fungicides.

Keywords: *Air-borne fungi, Biodeterioration, Buddhist paintings.*

Introduction

The classical paintings of Ajanta Caves near Aurangabad are based on the life of Lord Buddha, Jataka and Avadana stories, scenes from Buddhist plays and mythology. These glorious and classical painting by anonymous painters came into existence during the period of Gupta-Vakataka rule (450 A.D.). However, it is observed that day-by-day, these painting are being deteriorated which may be attributed to the physical and biological factors such as microbes and there is need to preserve them in different ways. Microorganisms are known to play an important role in the deterioration of organic matter and the aim of the present study was to find out type of micro fungi in the air and inside the caves responsible to deteriorate the paintings.

Plates containing Martin's rose bengal agar medium and 2 per cent protein (casein), citrus pectin and CMC 9caboxymethyl cellulose) separately were exposed for two minutes at the entrance of important cave Nos. 1, 2, 10 and 16. The plates were also exposed in the horseshoe valley and on the debris of the litter. Simultaneously Martin's rose Bengal agar plates containing 1 per cent carbendazim (Bavistin) and 2 per cent sulphur were also exposed to find out whether these fungi can be controlled with fungicides. Plates were exposed in triplicate at each spot.

A total of 27 fungal species were recorded on the plates exposed in the caves and in the premises of the caves. Maximum numbers of fungal species were seen in the caves No. 10 while the number was 7 to 11 in the cave Nos. 1, 2 and 16. On the leaf litter debris in the premises, 16 species were recorded whereas 15 species were noted in the valley where the caves are situated in the horseshoe valley. The dominant species occurring on the plates (more than 20 per cent colonies) were *Aspergillus niger, Fusarium oxysporum, Cladosporium oxysporum, Penicillium funiculosum, Curvularia lunata, Aspergillus fumigatus* and *Phoma herbarum.* Importance of air spora in the Biodeterioration has been already suggested by Gregory (1973). Most of these fungi have also been recorded from the soil and rhizosphere of forest trees around the Ajanta caves (Gangawane and Deshpande, 1972).

In order to find out the degradation of protein, pectin and cellulose, plates containing these substrates were also exposed. Fungi growing on these plates can be considered as biodegradant. Because it is known that paints used for the painting had been prepared from herbal or plant material and these substrates are major constituents of the paintings. Results indicate that of 12 species occurred on the plates containing protein, 7 on pectin and 6 on CMC. *Aspergillus niger, Cladosporium oxysporum,* and *Fusarium oxysporum,* occurred on all the three substrates while *Chaetomium, Hormiscium* sp. and *Phoma* sp. Were specific to CMC and *Helminthosporium* sp. and a sterile black mycelium were specific to protein substrate. All the species are known to produce proteolytic, cellulolytic and pectolytic enzymes (Gangawane and Deshpande, 1972, 1975; Sinha, *et al*, 1979; Cherry, 1982; Bose and Nandi, 1985; Gangawane, 1986). Inhibitor of fungal species on the plates containing carbendazim and sulphur may be helpful in the preservation of paintings in different ways. These are the broad-spectrum fungicides in the control of many fungal species (Saler and Gangawane, 1980).

Acknowledgment

Thanks are due to Capt. P.N. Kamble, Superintendent, Archaeological Survey of India, Aurangabad Circle, Aurangabad (MS) for their encouragement and keen interest in the work.

References

Bose, A. and Nandi, B., 1985. *Seed Res.*, 13(92): 19–28.

Cherry, J.P., 1982. *Phytopathology*, 73: 317–321.

Gangawane, L.V. and Deshpande, K.B., 1972. *Marathwada Univ. J. Sci.*, 11: 1–4.

Gangawane, L.V. and Deshpande, K.B., 1975. *Marathwada Univ. J. Sci.*, 14: 65–69.

Gangawane, L.V. 1986. *Ind. Bot. Reptr.*, 5(1): 110–112.

Gregory, P.H., 1973. *The Microbiology of the Atmosphere*. Leonard Hill Publication, London.

Saler, R.S. and Gangawane, L.V., 1980. *Pesticides*, 14(1): 22–26.

Sinha, V.K., Nagar, M.L. and Chauhan, S.K., 1979. *J. Ind. Bot. Soc.*, 58(1): 94–97.

Sustainable Environmental Management *Pages* **22–27**
Edited by: **Dr. L.V. Gangawane & Dr. V.C. Khilare**
Published by: **DAYA PUBLISHING HOUSE**

Chapter 5

Environmental Protection: A Case Study on the Highly Polluted Major Lakes of Hyderabad, Andhra Pradesh

S.V.A. Chandrasekhar

Freshwater Biological Station, Zoological Survey of India, 1-1-300/B, Ashoknagar, Hyderabad – 500 020, India

ABSTRACT

More than 400 years old city, Hyderabad had a number of water-bodes and some of them vanished and some are on the verge of disappearance. The city is representing a typical case of environmental degradation with most of the present big and small lakes getting polluted, out of which Hussainsagar, Saroornagar and Mir Alam lakes are the major ones. The morphometry of these lakes is also getting reduced due to encroachments and other unauthorized activities. Some remedial measures to protect these natural heritages are discussed in this paper.

Keywords: *Hyderabad, Water bodies, Environment, Pollution.*

Introduction

Hyderabad, the capital of Andhra Pradesh has five major water bodies viz., Hussainsagar, Saroornagar lake, Mir Alam lake, Osmansagar and Himayatsagar in addition to a number of small water bodies. Out of the five, first three are of polluted and other two are unpolluted and drinking water sources to the city. Historically, these impoundments had been created mainly to meet the drinking and irrigational water needs, formed a designed water conservation system in the region. During the second half of the 20th century, particularly after 1956 the Hyderabad city has became the capital of Andhra Pradesh, the city witness to three phenomena of great ecological consequences: first the unprecedented population growth and urbanisation, second industrialization and third intensive agriculture based on large scale inputs of synthetic fertilizers and insecticides due to this the environmental degradation has become inevitable. The destructive factors include the pollution from the untreated domestic sewage, solid waste and industrial effluents entering particularly in the catchment areas.

Many of the smaller water bodies have become destroyed due to dumping of waste, garbage and silt. The urbanization and industrialization in the lake catchments in particular had its adverse impact on three major water bodies, viz., Hussainsagar, Mir Alam and Saroornagar lakes and these three water bodies are typical examples of this obvious event.

Brief Description of the Lakes

Hussainsagar

This Lake situated between the twin cities of Hyderabad and Secunderabad, was excavated in 1562 AD mainly to store drinking water brought from Musi river by Balakpur canal. However, with the passage of time the lake lost its importance as a source of potable water, nevertheless, it was extensively used for washing, bathing and recreation. The lake gets its pollution load from surrounding industrial areas and settlements. Its water spread area is 27.5 sq km., length 3.2 km, width 2.8 km and maximum depth is 12.5m.

Mir Alam Tank

Constructed in 1806, it was one of the oldest tanks created on Musi river system and the only multi-arch (21 arches) dam of its kind in the world. It is about 7 km south-west of the city. The water body was used as a source of drinking water up to 1960 by the Rajendranagar municipality. Presently is serves as the main source of water to the Nehru Zoological Park and its surroundings. Its surface area is 1.7 sq km., maximum depth 13.41m, and catchment area of 16.5 sq. km. The catchment is made of rocky undulating terrain.

Saroornagar Lake

Impounded in 1626 AD Saroornagar lake was meant for agricultural and drinking purposes and today is one of the major aquatic ecosystems about 8 km on the eastern side of Hyderabad city. Its water-spread is 35 ha with a maximum depth of 6.1m. In the last two decades, due to growing urbanization the catchment has undergone drastic modification with the consequent effects on the morphometry and limnology of the lake ecosystem. Scientifically speaking, the lake has got aqua cultural, ecological and recreational potential.

Ecological Benefits

The water bodies as a whole, bestow the following incalculable ecological benefits in financial terms to its immediate environment.

1. Charging of groundwater table
2. Aquaculture
3. Flood regulation

4. Condition of the climate
5. Sustenance of biodiversity
6. Sustenance of colloidal water for vegetation
7. Recreational sites
8. Silt as a source of soil for gardens
9. Water use by lake dependent communities like washermen and small dairy farmers
10. Community asset with cultural and educational value.

As the urbanization at the catchment areas of these lakes several suggestions were given to the government to protect these water bodies and the delay in the decision making process led to rapid urbanization of the lake catchments with a number of authorised and unauthorized colonies in the catchment and hence the manifestation of pollution and destruction occurred. Fish Kill which is the result of manifestation of eutrophication has becoming a regular phenomena particularly in the last decade of the century in Hussainsagar as well as Saroornagar lakes in the case of species like *Notopterus notopterus* and *Cyprinus carpio* respectively.

Some voluntary Non-Government Organisations viz., Society for Preservation of Environment and Quality of Life (SPEQL) and a scientific Association, Indian Association of Aquatic Biologists (IAAB), Hyderabad made good efforts to protect these lakes through their 'Save the Lake Compaign' by enlightening the Government Authorities like Andhra Pradesh Pollution Control Board (APPCB), Municipal Corporation of Hyderabad (MCH), Environmental Protection Training and Research Institute (EPTRI) and Hyderabad Urban Development Authority (HUDA). Several efforts have also been made to educate the public through visual media.....*etc.*, under the 'Save the Lake Campaign'.

Water Quality

There are a few studies on the limnological aspects on these three lakes in the recent past on Hussainsagar (Malathi *et al.*, 2003), Saroornagar (Chandrasekhar 1997) and Mir Alam Lake (Anitha *et al.*, 2004).

Table 5.1

Parameter	*Hussainsagar Lake (1996-98)*		*Saroornagar Lake (1994-96)*		*Mir Alam lake (1998-2000)*		*Desirable IS 10500 1991*	*Permissible Limits IS 10500 1991*
	Ranges	*Mean Values*	*Ranges*	*Mean Values*	*Ranges*	*Mean Values*		
pH	6.0–9.0	7.4	7.0–8.3	7.85	7.0–8.9	8.6	6.5–9.0	No relaxation
Carbonates	10–80	44	0–145	14	–		–	–
Bicarbonates	180–470	277	250–350	357	150–415*	278*	–	–
Chlorides	196–414	277	300–600	417	80–320	230	250	1000
Total Hardness	190–720	347	450–650	521	150–410	165	300	600
Nitrates	5–80	38	9–76	53	1.5–8.0	3.6	45	100
Phosphates	3–140	57	0.5–4.4	2.43	0.1–0.9	2.3	–	–
B.O.D.	40–164	76	35–100	62	0.4–26	6.9	–	30
C.O.D.	132–420	266	80–240	212	11–118	46	–	250

*: Total alkalinity.

Ranges of the physico-chemical parameters of the three lake waters are given together with its mean values. The individual values of the parameters of the three lake waters have indicated its unsuitability for drinking waters, even though some values (Ranges/Mean values) are indicating positive values.

The higher values of chloride, Nitrate, Phosphate and particularly the effluents of Hussainsagar clearly indicate the pollution due to industrial wastes that are letting into the water body. The higher values of the other two water bodies are indicating the pollution due to domestic sewage from the surrounding colonies and higher phosphate values indicate the pollution due to the usage of detergents while washing the clothes. The Physico-chemical factors are showing that the lake water (s) are getting polluted due to untreated industrial effluents and domestic sewage (s).

Environmental Degradation

The agriculture has completely disappeared in the metropolitan limits particularly in the catchment areas of the lakes due to urbanization with a number of colonies surrounding the basins of Saroornagar lake and Mir Alam lakes. In the case of Hussainsagar lake, number of big and small sized industries have been developed in addition to the housing colonies in the lake vicinity. At present the Hussainsagar lake, is being used for recreational purpose and fish culture, the Saroornagar lake is being used for fish culture and Mir Alam lake waters are used as a source of water source for Nehru Zoological Park situated on the bank of the lake and also fish production.

All these three water bodies have high productivity as can be seen from the major commercial fish species available in these Lakes, viz. Catla catla, Labeo rohita, Notopterus notopterus, Cyprinus carpio, Cirrhinus mrigala, Channa punctata, Clarius batrachus, Heteropneustes fossilis, Mystus vittatus, Glossogobius giuris, etc. *The zoobenthos of these three lakes includes chyronomid larvae in Diptera and the malaco fauna viz.*, Bellamya dissimilis, B. bengalensis, Indoplanorbis exustus, Thiara tuberculata, T. scabra, Lymnaea accuminata, Gyraulus convexiuscullus, Cremnoconchus conius, Faunus ater and Gabbia stenothyroide. *Among the avian fauna, Pond Herons, Ducks, Egrets,* etc., *can be observed at the lakes.*

The three phenomena of ecological consequences as stated above viz. (a) Un-precedented population growth, (b) industrialization and (c) intensive agriculture based on large scale inputs of synthetic fertilizers and insecticides, have got adverse impact on the three water bodies manifesting in terms of pollution at ground as well as surface waters, biodiversity, destruction of catchments due to developmental activities followed by the reduction of water holding capacities of the water bodies. These degrading factors can be summarized as follows:

Pollution Due to Industrial Effluents/Domestic Sewage

Industrial effluents that are let into the water bodies particularly in the case of Hussainsagar that has got about 350 smalls/big industries on the lake basin is the main source of pollution. In the case of Mir Alam lake there are a few small industries on the lake basin whose wastes are letting into the lake. Domestic sewage from the surrounding housing colonies like Singareni which contains a load of nutrients, is main source of pollution in Saroornagar lake.

Silting

Silt that is generated through the construction of housing colonies..etc., not only at the catchment areas but also lake basins and 'idols' (that are made up of clay and Plaster of Paris) immersions during after the festivals like Ganesh Chaturthi, Vijaya Dasami.. *etc* reduces the depth of lake.

The chemicals that are being used as colours to different above idols like Zinc Oxide.. *etc.* are toxic to the biota and thus effects the biodiversity.

Encroachments

Due to lack of proper lake boundaries and survey records, unauthorized encroachments extending to the lake bed has become a common phenomenon to all these three lakes.

Washing Activities

Due to lake of Dhobi Ghats, the washermen are using the lake for washing the clothes by using the detergents that load phosphates to the water bodies, leads to the eutrophication.

Dumping of Garbage and Solid Wastes

These water bodies have become convenient places for dumping of wastes and garbage.

Remedial Measures

For proper rehabilitation of these three ecosystems measures like (a) Demarcation of lake boundaries after thorough surveys to avoid unauthorized encroachments, (b) modification of natural in-lets in the catchment-areas (c) regular de-silting of the lake particularly after the idol immersions every year, (d) laying of separate sewer lines should be made as a precondition for sanctioning the layouts particularly in the catchment-areas, (e) prevention of dumping of wastes in the lake with proper watch and ward systems, (f) developing the lake environs as recreational zones by construction of gardens, parks, water sports.. *etc.*, and establishing the educational centers for creating public awareness about the importance of eco-heritage. (g) Macrophytes like water hyacinths which act as de-polluters should be developed not only to reduce the nutrients like Phosphates and Nitrates but to remove pollutants. Periodical harvesting of weed biomass reduces the nutrients and improves the water quality. (i) Introduction of composite fish culture with select fish species to harvest biomass in all the layers to reduce the nutrient levels. (j) Proper locations for washing the clothes by constructing the Dhobi Ghats to prevent water pollution.

In the case of Hussainsagar lake, some steps had been taken like to prevent water pollution:

1. Fish species, *Cyprinus carpio* which act as depolluters, had been released into the lake in the year 1996, but no positive result could be seen.
2. Steps have been taken to divert the industrial effluents by a separate pipeline at Kukatpally Nalluh.
3. Sewage treatment plants have been installed to treat the domestic sewage through effluent ereatment plant with a well developed technology.
4. Lumbini Park with boating facility was constructed on the bank of the lake.
5. Water sports have been plying under the sports authority, the state government authority
6. The lake had been beautified by constructing necklace road as recreational purpose.

In the case of Saroornagar lake, the water body had been beautified by constructing a bund for recreational value. To avoid siltation, the Idol immersions have been decentralized area-wise to different water bodies, and so the siltation could be controlled up to certain extent. In spite of the above steps taken by the concerned authorities, proper achievement on depollution of these lakes could be attained. Steps are yet to be taken for protection of Mir Alam lake.

Acknowledgements

The author is grateful to the Director, Zoological Survey of India, Kolkata and the Officer-in-Charge, Zoological Survey of India, Hyderabad for encouragement and facilities.

References

Anitha, G., Chandrasekhar, S.V.A. and Kodarkar, M.S., 2005. *Rec. Zool. Surv., India,* Paper No. 235.

Chandrasekhar, S.V.A., 1997. Ecological studies on Saroornagar lake, Hyderabad. *Unpublished Ph. D. Thesis,* Osmania University, Hyderabad.

IS 10500 : 1991. *Indian Standard Drinking Water Specification*, Table 1.

Malathi, D., S.V.A. Chandrasekhar and M.S. Kodarkar, 2003. *Rec. Zool. Surv., India.*

Sustainable Environmental Management *Pages 28–32*
Edited by: **Dr. L.V. Gangawane & Dr. V.C. Khilare**
Published by: **DAYA PUBLISHING HOUSE**

Chapter 6

Some Facts Regarding the Mangroves of Maharashtra

N.S. Chavan, M.V. Gokhale and A.B. Telave

Department of Botany, Shivaji University, Kolhapur, 416004 MS, India

ABSTRACT

This article gives idea about some of the facts regarding the mangroves of Maharashtra. The points such as mangrove area, geomorphology, mangrove biodiversity, biomass and coastal fishery are discussed.

Keywords: *Mangrove ecosystem, Biodiversity, Geomorphology, Coastal fishery.*

Introduction

Mangrove ecosystem is one of the unique ecosystems in the world. It has its characteristic strategies location and very unique functions. These ecosystems are called as lifeline of the coastal region. Basically indirect uses are more than direct uses. It is one of the major food source and breeding sites for coastal fishery. Along the coast of Maharashtra there is 5039.63 ha. mangrove area (Bhosale, 2005), but it is under tremendous human pressure. The mangroves in Maharashtra fall under four different districts *viz.* Sindhudurg, Ratnagiri, Raigad, Thane and Mumbai. The estuaries in Sindhudurg district are comparatively smaller. Estuaries in Ratnagiri and Raigad district have comparatively wide spread. There was luxuriant occurrence of mangroves in Mumbai but presently one has to search for mangroves. The statement *i.e.* "The mangroves of Mumbai are vanished" is not totally wrong. In this paper it is tried to collect and discuss different causes of mangrove degradation.

Mangrove degradation can be studied at three different levels *i.e.* biodiversity, biomass and coastal fishery. Estuarine ecosystems are self-sustainable and studied indirectly on the basis of coastal fishery.

The amount and diversity in fishery resources is one of the important measures to study the level of mangrove degradation.

Estuarine ecosystem is unique assemblage of different environmental factors. It is a result of tidal action together with freshwater inflow and geomorphology of the region. Some of the environmental factors influencing mangrove forest are soil type, salinity, duration and frequency of inundations, accretion of substratum, strength of tidal exposure and shelters etc. Walsh (1974) has given five basic requirements for mangrove establishment. The important factors are aridity, tidal conditions, mineralogy, tropical temp, salinity, shallow shores water energy, sedimentation, geotectonic effects, fine-grained alluvium, large tidal range and muddy substrate. Bhosale (2005) considered man as one of the environmental factor interacting, affecting and exploiting mangroves and other related resources. Traditionally mangroves have been viewed as ecological and economic wastelands. In the past they have been regarded as suitable repositories for city waste and wide range of reclamation projects (Chapman, 1977).

Mangroves of Maharashtra are typical in that they fall at most of the places, riverine or fringe forest. As a matter of facts west coast of Maharashtra is a narrow strip of land. The rivers have a short coarse and they do not form big deltas. The estuarine limits may go upto 20 kms or so (Bhosale and Mulik, 1991).

Mangrove Area

Mangrove ecosystems include forest communities and open water surfaces such as network of canals, creeks sandy and muddy sediments without any forest, therefore the term mangrove is being interpreted in different ways (Blasco et. al 1998). All woody and grassy halophytic communities included in the intertidal zone are considered as potential mangrove areas. It is termed as mangrove land. Mangrove area is still wider concept than the other and always remains extremely vague. Its definition is not accurate enough for statistical purposes. It includes all mangrove and back mangrove plant communities including salt flats and an extended array of land use units resulting from mangrove conversion to other uses such as fish ponds, agriculture, salt pans etc. Therefore variation is always regarding their mangrove areas. According to Jagtap et. al (1994) mangrove area in Maharashtra is 210.17 hector. Untawale (1984), Blasco (1977), Sidhu, (1963), Khan, (1957), Qureshi (1957) reports 200-622 km ^{2} mangroves area for Maharashtra. Recently Bhosale (2005) has reported the area covered by mangroves in different districts of Maharashtra *i.e.* 5039.63 ha. Case study of Achra estuary of Maharashtra reveals that there is increase in area under mangroves from 1985 (Radhakrishnan, 1985) to 2004(Gokhale 2004). In Achra Radhakrishnan (1985) has reported 74.4 ha. mangrove area and 273 ha. total estuarine area. After that Bhosale (2002) reported 151 ha. mangrove area. The work of Gokhale (2004) reports 154.65 ha mangrove area and 56.16 ha. area under encroached mangroves. It indicates that there should be some increase in some mangrove areas further it may be suggestive that reclamation at some place may increase the extent of estuarine limits in other estuaries. Some times it is very difficult to decide wheather the land is newly formed or original mangrove land. Therefore on the coast of Maharashtra we can see the estuaries with heavy human encroachment as well as heavily spreading estuaries. In such cases the encroached mangrove land as well as new mangrove land support different mangrove species and both the areas are considered under mangrove area. Most of the reports are based on the area under mangroves and the consideration for this is different for different workers. Some workers consider the actual areas under canopy while some consider the total extent upto last mangrove occurrence. To estimate the total estuarine areas (the area upto high tidal reach) is therefore very essential. Comparative account of these areas for different estuaries will speak more about the rate of encroachment/mangrove conversion to other uses and the rate of estuarine spread.

Geomorphology

Geomorphology is the impression of land from developments within largely abiotic environment. Geomorphology developed as a part of geology but it is important from habitat ecology point of view. Geomorphological investigations throw light on the environmental changes over a range of time scale (Flenley, 1984). Geomorphology is a physical characterization of a system. It is determined like local geology, sea level change, tides, fresh water inputs, shoreline structures, water shade morphometery, natural disturbance and climate. Therefore Geomorphological features govern the development of a system (Kjerfve, 1990) ecology and geomorphologies are both characterized by having functional and historical aPproaches and both shares a concern with wide range of temporal and spatial scale

Both erosional and depositional processes characterize estuarine ecosystem. This ecosystem is unstable and constantly changing (Blasco et.al. 1998) it is observed that construction of erosion controlling banks diverts the estuarine currents towards the other banks resulting in erosion and loss of potential mangrove stands. Similarly the construction of bund disturbs the natural habitat level and influences the tidal spread. Bank and Bund construction also changes hydrodynamics and favors sediment deposition. It decreases the average depth of water column affecting hydrologic character. This process is evident in most of the estuaries of Maharashtra. Erosion from near by localities that is mountains, tablelands are also considerable from last few years. It is due to clearing forest for horticultural purposes. It also contributes in decrease of water column in river and estuary. Dredging of sand is another economic aspect of estuarine ecosystem. It is observed mostly near mouth. It also affects the pattern of water currents in estuarine ecosystem. Construction of banks, bunds require foreign material such as stones and soil from terrestrial origin these put the pressure on the fine grained alluvium resulting in hard compact sediments. Therefore in most of the encroached mangrove areas density of mangrove soil is higher than undisturbed areas. It affects root penetration and other microbial processes as well as organic matter transformation. Such areas show scrubby growth of mangroves. The areas show less frequency of tidal inundations therefore soil salinity increases. Though this phenomenon is controversial it haPpens in number of estuaries of Maharashtra. As the frequency of inundations is less there is no proper washing off of the salts coming from high tidal water. Salts get accumulated and increase soil salinity. As a result of this complex process we can observe dwarf/scrub forests of mangroves at upstream and landward of number of estuaries. Stunted forms of mangroves dominate this type of region and it is low in both nutrients as well as hydrological energy.

Mangrove Biodiversity

Prima-facie the biodiversity of mangroves in India is well studied. There are number of reports indicating the diversity and enlisting mangroves, their associates and other halophytic as well as algal species. Some of the reports are Kothari (2002), Jagtap *et al.* (1994) etc. The main problem in the field of biodiversity of mangroves is taking the suPport from previous reports. Therefore there is no report giving update information except Bhosale (2002). It can be explained by taking the example of critically endangered species on the coast of Maharashtra *i.e.* Xylocarpus Granatum. It has not occurred in very recent publication by Kathiresan and Qasim (2005). Therefore this aspect demands keen and thorough survey of each and every mangrove stand on the coast. The species, which are rare and endangered, are generally restricted to certain region and they have very few in number. This type of situation is observed in number of estuaries like Kolawal, Achra, Mithbav, and Rajapur etc. It is very essential to study the ecology of *X. granatum, Cynometra iripa, Sonneratia apetala, S. caseolaris* and *B. cylindrica* seedlings at micro level. As compared to other mangroves these species regenerate very rarely on the coast of Maharashtra (Telave, 2005 and Gokhale, 2004). Today's forests are the results of zonation and successional changes over long time. Therefore prediction about regeneration at any single site will be hasty. Constant observations on regeneration will solve this problem. It should be considered before undertaking the conservation plans. Mere plantations cannot give assurance for the survival. It is observed that the survival of planted seedling of number of mangroves is very less. One can obtain good results only after selecting proper sites.

Mangrove Biomass

There are no big deltas in west coast estuaries, whatever stands are there, are in the form of landward fringe or small islands. Also the rivers are small and depth is less. It favors constant contact of local inhabitants to mangroves for almost all estuaries of Maharashtra. Mangrove biomass is the important fuel for local people. Commercial trade is also observed in the form of stumps. As a result of this number of islands are denuded internally keeping the boundary with plants. Biomass conservation requires awareness and legal activities. Unfortunately the rare species like *Cynometra iripa, Xylocarpus granatum* and *Sonneratia apetala* are under this type of threat.

Coastal Fishery

Estuarine ecosystem is the cradle house of marine and coastal fishery. It harbors the 'fishery seed'. Different types of crabs, fishes, and molluskas are the important edible types of estuarine origin. In spite of that many marine fish species come over in this cradle to breed. Bhosale (2004) has made a survey on the coastal fishery of Malvan in Maharashtra and nearby areas. According to her biodiversity and biomass both are depleting very fast. It is due to heavy fishing and use of non-sustainable methods.

References

Bhosale, L.J., 2002. Categorization of mangroves of Maharashtra based on IUCN red list guidelines and germplasm preservation of threatened species. Final report submitted to Ministry of Environment and Forest, Government of India, New Delhi.

Bhosale, L.J., 2005. *Field Guide to Mangroves of Maharashtra*. Published by Shivaji University, Kolhapur; 315 pp.

Bhosale, L.J. and N.G. Mulik, 1991. Endangered Mangrove Areas of Maharashtra. In: *Proceedings of the Symposium on Significance of Mangroves,* Pune, March 1990.

Blasco, F., 1977. Wet coastal ecosystem. In: *Ecosystem of the World I*, (Ed. V.J. Chapman). Elsevier Scientific Publishing Company, Amsterdam.

Blasco, F., T. Gauquelin, M. Rasolofoharinoro, J. Denis, M. Aizpuru and V. Caldairou, 1998. *Mar. Freshwater Res.*, 49: 287–296.

Chapman, V.J., 1977. Wet coastal ecosystems. In: *Ecosystems of the World I.* Amsterdam, Elsevier, 428 p.

Flenley, J.R., 1984. Time scales in biogeography. In: *Themes in Biogeography* (Ed Taylor, J.A.), London, Croom Helm, pp. 63-106.

Gokhale, M.V., 2004. Studies on the mangrove environment of Achara estuary. *Ph.D. Thesis*, Shivaji University, Kolhapur, MS, India.

Jagtap, T.G., A.G. Untawale and S.N. Inamdar, 1994. *Indian Journal of Marine Sciences*, 23: 90–93.

Kathiresan, K. and S.Z. Qasim, 2005. *Biodiversity of Mangrove Ecosystems*. Hindusthan Publishing Corporation (India) New Delhi.

Khan, M.A.W., 1957. *Proceeding of Mangrove Symposium*, Government of India Press, Calcutta, pp. 97.

Kjerfve, B., 1990. Manual for investigation of hydrological processes in mangrove ecosystems. UNESCO/UNDP regional project '*Mangrove ecosystems in Asia and the Pacific*' RAS/79/002 and RAS/86/120.

Kothari, M.J., 2002. Mangrove diversity and its role for sustaining productivity of the N. W. Coast in India. In: *Proc. the National Seminar on Creeks, Estuaries and Mangroves: Pollution and Conservation*, pp. 226–233.

Qureshi, I.M., 1957. *Proceeding of Mangrove Symposium*, Government of India Press, Calcutta, p. 20.

Radhakrishnan, N., 1985. Studies on the mangroves along the central west coast: Achara (Maharashtra). In: *The mangroves, Proc. Nat. Symp. Biol. Util. Cons. Mangroves*, (Ed.) Bhosale, L. J. pp. 222–226.

Sidhu, S.S., 1963. *Indian Forester*, 89: 337.

Telave, A.B., 2005. Studies on starch and its distribution pattern in mangroves–with special reference to three species of *Sonneratia*. *Ph.D. Thesis*, Shivaji University, Kolhapur, MS, India.

Untawale, A.G., 1984. Proceeding of Asian Symposium on Mangrove Environment Research and Management, (Eds) E. Soepadmo, A. Rao and D. Macintosh. Percetaken Ardyas Sdn Bhd, Kualalampur, Malayasia, pp. 57.

Walsh, G. E., 1974. In: *Ecology of Hahophytes*, (Eds) Reimold, R.J. and Queen, W.H. Academic Press, New York, pp. 51.

Sustainable Environmental Management *Pages 33–51*
Edited by: **Dr. L.V. Gangawane & Dr. V.C. Khilare**
Published by: **DAYA PUBLISHING HOUSE**

Chapter 7

Emperical Relations for Eco-Labeling of Agricultural Farms

V.A.S. Daniel and S.B. Wagh

Department of Environmental Science, Dr. Babasaheb Ambedkar Marathwada University, Aurangabad – 431 001 (M.S.)

ABSTRACT

Nowadays eco-labeling of organic and chemical farms has become essential to maintain eco-friendly environment. For this different parameters have been registered in different countries. In India it is found that respiration rate, total count of soil microbes, percentage of organic carbon, insect diversity index, the average harvest diversity, average multiple cropping index etc. was noted to be more in organic farms. However more studies for long duration are required to bring definite conclusion.

Keywords: *Eco-labeling, Organic farming, Organic carbon*

Introduction

What are Eco-labeling Programmes?

"Eco-labeling" is a voluntary method of environmental performance certification and labeling that is practised around the world. An ecolabel is a display of written, printed, or graphic material on the packaging of a product giving information of environmentally preferable products based on an environmental impact assessment of a product compared to other products in the same category. An important feature of this impact assessment is that it *is* not limited to the environmental impacts from use and/or disposal of the product It also includes impacts from production of the product. The impact assessment is done by a third-party, either public or private. Thus the roots of ecolabeling can be found in growing global concern for environmental protection on the part of governments, businesses and the public.

There are many different voluntary (and mandatory) environmental performance labels and declarations. The International Organization for Standardization (ISO) has identified the following three broad types of voluntary labels, with ecolabeling fitting under the Type I designation:

Type I–a voluntary, multiple-criteria based, third party program that awards a license that authorizes the use of environmental labels on products indicating overall environmental preferability of a product within a particular product category based on life cycle considerations

Type II–informative environmental self-declaration claims.

Type III–voluntary programs that provide quantified environmental data of a product, under pre-set categories of parameters set by a qualified third parry and based on life cycle assessment, and verified by that or another qualified third party.

Understanding a Few Terms in Eco-labeling

General Claim

Claims used on products that are not independently verified are general claims; the manufacturer often places them on the product.

Logo or Seal

An identifying mark of the eco-label that is affixed to a product or item, a logo or seal is usually trademarked and indicates association with or approval by the owner of the mark.

Certifier

A certifier evaluates a company wishing to put a particular eco-label on its products and verifies that the products conform to the standards behind the label.

Accreditor

An accreditor develops a set of standards behind an eco-label and then authorizes or accredits a certifying organization to verify that any company using the eco-label on its products is following standards.

Brand

A brand is a printed mark made to attest manufacture or quality or to designate ownership

Eco-labeling is only one form of environmental labeling seen in markets today. Two other common forms are government-mandated labels and self-declarations (Kuhre, 1997). Examples of government mandated environmental labels are the fuel efficiency ratings required on new automobiles, the energy use guides required on household appliances, and environmental hazard warnings required on pesticides and products containing CFCs or toxic substances. Examples of self-declarations are manufacturer claims about recyclability, recycled content, solid waste reduction, biodegradability, and non-use of certain chemicals (*e.g.*, no phosphates). Two key features differentiate eco-labeling from these other forms of environmental labeling. Unlike government-mandated labels, ecolabels are voluntary. Unlike self-declarations, ecolabels involve standard setting and enforcement by a third-party. Without guiding standards and investigation by an independent third party, consumers may not be certain that the companies' assertions guarantee that each labelled product or service is an environmentally preferable alternative. This concern with credibility and impartiality has led to the formation of both private and public organizations providing third party labeling. Thus, an ecolabel is like a seal of approval because it is a signal of high standards as well as a signal that products meet standards.

An ecolabeling organization performs three key tasks: standard setting, certification, and marketing. Standard setting determines the environmental standards a product must meet to qualify for the ecolabel. Certification determines whether a given product meets those standards. Marketing develops customer awareness of and trust in the claim.

Major Benefits of Eco-labeling

Providing Information to the Consumer

Eco-labeling is an effective way of informing customers about the environmental impacts of selected products, and the choices they can make. It empowers people to discriminate between products that are harmful to the environment and those more compatible with environmental objectives. An eco-label makes the customer more aware of the benefits of certain products, for example, recycled paper or toxic-free cleaning agents. It also promotes energy efficiency, waste minimization and product stewardship.

Promoting Economic Efficiency

Eco-labeling is generally cheaper than regulatory controls. Empowering customers and manufacturers to make environmentally supportive decisions keep the need for regulation kept to a minimum. This is beneficial to both government and industry.

Stimulating Market Development

When customers choose eco-labeled products, they have a direct impact on supply and demand in the marketplace. This is a signal which guides the market towards greater environmental awareness.

Encouraging Continuous Improvement

A dynamic market for eco-labeled products encourages a corporate commitment to continuous environmental improvement. Customers can expect to see the environmental impacts of products decline over time.

Promoting Certification

Eco-labeling provides customers with visible evidence of the product's desirability from an environmental perspective. Certification therefore has an educational role for customers-and promotes competition among manufacturers. Since certified products have a prominent logo to help inform customer choices, the product stands out more readily on store shelves. Coveting the logo may induce manufacturers to re-engineer products so that they are less harmful to the environment.

Assisting in Monitoring

Another benefit of an official eco-labeling program is that environmental claims can be more easily monitored. Competitors and customers are in a better position to judge the validity of a claim, and will have an incentive to do so should a claim appear dubious.

Examples of Eco-labeling Programs

Blue Angel

Germany's Blue Angel was the world's first eco-labeling program, introduced in 1977. It relies on the voluntary cooperation of manufacturers to provide information about their products. It also relies on individuals and retailers to purchase more environmentally friendly products. Unlike other ecolabel programs, all Blue Angel products must be reclaimed by their manufacturers at the end of their useful lives because of a federal mandate. In 1991, Germany mandated a take-back program (of products and their packaging), which is credited with cutting 1 million tons of waste between 1991 and 1993.

The Blue Angel organization's Eco-Label Jury, composed of industry, environmental associations, trade unions, churches, and public authorities, scrutinizes and approves product groups twice yearly in light of developments in product design and environmental technology. Approved products are reviewed every two to three years to determine whether or not they should continue to be approved as part of the program.

Current criteria for the Blue Angel award include resource conservation, low greenhouse gas emissions, and fossil fuel efficiency, resulting in products with minimal climate impact. As applied to personal computers, the criteria used focus on power consumption, durability, and recyclability.

EU Ecolabel

The European Union's Eco-label program was developed in 1993 to foster the production and purchase of products that are more environmentally friendly than their peers. All countries with membership in the European Union can take part in this program. The logo is comprised of a flower with the EU's symbol–a circle of stars with the letter "E" in the center. Each awarded product must undergo a life-cycle analysis. At present, 10-20 per cent of EU products on the market have been manufactured in accordance with ecolabel criteria. Products that have been awarded Eco-label certification are reviewed after about three years, when acceptance standards may be raised.

Eco Mark

Japan's Environmental Association runs the country's Eco Mark program. This program was developed to encourage more environmentally friendly manufacturing and help consumers and retailers to identify products that do less environmental harm. The logo shows someone protecting the earth with her/his hands. Across the top are the words "Friendly to the Earth" while the product category is shown across the bottom. Products awarded the Eco Mark label must meet the following criteria: be less environmentally unfriendly in their manufacture, use, and disposal than their product peers and be more environmentally friendly that result in improved conservation.

Energy Star

Energy Star is a voluntary government program that was developed in 1992 by the U.S. Environmental Protection Agency to encourage energy efficiency in office equipment. Because it is a government program. Energy Star is unlike the other ecolabeling programs already mentioned. The Energy Star program focuses on energy efficiency to the exclusion of other environmental criteria. In addition to labeling energy-efficient products, the program has partnerships that encourage energy conservation, saving more than $1 billion in energy costs while reducing air pollution.

Environmental Choice

Canada's Environmental Choice program was designed in 1988 to assist consumers in identifying products and services that are less environmentally harmful than similar products. The logo is comprised of three doves (consumers, industry, and government); forming a maple leaf when combined, the doves are working together to improve Canada's environment. Criteria for the Environmental Choice ecolabel include: energy efficiency, low production of hazardous by-products and waste, recycled material use, product recyclability. Environmental Choice awards last for only one year. During that time, certified products are monitored for compliance with the guidelines relevant to their product categories. Following the one-year award, products are reviewed for renewal.

Green Seal

Green Seal was developed in 1992 to award a "Green Seal of Approval" to U.S. products that are environmentally superior to their peers. The logo, a green check mark over the globe, helps consumers and retailers to identify environmentally preferable products. Intended to help businesses, universities, and other organizations from having to wade through product environmental claims, Green Seal recommends products based on factors that include packaging, product performance, and environmental impacts.

NF Environment

NF Environment label was developed in France and is found on products that are less environmentally detrimental than (while performing on par with) their product peers. This program, founded in 1992, is voluntary and, like the others, was developed for businesses that want to make their products stand out for their environmental superiority. The logo shows a single leaf draped over a globe.

Eco-Labeling Programs Related to Agricultural Systems

Eco-labeling programmes for agricultural systems are related to products from eco-friendly systems of agricultural production. Eco-friendly agriculture can be defined as an approach to agriculture where the aim is to create integrated, humane, environmentally sustainable agricultural production systems. Maximum reliance is placed on self-regulating agro-ecosystems, locally or farm-derived renewable resources and the management of ecological and biological processes and interactions. Dependence on external inputs, whether chemical or organic, is reduced as far as possible.

Organic farming is one system of such agriculture practice based on dynamic interaction between the soil, plants, animals, humans, the eco-system and the environment.

Organic agriculture and processing is based on a number of equally important principles and ideas:

1. To produce high nutritional quality food in sufficient quality.
2. To interact in a constructive and life-enhancing way with natural systems and cycles
3. To encourage and enhance biological cycles within the farming system involving microorganism, flora and fauna, plants and animals
4. To help in the conservation of soil and water
5. To use, as far as possible, renewable on farm resources in locally organized agricultural systems
6. To work as far as possible, within a closed system with regard to organic matter and recycled, either on the farm or elsewhere

7. To provide all livestock with living conditions, which allow them to perform the basic aspects of their innate behaviour
8. To create a harmonious balance between crop production and animal husbandry
9. To minimize all forms of pollution that may result from agricultural practice
10. To completely abstain from the use of genetically engineered products as inputs or in any other form
11. To maintain the genetic diversity of the production system and its surroundings, including the protection of plant and wildlife habitats
12. To preserve and enhance traditional and indigenous knowledge, seeds, crop varieties and animal breeds
13. To allow everyone involved in organic production and processing a quality of life, to cover their basic needs and obtain an adequate return and satisfaction from their work, including a safe working environment.
14. To consider the wider social and ecological impact of the farming system
15. To produce non-food products out of renewable resources, which are fully biodegradable.

Organic farming has taken almost 50 years to become a real force in the marketplace. Even now, after 8 years of remarkably strong and steady growth, it still is only 1 or 2 per cent of the overall food economy. In India, every year about 3 lakh tonnes of organic agriculture products pass through third party eco-labeling programmes for the purpose of export. At present marketing of organic products in the domestic organic agriculture market is not in a large scale and the sale is limited to metros like Mumbai, Delhi, Kolkata, Chennai, Bangalore and Hyderabad. Third party certification in not very much involved in this domestic marketing of organic products which is mostly based on trust through individual initiatives of the farmers, non government organizations, some entrepreneurial traders etc. In Aurangabad, organic products are available in weekly organic bazaars organized by Institute for Integrated Rural Development (IIRD) in Aurangabad city. Organic products are directly sold in the market by organic producers with a weekly turnover of Rs. 5000 to 6000.

In order to provide quality assurance for marketing organic agricultural products, various eco-label programmes have been developed all over the world. Each program has its own set of standards for the products it labels or the goals it has set forth. In each case, these are designed to instruct growers in what they ne ed to do to qualify for that particular label. Each program requires an application and extensive documentation procedure from its participating growers.

Different eco-label programmes for agricultural products found in major countries are described below:

National Programme for Organic Production, India

India's eco-labeling programme came into force in the year 2001 under the National Program for Organic Production. The programme is developed and implemented by the Government of India through its Ministry of Commerce. Under the National Programme for Organic Production (NPOP), documents like National standards, Accreditation Criteria for accrediting, inspection and certification agencies, Accreditation Procedure, Inspection and Certification Procedures have been prepared and approved by the National Steering Committee. These documents were prepared on the basis of the guidelines evolved by the representative international organization, International Federation for Organic Agricultural Movement (IFOAM), EU Regulations and Codex standards. Under this programme organic agricultural products cannot be exported unless inspected and certified by an Inspection and Certification Agency duly accredited by one of the Accreditation Agencies designated by the Government of India. Inspection and Certification agencies offering eco-labels and accredited to the Accreditation agencies (APEDA, Coffee Board, Tea board, Spices Board etc) in India are ECOCERT, SKAL, MO, APOF, INDOCERT.

The National Organic Program (NOP), USA

In December 1997, the United States Department of Agriculture (USDA) issued the National Organic Program (NOP), which established standards for organic agricultural products as well as a certification system and labeling requirements. Since there are differences in the standards of the 33 private certifiers and between state agencies, a federal organic standard was found necessary by consumers, environmentalists and the US organic industry. The evolution of a regulation on organic food started in 1990 when the Organic Foods Production Act (OFPA) was passed by the US Congress as part of the 1990 Farm Bill. The OFPA is a production practices regulation, which is ruled by the NOP. The OFPA authorizes the U.S. Department of Agriculture (USDA) to develop specific organic production and handling standards and permit use of a USDA seal on products that have been certified by a federally accredited certifier to meet those standards. Imported agricultural products may be sold in the United States if USDA-accredited certifying agents certify them.

In lieu of USDA accreditation, a foreign certifying agent may receive recognition when USDA has determined, upon the request of a foreign government, that the foreign certifying agent's government is able to assess and accredit certifying agents as meeting the requirements of the USDA National Organic Program.

Jas Organic Certification Programme, Japan

JAS (Japan Agricultural Standard) law specifies labeling of agricultural products and other related regulation governed by the Ministry of Agriculture, Forestry and Fisheries of Japan MAFF). On June 1, 2000 a revised JAS law has been implemented and new legal organic certification program has been newly launched.

EU Eco-labeling Programme

The EU eco-labeling programme was established by the European Economic Community (EEC) Council Regulation No. 2092/91 of June 91 on "organic production of agricultural products and indications referring thereto on agricultural products and foodstuffs". Regulations of the European Economic Community are directly applicable as the law of the land in the member states. EEC regulations are self-executing.

Third Party Eco-labeling

ECOCERT SA

ECOCERT SA is an inspection and certification body accredited to verify the conformity of organic products against the organic regulations of Europe, Japan and the United States. The ECOCERT certification mark is one of the leading international organic certification marks, enjoying a good reputation and trusted by both consumers and the organic industry. ECOCERT SA, has its operational office in Germany and currently performs such inspection and certification services in about 70 countries outside the EU, on all continents. In various countries ECOCERT SA has established offices in order to provide services more efficiently. ECOCERT is represented in France, Belgium, Italy, Portugal, Germany, and Spain to certify European companies.

ECOCERT SA is accredited according to ISO Guide 65 (equals to European Norm EN 45011) by COFRAC. COFRAC is member of EA (European Accreditation) and IAF (International Accreditation Forum) and thus an internationally recognised accreditation body. In the United States, ECOCERT is accredited to the NOP standard by the USDA and in Japan to the JAS organic standard by MAFF. In India ECOCERT is accredited to the NPOP of the Indian Government. As far as needed in countries with national organic legislation ECOCERT aims to obtain accreditation according to the respective national requirements.

SKAL International

Skal International is an inspection and certification organisation, which certifies organic products, processes and inputs besides, sustainable forest/wood and textile. Skal International operates worldwide and is accredited according to various national requirements.

Skal International is the only inspection and certification organisation that may issue the certification mark, EKO quality symbol, outside the European Union. The consumer can recognize an organic product, inspected and certified by Skal International, by the EKO quality symbol which is used when raw materials originate from organic cultivation and are processed using organic methods. EKO is a certification mark. It is neither a cultivation mark nor a trademark.

Soil Association

The Soil Association's organic symbol is UK's most recognizable trademark for organic produce. Products labeled by the Soil Association symbol are produced and processed to strict and rigorous environmental and animal welfare standards.

Besides setting standards to ensure the integrity of organic food and other products, Soil Association also undertakes inspection of producers, processors and suppliers and certifies such organic agriculture products.

Institute for Marketecology (IMO)

The Institute for Market Ecology (IMO) is one of the first and most renowned international agencies for inspection, certification and quality assurance of eco-friendly products. Its activities are accredited by the Swiss Accreditation Service (SAS) according to EN 45011 (ISO 65), which is the international standard for certification. Since more than 20 years, IMO has been active in the field of organic certification and in the sectors of natural textiles, sustainable forestry, and social accountability monitoring.

IMO offers not only certification according to Regulation (EEC) N° 2092/91 for organic production, the IVN Standards for natural textiles or FCS principles for sustainable forestry, but also co-operates with any existing label schemes on the market as per need of the customer. IMO also co-operates with the popular private label Naturland (IFOAM accredited) and conducts Naturland inspections worldwide. The range of operations reaches from small organisations to large plantations and also includes collection of wild products and traditional production systems. IMO certifies all types of agricultural products, from traditional produce such as coffee, tea, spices, cocoa, nuts, fruits, vegetables, cereals or pulses to cotton, dairy products, honey or fish and seafood. This also includes all types of food processing, product manufacturing and international trading activities.

The IFOAM Organic Guarantee System

IFOAM's Organic Guarantee System unites the organic world through a common system of standards, verification, and market identity. It is the practical realization of IFOAM's commitment to harmonize an international guarantee of organic integrity. The core of the Organic Guarantee System (OGS) is the IFOAM Accreditation Program. Beyond IFOAM, the International Organic Accreditation Service (IOAS) and IFOAM Accredited Certifiers are key components of the OGS.

IPOAM Accreditation Program

The IFOAM Accreditation Program is a service offered to certification bodies. It is administered for IFOAM by the International Organic Accreditation Service (IOAS), which is an independent organization. IFOAM Accredited Certification Bodies (ACBs) nust demonstrate compliance with the IFOAM Norms.

IFOAM Seal

The IFOAM Seal is a market- oriented mark of compliance with the IFOAM Accreditation 'rogram. The seal may be used on products that are certified by IFOAM ACBs. This lark ensures wholesalers, retailers, and consumers that a product and its producer are rganically certified within the IFOAM Organic Guarantee System. IFOAM Seal use on roducts is implemented through a contract signed between an ACB and the IOAS, and a corresponding contract between an ACB and its certified parties.

International Organic Accreditation Service

Although it operates as an independent body, the IOAS is a key component of the Organic guarantee System. IOAS is licensed by IFOAM to administer the Organic Guarantee–stem. It accepts and reviews accreditation applications, conducts site evaluations, and grants IFOAM accreditation to compliant applicants. IOAS also administers the IFOAM Seal, and together with IFOAM, it promotes the IFOAM Accreditation Program.

IFOAM Accredited Certification Bodies

ACBs made significant progress toward implementing a Multilateral Agreement (MLA) amongst them. The MLA creates multilateral equivalence at the level of the Accreditation Criteria and the IFOAM Basic Standards. Implementation of the MLA streamlines certificate acceptance, and thus helps to support and ensure orderly market transactions and trade. As an initiative that is administered by the ACBs, the MLA also supports the objectives of the IFOAM Organic Guarantee System

Other Agricultural Ecolabels

"Certified organic" has become the most widely known ecolabel, with clearly defined soil and crop management programs, especially the avoidance of synthetic pesticides, fertilizers, sewage sludge or biosolids, genetically engineered organisms and ionizing radiation. However there are other forms of eco-labels in agriculture, which are intermediate between conventional and organic production. An important difference between organic label and other eco-label is that organic fanning is more fanning system-focused whereas ecolabeling is more of a consumer-oriented, market-driven concept. For example, the specific crop production farm or site is important in organic farming because this means that a produce should be traceable back to the field in which it was grown.

Besides, the large number of existing ecolabels does not currently have a single, comprehensive standard as compared to organic certification programs. Some ecolabels already have clearly defined standards that focus on reduced pesticide and fertilizer use, PM and other sustainable practices. Since ecolabels do not generally prohibit pesticide use >r require a minimum pesticide residue standard (5 per cent of allowable EPA standards, under JSDA National Organic Standards), there is no transit! m period after initial certification by some ecolabel programs. In contrast, there is a 3-year transition period from conventional to certified organic production.

Some of the eco-labels developed are by Stemilt's Responsible Choice Program in Vashington, the Massachusetts-IPM Partners With Nature program, the Core Values program for apple fanners in the Northeast, Wegmans Food Stores using standards created y the New York-IPM Program, and California Clean Growers®. The stated goals of each program are basically the same: to produce high quality agricultural products in such a way lat the environmental impact from fanning on the surrounding ecosystem is minimized rid economic competitiveness is maintained. These eco-labels have no formal inspection id certification processes but an inbuilt point system for each set of guidelines and quality assurance corresponds to whether an included guideline was followed that is specific to at crop. Some other examples of ecolabels are "Protected Harvest," supported by the World Wildlife Fund, the "Healthy Grown" potatoes produced under a bio-TPM management system that measures and restricts pesticide use. Another ecolabel, "Food Miles" takes into account food miles (the distance food travels from where it is grown to here it is distributed, purchased, and consumed and the amount of food product transported).

Wegmans's Food Stores is another example where the food store has taken on the extra costs of providing environmentally labeled food. New York IPM guidelines are currently being used in the marketplace at Wegmans Food Stores through the collaboration of several different parties, which choose the specific growers to use, collect appropriate data, enforce and collect grower records, and inspect and make sure that the product is grown using IPM techniques. Besides, the store has also instituted a consumer education program within the store and over various media channels and includes brochures, short in-store videos, and employee training.

Overview of Different Eco-labeling Standards

One of the main objectives of standardization is usually that everybody adheres to the same standards, *i.e.* the same procedures or product specifications. This eases logistical procedures, facilitate trade, prevent consumer deception and improve quality.

Standards are defined by ISO as documented agreements containing technical specifications or other precise criteria to be used consistently as rules, guidelines or definitions, to ensure that materials, products, processes and services are fit for their purpose.

From this definition it becomes clear that standards are not only used for standardization, but also as "guidelines", *i.e.* for capacity building.

Some of the general claims on labels are, pesticide free, safe and natural, bird friendly Eco-OK, Fair Trade Certified, cruelty free, environmentally friendly, humane treatment 01 animals etc. How exactly these labels guarantee really the claim of the label is not often clear because most often the standards are not meaningful or their label or logo is not consistent in meaning from product to product.

What Makes a Good Eco-Label

1. Attainable but strict standards
2. Should be verifiable
3. Meaningful criteria

4. Requirements that are updated and encourage progress
5. Reputable administration.

Ecolabeling programs vary substantially in terms of the comprehensiveness of their environmental standards. Some ecolabels concern themselves with a single environmental impact within a single stage of the lifecycle of a product like the EPA Energy Star program which focuses only on energy conservation in the use of computer equipment. In contrast, ecolabeling programs like Green Seal and those of many European countries consider multiple environmental impacts throughout the stages of the lifecycle of a product.

Ecolabeling standards based on reducing environmental impacts throughout the full life cycle of a product use a method known as Life Cycle Assessment (LCA). LCA is defined 15 involving four sets of tasks (U.S. EPA 1993c). The first task is to define what institutes the life cycle of a product including extraction of raw materials, manufacturing. distribution, product use, and disposal. The second step involves an inventory of environmentally significant inputs (*e.g.*, energy, water) and outputs (*e.g.*, emissions to air and water, solid waste) throughout the various life cycle stages. The third step is to assess the impacts of environmental inputs and outputs on ecosystems, human health, and natural resource stocks. Of all these steps, this third step is the most controversial because there is still great scientific uncertainty about the fate and effects of various pollutants. The final step is to evaluate options for reducing environmental impacts throughout the product's life cycle.

Product standards are specifications and criteria for the characteristics of products. Process standards are criteria for the way the products are made. Social and environmental standards in agriculture are essentially process standards. Standards are thus rules for production for organic agriculture.

Process standards can be further divided into *management system standards* and *performance standards*. Management systems standards set criteria for management procedures, for example for documentation or for monitoring and evaluation procedures. They do not set criteria for the performance of the management system in terms of what actually happens in the field or the packing station. Performance standards, in contrast, set verifiable requirements for factors such as the non-use of certain pesticides, or the availability of sanitary services.

Setting international standards has proven to be very difficult due to the variety of circumstances that exist around the world. This is especially true for agricultural practices, which have to respond to differences in climate, soils and ecosystems, and are an integral part of cultural diversity. In response to this diversity, international environmental and social standards are often normative standards, *i.e.* generic standards or guidelines to be used as a framework by local standard-setting or certification bodies to formulate more specific standards. It has to be noted that environments' and social standards in agriculture usually do not have the purpose of standardization per se, but are developed to improve environmental and social sustainability in the variety of existing farming and agro-trade systems.

In most ecolabeling programs, private or public, the standard setting process is done by a Standards Committee and is a lengthy process. It involves the following process

Standards Setting Process

1. First a product category is identified by the ecolabeling organization
2. The next step is to develop a description of some or all the stages of a product's life cycle and the kinds of environmental impacts associated with each stage. In practice, it is impossible to examine all impacts, so most programs try to identify those impacts, which differ the most

across different companies' products. Standards are then proposed for reducing these environmental impacts.

3. These standards are made available for public review and comment. The standards are revised to reflect public comment and then finalized. A scientific review panel and an appeals process is also a part of the standard setting process.
4. Periodic review is undertaken to ensure that standards reflect technological progress

Different types of Standards

1. International Standards *e.g.* Codex Alimentarius, IFOAM Basic Standards
2. Regional Standards *e.g.* EU Council's (EEC) N° 2092/9 i
3. National Standards *e.g.* USDA organic standards. Japan Agricultural Standards, National Standards for Organic Production (India)
4. Certification Standards *e.g.* Organic detailed standards of certification agencies

Codex Alimentarius

"Foods described using the term organic are a product of organic farming, which is a system of farm design and management practices that seek to create ecosystems, which achieve sustainable productivity and provide weed and pest control through a diverse mix " of mutually dependent life forms, recycling plant and animal resides, crop selection and rotation, water management, tillage and cultivation. Soil fertility is maintained and enhanced by a system, which optimizes soil biological activity as the means to provide a balanced nutrient supply for plant and animal life as well as to conserve soil resources. Pest and disease management is attained by means of encouragement of a balanced host predator relationship, augmentation of beneficial insect populations, biological and cultural control and mechanical removal of pests and affected plants." Definition by Codex Alimentarius FAO/WHO.

Codex Alimentarius means "food code" and is the compilation of all the Standards, Codes of Practice. Guidelines and Recommendations of the Codex Alimentarius Commission. The Codex Alimentarius Commission (Codex) is the highest international body on food standards. The Commission is a subsidiary body of the UN Food and Agriculture Organization (FAO) and the World Health Organization (WHO) set up in 1961 and open to all countries that are members and associate members of the FAO and the WHO. The Commission currently has 165 member countries, representing more than 98 percent of the world's population. It meets every two years.

The Codex system was set up to protect the health of consumers, ensure fair practices in international food trade and to coordinate all international (bod standards work Codex standards are based on sound science and are accepted as the benchmarks against which national food measures and regulations are evaluated within the Uruguay Round Trade Agreements. One of the principal purposes of Codex is the preparation of food standards. Codex adopts international recommended standards, guidelines and codes of practice after thorough consideration by all Codex member countries. The Codex Alimentarius contains nore than 200 standards. There are general standards or recommendations for: food labeling; food additives; contaminants; methods of analysis and sampling; food hygiene; nutrition and foods for special dietary uses; food import and export inspection and certification systems; residues of veterinary drugs in foods; and pesticide residues in foods. Prevention of food contamination is the best approach to prevent outbreak of food borne illnesses and the Codex

Alimentarius has established international codes of practice on food hygiene and other guidelines for good food product) on and handling practices.

The work of the Codex Commission goes beyond means of removing trade barriers. It also encourages countries to adopt ethical practices. The Code of Ethics for International Trade in Food, for example, calls on parties to stop dumping poor-quality or unsafe food on to international markets.

An increasing number of countries are aligning their national food standards, or parts of them with those of the Codex. This is particularly so in the case of additives, contaminants and residues. The World Trade Organization (WTO) Agreements on the Application of Sanitary and Phytosanitary Measures (SPS) and on Technical Barriers to Trade (TBT) encourage the international harmonization of food standards on the basis of Codex standards.

The IFOAM Norms

The IFOAM Basic Standards provide a framework for certification eco-labeling programmes world wide to develop their own regional and national standards. The "Basic Rules of IFOAM for Organic Agriculture" was first published in mid-1970s. Since then they have been used as guidelines for developing standards taking into account the particular landscape, climate and cultivation techniques of specific regions. The IFOAM Basic Standards for Organic Production and Processing (IBS) and the IFOAM accreditation Criteria for Bodies Certifying Organic Production and Processing (IAC) constitute the IFOAM Norms. The IFOAM Norms are periodically revised and expanded by IFOAM in consultation with a broad group of its members and other stakeholders.

The Norms are the basis for IFOAM's Organic Guarantee System. The relevant IFOAM policies and procedures explain how these Norms are developed, revised, approved, interpreted and how they function as a benchmark for other standards. IFOAM Basic Standards deal with guidelines for conversion to organic agriculture, crop production management, animal husbandry, storage and transportation of products, processing of products, labeling and consumer information. The standards also have guidelines relating to labour welfare, rights of children small farm holders rights and indigenous people's rights.

(PEC) N° 2092/91 on Organic Production

This regulation is one of the international standards that were developed for consumer protection by legally defining and making public the basis of organic food label claims Besides it was considered that it would provide product norms for a unified European market. The standards were drafted using already privately developed normative systems of organic associations. The "principles of Organic Production at Farm Level' correspond to the "Basic Rules for Organic Agriculture" of IFOAM (International Federation of Organic Agricultural Movements).

"Organic production methods constitute a specific form of production at farm level."

"Organic production methods entail significant restrictions on the use of fertilizers and pesticides, which may have detrimental effects on the environment or result in the presence. of residues in agricultural produce". (EEC) N° 2092/91 on Organic Production

Regulations are related to crop production, pest, disease and weed management, animal husbandry, etc. A feature of the principles set out by EEC states that the conversion period for farms should be two years before transition to organic.

United States Department of Agriculture (USDA) Regulation

The regulations prohibit the use of genetic engineering, ionizing radiation, and sewage sludge in organic production and handling. As a general rule, all natural (non-synthetic) substances are allowed in organic production and all synthetic substances are prohibited. The National List of Allowed Synthetic and Prohibited Non-Synthetic Substances, a section in the regulations, contains the specific exceptions to the rule. USDA regulations have the following standards:

Production and handling standards address organic crop production, wild crop harvesting, organic livestock management, and processing and handling of organic agricultural products.

Organic crops are raised without using most conventional pesticides, petroleum-based fertilizers, or sewage sludge-based fertilizers. Animals raised on an organic operation must be fed organic feed and given access to the outdoors. They are given no antibiotics or growth hormones.

Labeling standards, which are based on the percentage of organic ingredients in a product. Products labeled "100 percent organic" must contain only organically produced ingredients. Products labeled "organic" must consist of at least 95 percent organically produced ingredients. Products meeting the requirements for "100 percent organic" and "organic" may display the USDA Organic seal.

Processed products that contain at least 70 percent organic ingredients can use the phrase "made with organic ingredients" and list up to three of the organic ingredients or food groups on the principal display panel, but cannot display the USDA seal on the package.

Processed products that contain less than 70 percent organic ingredients cannot use the term "organic" other than to identify the specific ingredients that are organically produced in the ingredients statement.

Certification standards establish the requirements that organic production and handling operations must meet to become accredited by USDA-accredited certifying agents. The information that an applicant must submit to the certifying agent includes the applicant's organic system plan. This plan describes (among other things) practices and substances used in production; record keeping procedures, and practices to prevent commingling of organic and non-organic products. The certification standards also address on-site inspections.

Farms and handling operations that sell less than $5,000 a year in organic agricultural products are exempt from certification. They may label their products organic \f they abide by the standards, but they cannot display the USDA Organic Seal. Retail operations, such as grocery stores and restaurants, do not have to be certified.

Accreditation standards establish the requirements an applicant must meet in order to become a USDA-accredited certifying agent. The standards are designed to ensure that all organic certifying agents act consistently and impartially Successful applicants will employ experienced personnel, demonstrate their expertise in certifying organic producers and handlers, and prevent conflicts of interest and maintain stnct confidentiality.

JAPAN Agricultural Standards

(Notification No.59 of the Ministry of Agriculture, Forestry and Fisheries of January 20, 2000)

The purposes of this standard are to establish the criteria, etc. of production methods for the organic agricultural products.

Principles of Production of Organic Agricultural Products

The principles of the production of the organic agricultural products are as follows.

1. To sustain and enhance the natural recycling in agriculture, the productivity of the farmland derived from the soil properties shall be generated by avoiding the usage of the chemical synthetic fertilizer and agricultural chemicals, and the organic agricultural products shall be produced in fields adopting such cultivation management method as reducing the load derived from the agricultural production on the environment as much as possible
2. In collection fields (meaning the field for collecting the agricultural products growing spontaneously, being the same hereafter), to collected the agricultural products by such methods as affecting no damage for preserving the ecosystem of the collection fields.

Criteria of Production Methods: Highlights of the criteria of the production methods are as follows:

Conditions of Fields, etc.

1. To clearly divide the field so as to protect it from the drifting fertilizer, soil improvement materials, or agricultural chemicals. In the paddy field, the necessary measures shall be taken to prevent the prohibited substances from contaminating the agricultural water.
2. The criteria of manuring, sowing and planting, controlling noxious animal and plant must be based on the cultivation at least 3 years before the first harvesting of perennial plants (except for the pasture grass) and at least 2 years before the sowing or planting of the other plants than perennial plants (in the case of newly developed field or the field which has not been used for cultivation, prohibited substances must not be used at least 2 years, and these criteria must be based on the cultivation at least one year.
3. The collection field shall be defined as a prescribed section protected from the drift of the prohibited substances from the circumference and utilizing no prohibited substances for 3 years or more before collecting the agricultural products.

Manuring Practice in Fields, etc.

1. The productivity of the farmland shall be preserved and promoted only by applying the compost derived from the remainders of the agricultural products produced in the said fields, etc. and methods effectively utilizing biological functions of the organism inhabiting and growing in the fields or in the circumference (in cases where the productivity of the farmland cannot be preserved and promoted only by the methods utilizing the biological functions of the organism inhabiting and growing in the said fields or in the circumference, utilize only the fertilizers and (he soil improvement materials given as attachment.

Seeds and Seedlings to be Sown or Planted in Fields

1. To utilize seeds and seedlings complied with the criteria of conditions of the fields, etc., the criteria of the manuring practice in the fields, etc., the criteria of the control of noxious animal and plant in the fields, etc., and the criteria of the management concerning the transportation, the selection, the processing, the cleaning, the storage, the packaging, and other processes. This is not applicable to cases of being hard to obtain them in the ordinary means.
2. To be produced without using recombinant DNA technology (meaning technology preparing the recombinant DNA by connecting DNA through the breakage and reunion using enzyme, transferring it into live cells, and proliferating it; being the same hereafter.)

Control of Noxious Animal and Plant in Fields, etc.

1. To be executed only by the cultivation method (to control noxious animal and plant by intentionally executing works generally performed as parts of the selection of crop lists and variety, the adjustment of the cropping time, and other cultivation management of the agricultural products so as to suppress the emergence of noxious animal and plant), physical method (to control noxious animal and plant by methods using light, heal, sound, etc, or manual or mechanical methods), biological method (to control noxious animal and plant by introducing microorganisms suppressing the proliferation of microorganisms being the cause of diseases, predators of noxious animal and plants repelling noxious animal and plant, or plants having effects of suppressing the emergence of noxious animal and plant, or by improving the environment suited for growing them), or an appropriate combination of these methods (in cases of being critical or seriously risky for the agricultural products and being impossible of effectively controlling noxious animal and plant in the fields, etc., only by an appropriate combination of these methods, use the agricultural chemicals noted in the attachment.

Management Concerning Transportation, Selection

1. In the transportation, selection, processing, cleaning, storage, packaging, and other processes, control in such a manner so as the product is not being mixed with other agricultural products than the organic agricultural products.

National Standards for Organic Agriculture (India)

The national standards were developed by the Steering committee for the National Programme for Organic Production. Each section of the document contains general principles; recommendations and specific standards have been set for

Crop Production

1. Conversion period, bio-diversity in crop production, fertilization policy, pest and weed management, contamination control, soil and water conservation, collection of non-cultivated material of plant origin and honey

Animal Husbandry

1. Brought in animals, breeds and breeding, mutilations, animal nutrition, veterinary medicine, transport and slaughter, bee keeping

Food Processing and Handling

1. Pest and disease control, ingredients, additives and processing aids, processing methods, packaging,

Labeling

Storage and Transport

Products that are permitted, restricted and not allowed for use in fertilising and soil conditioning, for plant pest and disease control, feed materials for animal nutrition, products authorized for cleaning and disinfection of livestock buildings, besides approved ingredients and processing aids are given in detail as annexures.

Procedures for Eco-labeling of Farms

Eco-labeling of farms involves detailed procedures for inspection and certification aiming at establishing a mutual partnership between the producer and certifier, besides efficient and objective fact finding, decision making and control of commercial transactions. The procedure is designed for the benefit of the operators involved in organic agriculture as well as that of the consumers.

Steps Towards Certification

The procedure may be divided into the following steps:

1 Application
2. Cost Estimation and contract
3. Inspection
4. Reporting and certification decision-making
5. Final payment and certification
6. Transactions and Follow-up

Application

Applications should be addressed to the certification agency providing the eco-label which would supply all details of the procedures to follow. As first step information about the farm has to be provided to receive a reliable cost estimate.

Contract and Cost Estimate

Based on the information supplied, an inspection contract and a cost estimate would be received from the agency for approval. The inspection program, scopes, time-table, location and the technical details of the procedures are then agreed upon.

Inspection

The agency's experienced inspectors will be appointed to conduct the inspection visit. The inspector may be locally based or may be selected from a team of international inspectors. They will be selected according to the inspection program, the type of operation, language skills or knowledge of standards other than the EC organic rule.

Details such as an exact inspection time table and notification will be arranged by the inspector himself, in accordance with the detailed assignments the inspector receives from the certification agency. The inspection includes introduction of the inspector himself and the inspection program: visits of fields and producers, inspection of premises, stocks and processing units, updating of documents, checking of bookkeeping records and final discussion. Before the end of the visit, the inspector will summarise the deviations found from the standards. Fact finding and verification during inspection are reported and submitted together with all relevant documentation to the certification agency office by the inspector.

Certification and Reporting

The aim of certification is decision-making and reporting. The certification staff will retrieve findings of the inspection report compare them with the requirements as laid down n the EC regulation 2092/91 or additional standards as relevant and make decision according to approved procedures. The concluded deviations, conformity and improvement measures will be clearly stated in the certification protocol which will be sent to the producer together with the inspection report.

Once the producer agrees to this decision, the certification agency is ready to issue a certificate stating the conformity of operation. An appeal may be filed against any decision during the certification process and it should be submitted in writing to the certification agency office.

Payment

Services will be invoiced by the certification agency, 70 per cent of the estimated inspection costs must be paid prior to the assignment of an inspector. The final payment of the balance will be invoiced together with the certification decision and payment is to be received before the issuing of a certificate.

Transactions and Follow-up

Before importation into the EC can take place, importers have to apply for import authorisation to the competent authorities in the country where the importer is based. For this purpose it is necessary that the inspection report needs be submitted to the relevant authorities. The certification agency also helps to present the import application and/or provide additional required documents.

To comply with ongoing quantity checks and legal requirements, transaction certificates will be issued to the importer for each importation of certified commodities into the EC. Based on information supplied by the exporter, the certificate will usually be sent to the importer.

Follow-up of inspection and certification are conducted and it consists of updating relevant information of the project in the agency's files It includes the follow-up of improvements id conditions by the operations to continue compliance with the EC standard.

Case Study: Eco-labeling through ECOCERT International Programme

The Ecocert International inspection plan (Ecocert Third Country Certification) is based on

1. Articles 8 and 9 and relevant provisions of the Council Regulation (EC) 2092/91, as amended, particularly on annex III General Provisions (GP) and Specific Provisions (SP)
2. International standard regulations for agricultural and/or livestock products, foods or non edible goods (Codex Alimentarius, IFOAM, etc.)

National organic legislation for agricultural and/or livestock products, foods or non edible goods (USA, Turkey, Costa Rica, India, etc.)

In context of above parameters the eco-lebelling of farms were conducted during June 2000 to December 2003 for organic and chemical farms. It was seen that respiration rates for both irrigated and non-irrigated organic plots were about five to six times greater than conventional chemical plots. Whereas there was no major change in soil respiration. There was no significant difference of microbial count between irrigated and non-irrigated plots. The organic plots had total count of about 1.7 times more when compared to chemical farms.

The percentage of organic carbon in organic farms was two times higher than chemical farms. The reciprocal Simpson's index for insect diversity showed continuous increase of organic farms. In case of conventional chemical farms, the farmer applies biocides which totally destroys all insect population to zero. Moreover the insect predators are gradually eliminated. But in case of organic plots the number of individuals in almost constant. The average harvest diversity index of organic farms was about 2.5 times than that of the conventional farms. Average multiple cropping index in case of organic have a very gradual increase while in case of chemical farming. There was gradual decrease during the three years of study. However it is difficult to conclude on any particular difference between organic and chemical plots during the three years study (Daniel, 2005).

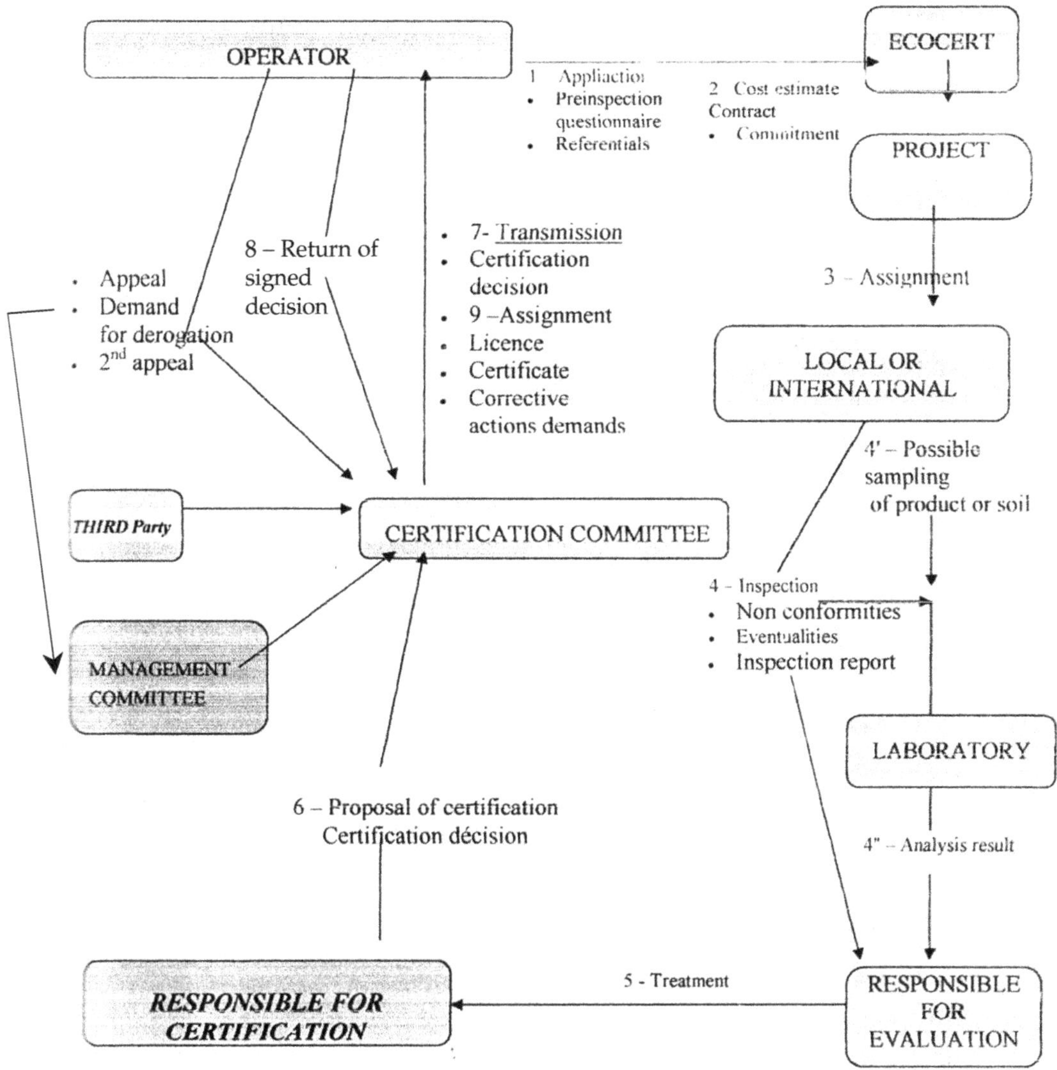

References

Daniel, V.A.S., 2005. Study of imperial relations for eco-labeling of farms. *Ph.D. Thesis*, Dr. B.A. Marathwada University, Aurangabad.

Kuhre, W. Lee, 1997. *ISO 14020s: Environmental Labeling-Marketing*. Upper Saddle River, NJ: Prentice Hall.

U.S. Environmental Protection Agency, Office of Pollution Prevention and Toxics 1993a. Evaluation of Environmental Marketing Terms in the United States. Washington, D.C. EPA 741-R-94-003, February, 1993a.

Sustainable Environmental Management *Pages 52–63*
Edited by: **Dr. L.V. Gangawane & Dr. V.C. Khilare**
Published by: **DAYA PUBLISHING HOUSE**

Chapter 8

Impact of Distillery Effluents on Water and Soil Quality

Jayashree Deshpande[1] and V.S. Lomte[2]

[1]*PG Department of Environmental Science, SBES College of Science, Aurangabad – 431 001, MS, India*
[2]*Department of Environmental Science, Dr. B. A. Marathwada University, Aurangabad – 431 004, MS, India*

ABSTRACT

There are nearly 166 distilleries with total production of 500 million liters of alcohol per year. However, the effluents produced cause different environmental effects. Therefore, impacts of these effluents on various environmental parameters such as soil, water and microbes have been reviewed. In addition it is proved that nitrogen-fixing microorganisms are very useful for the treatment of effluents to maintain a quality environment.

Keywords: *Distillery effluents, Water pollution, Soil pollution, Azotobacter chrooccocum.*

Introduction

There are nearly 166 distilleries with a total production of 500 million litres of alcohol per year. It is also estimated that about 12 to 14 litres of waste water is released for every litre of alcohol. The distillery waste water is also known as spent wash, spillage or vinase. These are from the molasses based distillery. The population equivalent of distillery waste water in India is nearly 7 times more than the entire population. Most of the distilleries are not treating waste water because of non-availability of a sure process under Indian conditions (Handa and Seth, 1990).

Farooqui *et al.* (1998) studied the characteristics of granular sludge developed on cane sugar mill waste, including the effect of temperature on methanogenic activity of the sludge granules, yield coefficient of methanogenes, adaptation of the sludge to different wastes, the effect of unfed storage of granular sludge at room temperature (25–30°C) on its methanogenic activity and physical characterization of the sludge.

Sludge from UASB reactors operating at three different distilleries were analysed for total solids, volatile suspended solids, polymers, sedimentation characteristics, methanogenic activity and methanogenic population. The sludge from Associated Distillery, Hissar showed highest methanogenic activity even though its volatile suspended solid content was lowest (49.9g L^{-1}) while lowest activity was found in sludge from Panipat Distillery having highest volatile suspended solid contents (97.5 g L^{-1}). Acetoclastic methanogenic population was also higher in the sludge from Hissar distillery. The sedimentation characteristics of the sludge were dependent on their proximate composition (Sharma and Singh, 2000).

Recent studies by Senthil Kumar *et al.* (2001) on pollution studies of sugar mill effluent showed that most of the physico-chemical parameters like colour, odour, total solids, COD, BOD, alkalinity and fluoride were found to exceed than the ISI prescribed permissible values while pH, phosphate and sulphate were found within permissible limits. The concentrations of toxic metals such as Cd, Cu, Fe, Hg, Mn, Ni, Pb, and Zn were also determined by inducting plasma emission spectrophotometry (ICP–AIS). The results showed that Fe, Hg, Mg, Mn and Pb content were higher than the permissible limits.

Distillery Wastes and Water Quality

Effect of distillery effluents on river Wainganga has been studied in detail by Chauhan (1991). The study was carried out before and after closure of factory. Addition of effluent to the river caused toxic condition by increasing BOD, COD and TSS alongwith decrease in DO. This was due to the addition of distillery effluent at high concentration. Chauhan (1991) again studied the impact of distillery wastes on aquatic environments. pH of the distillery effluents were always highly acidic, BOD increased upto 836 per cent in the river. Due to heavy organic loading, only tolerant species of organisms survived. Diatoms with 54 species contributed 69–97 per cent to the total spp. *Navicula pygmea* was recorded between 2.49 to 3.24 per cent. Ionic concentration showed lasting effect on the river after closure of distillery. Ammonical nitrogen rapidly assimilated with optimum growth of algae in October, November and December. Abou-EL-Naga *et al.* (1999) made the survey in South Cairo to evaluate the effect of industrial activities on chemical pollution of irrigation water. Concentration of contaminants varied depending on several factors, which are distant from the polluting sources near the industrial complex. Wang *et al.* (1999) showed that sugar mill waste water did not pollute the ground water.

Distillery Wastes and Soil Quality

Some literature is available on the effect of distillery waste on soil characteristics. Kaushik *et al.* (1997) studied the effect on the enzymatic changes in salt affected soils following sugar mill effluent irrigation. The effluents stimulated dehydrogenase and invertase activity but there was no significant change in the enzyme urease. Long term irrigation of soils for 12 years caused more stimulation of dehydrogenase and invertase activities as compared to short term irrigation.

Lugo Lopez *et al.* (1981) noted that use of industrial residue can control the soil erosion. They found that there was marked response to low organic matter sources like sugarcane trash and filtered press cake. In addition, their laboratory study also indicated that an excellent stable soil structure developed after treatment with residues like black strap molasses and rum distillery slops.

Velloso *et al.* (1982) applied the distillery residues to the soil and studied the leachates. It was noted that the nitrate leachate decreased while the ammonium leaching increased with increasing levels of residues applied. Leaching of nitrate was about 30 ppm less and leaching of ammonium about 4 ppm, which was more than the control.

The potential utilization of wastes from red wine distilleries as crop fertilizer was studied in pots. Application of fresh sludge resulted in a brief period of nitrate immobilization in the soil, which was less marked with mature sludge. This matured sludge also increased dry matter yields of *Lolium multiflorum* cv. *tetrone* (Morisot, 1986). Sweeney and Graetz (1988) noted the chemical and decomposition characteristics of anaerobic digester effluents applied to soil. They observed that the molasses residue effluent sample had a greater K concentration on an oven dry basis than swine or bovine waste material. Similarly, Leon *et al.* (1989) recorded that distillery effluent had a high salinity and can be applied for irrigation in very permeable soils if mixed with water of low salinity where the previous crop and chemical characteristics of the soil are known. Stehlic (1986) also noted that distillery wastewater gave bad results even if diluted with 5 parts clean water. Vantour *et al.* (1987) used the effluents from alcohol factories as amendments for saline soil. It was seen that the application of these effluents reduced the total phosphorous content. Somwanshi and Yadav (1990) also noted the effect of spent wash on soil chemical properties and composition of leachates. It was noted that addition of concentrated spent wash results in the increase of salinity of both, soil and ground water. Later, Chapman *et al.* (1995) studied the removal of soluble organic carbon from winery and distillery wastewater by application to soil. They proposed a logistic model to remove water-soluble carbon from synthetic winery and distillery waste water containing 14 C–labelled lactic acid and glycerol after their application to the top soils of a brown earth. It was noted that oxidative microbial decay was responsible for the removal of 14 C of the solution. Correlations between some chemical properties of soil were studied in sugarcane soils irrigated with distillery effluents by Machado-de-Armas *et al.* (1994). They found that pH was strongly correlated with Mg and Na cations. Thus, the irrigation of distillery effluents increased soil Mg and Na contents. Pande *et al.* (1995) carried out studies on the use of distillery waste as fertilizer. They found that techniques applicable to vinesse increased the yield of sugar beet, potato and other vegetables by 20 per cent but had adverse effects on legumes and no effect on rice.

Qzair Aziz *et al.* (1994) conducted an experiment on the impact of refinery effluents at the field level. Soil irrigated with effluent showed no significant change in pH, total organic carbon, calcium, water-soluble salts, cation exchange capacity and SAR.

Wang *et al.* (1999) also showed that waste water from sugar mills in China do not pollute the soils at the rate of 100 litre/ha. In general they found that the waste water was highly suitable for irrigation of sugarcane. Dhaliwal and Singh (2000) indicated that distillery effluent caused lowest surface tension, higher viscosity and density than the ordinary water. But due to the acidic nature, CD could not be used for irrigation in the fields, whereas SW was not found to have any harmful effects on its flow behaviour in soils.

Suresh Chandra *et al.* (2002) studied the effect of potassium salts and distillery effluents on carbon mineralization in soil. Addition of effluents with salts stimulated C–mineralization but a decline was noticed at higher concentrations. Bhat *et al.* (1997) noted the effect of distillery effluents on soil properties. They noted accumulation of soluble C1 and SO_4 occurred in the soil in all treatment. Gypsum reduced the accumulation of C1 and SO_4 when compared with other treatments.

Distillery Waste and Soil Organisms

Juwarkar and Dutta (1990) showed that population of bacteria, fungi, actinomycetes and N–fixing bacteria were decreased due to the treatment of diluted and raw distillery waste. The growth of *Rhizobium* and *Azotobacter* were also reduced after raw waste water application. Interestingly, distillery effluents had no effect on *Trichoderma harzianum*. Pre-treated seed with *Macrophomina phaseolina* and a fungal suspension of *Trichoderma harzianum* indicated the DSI was affected by the addition of effluents. Hence, it was concluded that *Trichoderma herzianum* can be used for the biological control of *Macrophomina phaseolina* even in the presence of distillery effluents. Kumar *et al.* (1990) studied the growth of six unidentified bacterial strains from distillery waste at temperature 30–45°C and pH 3–10. All these strains showed good growth at pH 7. The gradual decrease of COD was also observed (15000 mg/L). Goyal *et al.* (1995) applied distillery waste water to a field soil and found increased soil microbial biomass antihydrogenase activity. Smith (1997) found that soil samples contaminated with Cu and Zn from past applications of pig slurry and with Cu due to treatment with whisky distillery waste were assessed. *Rhizobium* effective in N–fixation was present in all soils supporting the host plant irrespective of metal concentration in soil which increased to 300 mg Cu/kg and 2000 mg Zn/kg. The work demonstrated that nodulation and effective N–fixation by white clover occurred in sludge treated soils above the current UK maximum permissible concentration of heavy metals. Goyal *et al.* (2002) observed soil microbial properties and growth of Mungbean grown in the soil treated with distillery spent wash. The amount of microbial biomass in untreated soil declined slowly over 90 days period. On the other hand, the microbial biomass increased significantly with increasing rate of spent wash amendment and reached maximum peak after 30 days. Thus, it was concluded that application of spent wash to soil improves soil microbial properties and plant growth up to a certain level but repeated application may lead to soil salinity and toxicity to plants.

Patil (1998) studied the occurrence and tolerance of algae in distillery wastewater. He showed that algal diversity along with the physico-chemical characteristics of wastewater from pond and lagoons of distillery spent wash, located around Pravaranagar distillery varied. Eleven taxa belonging to Cyanophyceae, Euglenophyceae, Chlorophyceae and Bacillariophyceae were recorded. The genus *Oscillatoria* could tolerate upto 60 per cent spent wash. Kothari *et al.* (1997) noted the effect of petroeffluent on mycorrhiza in soil. The rhizosphere of six crops showed efficient mycorrhization with vesicular arbuscular mycorrhiza (VAM). The rhizosphere was rich in VAM species like *Aculospora appendicula, A. bereticulata, A. foveata, A. sporcaria, Glomus aggregatum, G. deserticola, G. heterosporum, G. microcarpum* and other unidentified VAM species.

Effect on the earthworm *Megascolex pumilio* was studied by Kailasam *et al.* (2001). Addition of distillery effluent was found to change the soil quality like pH, N, K, BOD, COD and TDS. However, the effluent was found to be toxic to this earthworm. This suggests that the improvement of the quality of effluent should be done by effective treatment.

Distillery Waste and Plants

There are some reports on the application of spent wash, spent slurry and press mud composts on maize. Patil and Shinde (1995) used these at the rate of 1.5 per cent for maize cv African Tall and noted that dry matter yield was significantly increased by FYM, spent slurry (3:1 and 5:1) and spent wash (5:1) compost. However, the treatment supplied with spent wash liquid only, DM yield/pot was significantly decreased. Similarly the effect of distillery effluents on some physiological aspects in maize was studied by Ramana *et al.* (2002). The field experiment was conducted in India for two years. They used raw spent wash (RSW), biomethanated spent wash (BSW), lagoon sludge (LS), recommended

NPK + FYM and control. It was found that application of distillery effluents increased leaf area, chlorophyll content, nitrate reductase activity, total dry–weight and grain yield. In addition application of distillery effluent on maize crop and soil properties was done by Singh and Raj Bahadur (1998). The effluent having BOD of 4620 mg/litre was applied for pre- sowing irrigation. Twelve presowing irrigations with this effluent had no adverse effect on the germination of maize but improved the growth and yield. The pH and electrical conductivity of the soil was slightly increased. Moreover, soil organic carbon, nitrogen, P and K content also increased significantly with the increase in the number of pre sowing effluent irrigation. At the same time the infiltration rate and the bulk density of soil were considerably reduced.

Effect of distillery waste water on nodulation and nutritional quality of groundnut was studied by Juwarkar *et al.* (1990). They showed that *Rhizobium* population and protein and fat content of groundnut were adversely affected by the irrigation with distillery waste water. Field study of the crop productivity on agricultural soils with reference to refinery effluent treated soils was studied by Qzair *et al.* (1994). They indicated that the effluent increased all the growth and yield parameters of 3 different wheat varieties. This was due to the improvement of physico-chemical properties of soils. The existence of *Lens culinaris* L. under toxic effect of distillery effluent was observed under various concentrations by Verma and Verma (1995). The optimal concentrations of effluents were found to be beneficial for better growth of this crop. Disposal of large quantity of distillery effluents by different methods to irrigate the crops was studied by Rajannan *et al.* (1998). It was noted that the effluents have high nutritional values because of the presence of organic matter as compost. The possibility of saving P and K fertilizers is also assumed with the improvement of other soil properties including soil enzymes. Further, the impact of various concentrations of biologically treated distillery effluents on the growth of *Vigna radiata* was studied by Subramani *et al.* (1999). Simultaneously, *Ceratophyllum demersum* was also grown in distillery effluent for 5 days. The treated effluent was used for irrigation of *Vigna radiata* at various concentrations. *Ceratophyllum* was found to be highly capable of removing pollutants from the effluent and therefore, the crop *V. radiata* was found to give more yield. Interestingly, Dhaliwal and Singh (2000) studied the flow behaviour of effluent in soil. It was seen that due to acidic nature of soil, CD could not be used for irrigation in the fields whereas sewage water was not found to have any harmful effect on the flow behaviour of soil. Recent studies by Misra and Pandey (2002) showed that seeds of *Cicer arietinum* exposed to the distillery effluent (10–15 per cent) and concentration of leachates of flash light factory sludge (5 per cent) was found to be beneficial for the growth of root and shoot. However, 100 per cent concentration was inhibitory.

Some studies on the germination of various crop plants in the distillery effluent are available. Shinde *et al.* (1988) found that distillery effluents were inhibitory for the germination of gram (*Cicer arietinum*) at the concentration more than 50 per cent. Rani and Shrivastava (1990) noted the significant reduction in the germination of *Pisum sativum* indicating harmfulness of the effluent. Subramani *et al.* (1995) found that water hyacinth was more efficient in removing harmful effects of effluents and the effluent was found to be stimulatory for germination and seedling growth of green gram (*Vigna radiata*).

Reports on the effect of distillery effluent on seed germination in some vegetable crops are also available. Ramana *et al.* (2002) studied effect of different distillery effluent concentrations (5–100 per cent) on seed germination, speed of germination, peak values and germination value in tomato, chilli, bottle gourd, cucumber and onion. The distillery effluents did not show any inhibitory effect on germination of tomato at low concentration. On the other hand at higher concentrations (75–100 per cent) complete failure of germination was observed in all the vegetable crops used. Therefore, it is suggested that the effect of distillery effluent is also crop specific and due care should be taken before using distillery effluent for pre-sowing irrigation purposes.

Distillery Wastewater Treatment

It is essential to release any industrial effluent only after treatment for controlling pollution. These are physico-chemical and biological treatment processes. Environmentalists have done some researches on the treatment of different industrial wastes and many industries are using these methods to release the treated effluents in the environment. Juwarkar *et al.* (1987) mixed the paper mill waste water with chlorination and hypochlorite waste water at various proportion and found that SAR of waste water was reduced from 18.8 to 5.3. This treated wastewater was used to irrigate soil samples. All the soil samples equilibrated with PMW and crossed the ESP limit alkali soil. Similarly soils of loamy sand, sandy loam, sandy clay loam and loam texture with good drainage did not exhibit salinity problem on the water had electrical conductivity of 2000 to 3000 µmhos/cm. Saxena *et al.* (1986) treated the tannery effluents with chlorides and carbonates of sodium, ammonia, calcium, sodium dichromate, myrobalam extract and sulphuric acid. These treated effluents were used for irrigation purposes in dry areas. The effluents stimulated the growth of many crop plants but also had toxic chemicals, which have retarded the germination.

Kumar *et al.* (1997) noted that anaerobically digested molasses spent wash is a dark brown recalcitrant effluent which has high COD and high pollution potential. Bacterial enrichment cultures were set up to obtain isolates capable of distillery effluents decolourisation and bioremediation. Two-gram negative aerobic bacterial cultures isolated from soil samples were collected at Associated Distillery Ltd., Hissar, India. Both the cultures grew well on medium containing distillery waste supplemented with glucose as readily available C source. These bacteria gave maximum decolourisation (36.5 per cent and 32.5 per cent) and COD decrease (41.0 per cent and 39.0 per cent) after 8 days. Again, Kumar *et al.* (1997) studied the soil samples from the same distillery containing cane molasses. These were incubated with diluted anaerobically digested spent wash and a facultative anaerobic pure bacterial culture. The bacterial culture L2 achieved 31 per cent decolourisation and 56 per cent COD decrease in 7 days. By this, they showed that such culture of bioremediation of digested spent wash offers a realistic approach for decreasing the pollutant potential of the recalcitrant waste before disposal.

Reboll *et al.* (2000) studied the use of sewage waste water on *Citrus* trees irrigation during three different seasons. Many biochemical changes in *Citrus* leaves were observed when compared with the irrigation of normal ground water but the soil elements were within the optimal ranges. Rohella *et al.* (2001) studied the removal of colour, turbidity and COD from pulp and paper mill effluents using polyelectrolytes for pollution control and its safe disposal. The effluent was treated with 0.2 ml/L of Rishlyto 80 L for the removal of colour, turbidity and COD. It was noted that this method is techno-economically viable. Decolourisation of textile effluent was tried by Sharma and Arora (2001), as many dyes and pigments are used in processing of textiles. Many of these dyes are potential hazards and some of them are even carcinogenic. They used the methods such as coagulation, ion exchange, membrane filtration and surface adsorption. These methods reduced BOD, COD, foul odour and taste.

Biological treatment of industrial effluents has been observed to be more useful in removal of colour, oil etc. There are some reports available on this aspect. Ramana *et al.* (2000) used *Acetobacter xylinum* for the production of cellulase membrane in the wastewater. The carbon and nitrogen sources were added in the wastewater. The bacterium was able to utilize a wide range of protein and nitrogen sources. Sodium dodecyl sulphate polyacrylamide gel electrophoresis (SDS–PAGE) analysis of pellicle proteins revealed electrophoretic bands of molecular masses in the range of 116–20 Kda. Furthermore, the strain can be useful for removal of various nitrogenous and carbon substrates present in waste water. Srinivasan and Murthy (2000) used *Trematis versicolor* for removal of the waste water colour. This organism is a white rot fungus and has the ability to degrade lignin. The fungus can produce non-specific enzyme system, which oxidizes the recalcitrant compounds present in the effluent. The fungus could decolourise the effluent from pulp plant and dyeing industry. Kouser and Singara Charya (2002) applied four fungi (*Aspergillus niger, Curvularia lunata, Fusarium oxysporum* and *Mucor mucedo*) isolated from textile and dye contaminated soils for colour removal of textile and dye wastewater. The physical, chemical and biological characteristics were analysed and related to the process of decolorization. *Aspergillus niger* and *Mucor mucedo* were resistant in the soils and also efficient in the decolorization and enzyme production. Other two fungi were not so efficient in colour removal and also enzyme secretion. Chauhan and Thakur (2002) treated the pulp and paper mill effluent by *Pseudomonas fluoroscens.* This bacterium degraded 4 chlorophenol through an oxidative route as indicated by oathring cleavage. The strain produced significant reduction in colour (75 per cent), phenol (66 per cent), chemical oxygen demand (79 per cent) and lignin (45 per cent). However, there was increase in chloride content (92 per cent) after 15 days in bioreactor. Similar observations were made by Jeyaramraaja *et al.* (2001) when they used *Aspergillus fumigatus* to declorize the paper mill effluent in a bioreactor.

Researches on mechanical and chemical treatment of distillery effluents can be revealed as below. Gadre and Godbole (1986) used two bench scale up flow anaerobic filters for anaerobic digestion of waste. The results showed that there was substantial reduction in pollution load with simultaneous production of biogas. Essentially the in-built buffering capacity of anaerobic filter was more useful. Routh and Dhaneshwar (1986) also showed that anaerobic treatment of spent wash in immobilized cell reactor was more efficient than conventional methods. The advantage was first acidification and second methanogenesis phase. Ramteke *et al.* (1989) used pyrochar from paper mill sludge for removal of colour of distillery effluent. They observed that granular and powdered activated pyrochar which can remove colour and COD. Berchmans and Vijayvalli (1989) used electro-chemical oxidation technique for removal of BOD and decolourisation. For this, performance of hybrid anaerobic reactor using distillery effluent was assessed by Bardiya *et al.* (1995). The process involved was diphasic in nature with hybrid reactor design for methane digester. It was also seen that optimum biogas production of 10 kg/m^3/day in hybrid reactor. Patil and Kapdnis (1995) studied the decolourisation of spent wash pigment melanodin by chemical and biological methods. Spent wash from an anaerobic digester was treated with hydrogen peroxide, calcium oxide, and soil bacteria, at 144 hours of incubation with treatment at varied concentration of hydrogen peroxide, the maximum decolourisation and COD reduction was 98.67 and 88.40 per cent respectively. Vaidyanthan *et al.* (1995) made an attempt to determine the bio-kinetic coefficients for a two stage anaerobic digester in a bench scale model using distillery spent wash and application of these coefficients for the design of digester. The distillery spent wash treatment by anaerobic filter has also been studied by Ilyas *et al.* (1998). They concluded that conservation of fresh water and preservation of environment, imparting greater efficiency and stability to the reactor. Handa and Seth (1990) showed that among the various treatment alternatives, anaerobic digestion appeared to be most attractive and produces a needful by-product biogas.

Daryapurkar and Chakrapani (1999) showed that yeast sludge is most important waste stream generated from alcohol distillery for which no useful applications are still practiced in India. Bhagatkar (1999) studied the aerobic bio-oxidation of post anaerobic distillery effluents. The study evaluated the biological kinetic parameters like K, KS, Y, Kd, Umax for an activated sludge system using post

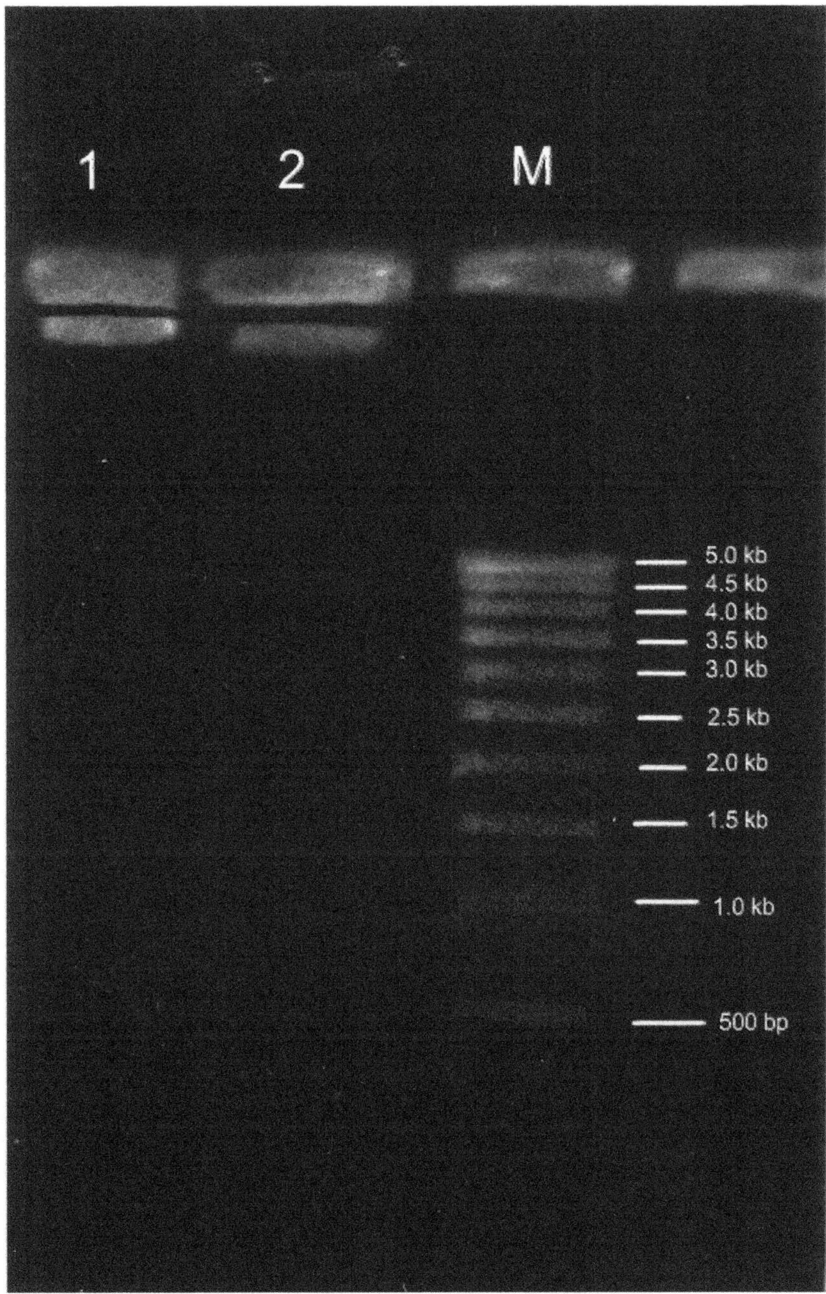

1: Azo 1 isolate; 2: Azo 7 isolate; M: Marker

Figure 8.1: Plasmid Profile of *Azotobacter chroococcum* Isolates of Distillery Effluent Treatment and Colour Removcal Process

distillery effluents. A bench scale continuous reactor of 6-9 liters was also developed. Akunna and Clark (2000) reported the performance of granular bed anaerobic baffled reactor used in the treatment of whisky distillery waste water. It was found to be very effective. It also combined the advantages of baffled reactor system and up-flow of anaerobic sludge blanket systems. Acidogens were mostly non-granular while methanogenes were granular. Recent studied by Mali (2002) showed aerobic composting of spent wash. The composting process is completed within 20–22 days while press mud and spent wash absorption ratio is 1: 1.7 for fresh press mud.

During this biotechnological era, biological treatment of wastewaters is gaining much importance. Seenappa and Rao (1995) used earthworm *Eudrilus eugeniae* for conversion of distillery waste into organic manure. Similarly Joshi and Kapadnis (1997) gave fungal and microbial biomass treatment to distillery waste. This helped in preparation of compost fertilizer. The fungal biomass of *Aspergillus* sp. and *Phialomyces* was developed on spent wash to remove the colour. The decolourised spent wash was then enriched with nitrogen content by growing mixed culture of *Azotobacter* sp. This compost fertilizer is used as other agro based waste products like press mud. A good compost fertilizer was obtained within 45 days. Interestingly Singh and Srivastava (1999) removed the basic dyes from aqueous solutions by chemically treated *Psidium guyava* leaves. The leaves were treated with formaldehyde and sulphuric acid. 2g/L leaves were sufficient for removal of methylene blue. Gonzalez *et al.* (2000) used white rot fungus, *Trametes* sp. to detoxify the distillery wastewater. Use of thermotolerant yeasts, *Saccharomyces cerevisae* and *Kluyveromyces marxianus* has been also made (Abdel Fattah *et al.* (2000). Inthorn *et al.* (2001) treated molasses pigment solution with *Chlorella saccharophila* and *C. vulgaris* and removed the colour to 30–33 per cent.

Removal of the effluent colour is very important aspect as major quantity is discharged into nearby rivers and river water also becomes coloured. In order to remove colour of distillery effluent two isolates of *Azotobacter chroococcum* were inoculated in the effluent in the conical flask. Surprisingly, the un-inoculated effluent showed high absorbance, while there is drastic reduction in the absorbance of inoculated effluent. One could also observe visually the clear colour in the effluents due to treatment of effluent with Azotobacter isolates which are also N– fixers. Efforts were made to find out the gene on the plasmids. The isolates were cured with acridine orange. But it showed that there was not much variation in absorbance of effluent between cured and uncured strains Figure 8.1. This showed that curing eliminated the smaller plasmid, and the gene may be present on larger chromosome (Deshpande, 2004).

References

Abdel-Fattah, W.R., Fadil, M., Nigam, P. and Banat, I.M., 2000. *Biotechnology and Bioengineering*, 68(5): 531–535.

Abou-El-Naga, S.A., El-Shinnawi, M.M., El-Swaaby, M.S. and Salem, M.A., 1999. *Egyptian Journal of Soil Science*, 39(2): 263–280.

Akunna, J.C. and Clark, M., 2000. *Bioresource Technology*, 74(3): 257–261.

Baradiya, M.C., Hashiya, R. and Chandna, S., 1995. *Journal of Association of Environment Management*, 22: 237–239.

Berchmans, L. and Vijayvalli, R., 1989. *Indian Journal of Environmental Health*, 31(4): 309.

Bhagatkar, M., 1999. *Journal of Indian Association of Environment Management*, 26(3): 177–182.

Bhat, S.N., Doddamani, V.S. and Lamani, B.B., 1997. *Karnatak Journal of Agricultural Sciences*, 10 (4): 982–986.

Chapman, J.A., Correll, R.L. and Ladd, J.N. 1995. *Australian Journal of Grape and Wine Research*. 1(1): 39–47.

Chauhan, A. 1991. *Indian Journal of Environmental Health*. 33(2): 203.

Chauhan, N. and Thakur, I.S., 2002. *Pollution Research*, 21(4): 429–434.

Daryapurkar, R. and Chakrapani, D., 1999. *Journal of Indian Association of Environment Management*, 26(3): 141–149.

Dhaliwal, R.S. and Singh, K., 2000. *Communications in Soil Science and Plant Analysis*, 31(5–6): 605–614.

Deshpande, J., 2003. *Ph.D. Thesis*, Dr. B.A. Marathwada University, Aurangabad.

Farooqui, I.H., Khursheed, A. and Siddiqui, R.H., 1998. *Indian Journal of Environmental Health*, 40(1): 58–66.

Gadre, R.V. and Godbole, S.H., 1986. *Indian Journal of Environmental Health*, 28(1): 54–59.

Gonzalez, T., Terron,M.C., Yague, S., Zapico, E., Galletti, G.C. and Gozalez, A.E., 2000. *Rapid Communications in Mass Spectometry*, 14(15): 1417–1424.

Goyal, S., Chander, K. and Kapoor, K.K., 1995. *Environment and Ecology*, 13(1): 89–93.

Goyal, S., Kapoor, K.K. and Chander, K., 2002. *Environment and Ecology*, 20(3): 640–642.

Handa, B.K. and Seth, R., 1990. *Journal of Indian Association of Environment Management*, 17: 44–54.

Ilyas, M., Isa, M.H., Farooqui, I.H., Khursheed, A. and Siddiqui, R.H., 1998. *Indian Journal of Environmental Health*, 40(2): 153–159.

Inthorn, D., Silapanuntakul, S. and Chanchitprecha, C., 2001. *Pollution Research*, 20(4): 691–696.

Jeyaramraja, P.R., Anthony, T., Rajendran, A. and Rajkumar, K., 2001. *Pollution Research*, 20(3): 309–312.

Joshi, R.D. and Kapadnis, B.P., 1997. *National Seminar on Biomass Productivity and Utilization* (6–8 Feb. 1997).

Juwarkar, A., and Dutta, S.A., 1990. *Environmental Monitoring and Assessment*, 15(2): 201–210.

Juwarkar, A.S., Bhalkar, D.V. and Subrahmanyam, P.V.R., 1987. *Indian Journal of Environmental Health*, 29(4): 313–321.

Juwarkar, A., Dutta, S.A. and Pandey, R.A., 1990. *Journal of Indian Association of Environmental Management*, 17: 24–25.

Kailasam, N., Subramaniam and Selvam, M., 2001. *Journal of Ecotoxicology, Environment Monitoring*, 12(1): 21–25.

Kaushik, A., Kadyan, B.R., Manchanda, H. and Kaushik, C.P., 1997. *Ecology, Environment and Conservation*, 3(2): 107–109.

Kothari, I.L., Joseph, U. and Patel, C.R., 1997. *Journal of Mycology and Plant Pathology*, 27(3): 275–278.

Kousar, N. and Singara Charya, M.A., 2002. *Indian Journal of Environmental Health*, 44(1): 65–70.

Kumar, S., Shukla, G.L. and Agarwal, P.K., 1990. Technical papers of the Fortieth Annual Convention of the Deccan Sugar Technologists' Association, Part 1, B 65–76.

Kumar, V., Wati, L., Nigam, P., Banat, I.M., McMullan, G., Singh, D. and Marchant, R., 1997. *Microbios*. 89(359): 81–90.

Leon, M., Berdan, I., Traviesco, L. and Sanchez, E., 1989. *Biotechnology Letters*, 11(3): 217–218.

Lugo-Lopez, M.A., Abruna, F. and Perez-Escolar, R. 1981. *Bulletin, Agricultural Experiment Station*, University of Puerto Rico, 226: 26.

Machado-de-Armas, J., Arzola-Pina, N. and Vega -Herrera F., 1994. *Centro-Azucar*, 21(1): 30–39.

Mali, D.S., 2002. *Journal of Ecotoxicology, Environment Monitoring*, 12(2): 101–104.

Misra, V. and Pandey, S., 2002. *Pollution Research*, 21(4): 461–467.

Morisot, A., 1986. *Agronomie*, 6(2): 203–212.

Patil, G.D. and Shinde, B.N., 1995. *Journal of the Indian Society of Soil Science*, 43(4): 700–702.

Patil, N.B. and Kapadnis, B.P., 1995. *Indian Journal of Environmental Health*, 37(2): 84–87.

Patil, S.S., 1998. *Journal of Indian Pollution Control*, 14(2): 89–93.

Qzair, A., Arif Inam and Siddiqui, R.H., 1994. *Indian Journal of Environmental Health*, 36(2): 91–98.

Rajannan, G., Parvinbanu, K.S. and Ramaswami, P.P., 1998. *Indian Journal of Environmental Health*, 40(3): 289–294.

Ramana, S., Biswas, A.K.,Kundu, S., Saha, J.K. and Yadava, R.B., 2002. *Bioresource Technology*, 82(3): 273–275.

Ramana, K.V., Tomar, A. and Singh, L., 2000. *World Journal of Microbiology and Biotechnology*, 16(3): 245–248.

Ramteke, D.S., Wate, S.R. and Moghe, C.A., 1989. *Indian Journal of Environmental Health*, 31(1): 17–24.

Rani, R. and Srivastava, M.M., 1990. *International Journal of Ecology and Environmental Sciences*, 16(2–3): 125–132.

Reboll, V., Cerezo, M., Roig, A., Flors, V., Lapena, L. and Garcia, P., 2000. *Journal of the Science of Food and Agriculture*, 80(10): 1441–1446.

Rohella, R.S., Choudhary, S., Manthan, M. and Murty, J.S., 2001. Removal *Indian Journal of Environmental Health*, 43(4): 159–163.

Routh, T. and Dhaneshwar, R.S., 1986. *Ind. J. Environ. Hlth.*, 29(2): 105–117.

Saxena, R.M., Kewal, P.F., Yadav, R.S. and Bhatnagar, A.K., 1986. *Ind. J. Environ. Hlth.*, 28(4): 345–348.

Seenappa, C. and Jagannath Rao, C.B., 1995. *Journal of India Association of Environment Management*, 22: 244–246.

Senthil Kumar, R.D., Narayanswamy, R. and Ramakrishnan, K., 2001. *Pollution Research*, 20(1): 93–97.

Sharma, J. and Singh, R., 2000. *Indian Journal of Microbiology*, 40: 203–205.

Sharma, J.K. and Arora, M.K., 2001. *Pollution Research*, 20(3): 453–457.

Shinde, D.S., Trivedy, R.K. and Khatavkar, S.D., 1988. *Pollution Research*, 7(3–4): 117–122.

Singh, D.K. and Srivastava, B., 1999. *Indian Journal of Environmental Health*, 41(4): 333–345.

Singh, Y. and Raj Bahadur, 1998. *Indian Journal of Agricultural Sciences*. 68(2): 70–74.

Smith, S.R. 1997. *Soil Biology and Biochemistry*, 29(9–10). 1475–1489.

Somawanshi, R.B. and Yadav, A.M., 1990. *Agricultural papers of the Fortieth Annual Convention of the Deccan Sugar Technologists' Association,* No. 1: 101–108.

Srinivasan, S.V. and Murthy, D.V.S., 2000. *Journal of Indian Association of Environment Management.* 27(3): 260–264.

Stehlik, K., 1986. *Scientia Agriculturae Bohemoslovaca,* 18(4): 291–306.

Subramani, A., Sundarmoorthy, P. and Lakshmanachary, A.S., 1995. *Pollution Research,* 14(1): 37–41.

Subramani, A., Sundramoorthy, P., Saravanan, S., Selvaraju, M. and Lakshmanachary, A.S., 1999. *Journal of Industrial Pollution Control,* 15(2): 281–286.

Suresh Chandra., Joshi, H.C., Pathak, H., Jain, M.C. and Kalra, N., 2002. *Bioresource Tecnology,* 83(3): 255–257.

Sweeney, D.W. and Graetz, D.A., 1988. *Journal of Environmental Quality,* 17(2): 309–313.

Sweeney, D.W. and Graetz, D.A., 1991. *Agriculture, Ecosystems and Environment,* 33(4): 341–351.

Vaidyanathan, R., Meenambal, T. and K. Gokuldas, 1995. *Indian Journal of Environmental Health,* 37(4): 237–242.

Vantour, A., Mena, M., Martinez-Cruz A. and Martinez, V., 1987. *Ciencias-de-la-Agricultura,* 30: 92–101.

Velloso, A.C.X., Nunes, M.R. and Leal, J.R., 1982. *Pesquisa Agropecuaria Brasileira,* 17(1): 51–55.

Verma, A. and Verma, A.P., 1995. *Journal of Indian Pollution Control,* 11(2): 151–154.

Wang, M,C., Tzeng, J.S., Chen, L.F. and Wang, P.L., 1999. Evaluation of soil treatment sites irrigated with wastewater from Nanching, Chishan and Kaohsiung sugar Mills. Report of the Taiwan Sugar Research Institute 165: 41–59.

Sustainable Environmental Management *Pages* **64–74**
Edited by: **Dr. L.V. Gangawane & Dr. V.C. Khilare**
Published by: **DAYA PUBLISHING HOUSE**

Chapter 9

Environmental Issues: Herbal Pharmaceutical Industries

***D.M. Dharmadhikari*[1] *and A.P. Vanerkar*[2]**

[1]*National Environmental Engineering Research Institute, Nehru Marg, Nagpur – 20*
[2]*National Institute of Miners' Health, JNARDDC Campus, Wadi, Nagpur – 23*

ABSTRACT

Herbal pharmaceutical industries generate huge volume of wastewater, which is highly concentrated in terms of COD, BOD, and SS in the range of 21960-26000 mg/L, 1200-15660 mg/L and 5460-7370 mg/L respectively. It is not feasible to treat this wastewater in a single stage treatment, instead a combination of primary physicochemical and secondary aerobic biological treatment is required to reduce the pollutants. Earlier studies have showed that inspite of two-stage treatment, the final treated effluent quality exceeds marginally on the higher side then the stipulated standard limit, and demanded further treatment to polish the effluent. Primary physicochemical treatment studies were carried out using conventional coagulants like Lime and Alum with three types of polyelectrolytes *i.e.* cationic, anionic and nonionic in nature. Results indicated that Alum 300 mg/L + synthetic polyelectrolyte (Oxyfloc-FL-11) as the best combination with respect to organics removals. Further this effluent was selected for aerobic biological treatment, which had an average COD, BOD, and SS removals were 6266 mg/L, 2867 mg/L and 637 mg/L respectively. Treated effluent from the aerobic system was selected based on optimum organic loading, HRTs and MLSS concentration and was subjected to Fenton's treatment (Advanced Oxidation Process) to achieve the standard promulgated by CPCB. The aerobic effluent had a COD of 288 mg/L, and BOD of 94 mg/L with SS concentration of 14 mg/L. Prior to the application of Fenton's reagent, pH, Contact Reaction Time (CRT), and chemical doses were optimized. Studies indicated a pH of 3.5, CRT of an hour and chemical dose were 200 mg/L of Ferrous sulphate and 6 ml/L of Hydrogen peroxide (30 per cent strength) to be optimized to achieve effluent removals. Fined treated effluent complied with the standards.

Keywords: *Herbal, Primary physicochemical, Aerobic biological treatment, Fenton's reagent.*

Introduction

India being an ancient country having Vedic origin even today considerable population believes in herbal medicines which is an origin of Ayurveda. Ayurvedic medicines (herbal medicines) manufacturing were predominantly practices in ancient Indian and references are quoted in many of our scriptures and ancient literatures. Herbal medicines are known as Ayurvedic medicines in India.

Of all the sciences India possesses the largest and most glorious tradition in the field of herbal medicines. Ayurveda, the science of long life, body and mind is said to have divine origin. Herbal medicines aim not only at the cure of diseases but also at preserving health of the normal population. Ayurveda has an integrated view of man, his body and mind as also of surroundings, which include the plants, minerals and animals. It uses mostly medicinal plants and minerals in diseases to bring back the body equilibrium.

Herbal pharmaceuticals and their use are as old as our civilization. The earliest recorded evidence of their use in Indian, Chinese, Greek, Roman and Syrian text dates back to about 5000 years. During the last 10 years the whole world, specially the developed one, has been swept by a green revolution with the realization of health hazards and toxicity associated with indiscriminate use of synthetic drugs and antibiotics, there has been a general realization in the West as well as East that anything in nature is safer as compared to synthetics.

According to the World Health Organization (WHO), as much as 80 per cent of the world population relies on traditional medicines. According to the WHO report, a considerable percentage of people both developed and developing countries use medicinal plant remedies and the number is on the increase, especially among younger generation (Pushpangadan *et al.*, 1997; Jaggi, 1984; Mukerji Subbarayappa; CSIR, 1987; Rajashekharan, 2002). In 1989 WHO adopted a resolution that herbal medicine is of great importance to the health of individuals and communities.

The countrywide scenario of herbal pharmaceutical industries reveals that the products are produced on batch basis. The process involves the cleaning of herbs (like flowers, stem, bark, nuts, fruits, resins, seeds, roots, leaves etc.) to remove dust and soil adhered to the material. They are thoroughly dried and passed through cutters or ball mills as per the requirement and are further subjected to extraction, fermentation, distillation, decoction preparation, percolation as per the requirement. During the manufacturing of herbal medicines huge volumes of waster is required for different process. Though the product is based on herbs, synthetic organic and inorganic chemicals are also used in the manufacturing process. Chemicals that are used include alcohol, sugar, gelatin, lactose, mineral salts, some trace quantities of metals in the form of oxides (Gold, Silver, Mercury, Cupper, Lead, Zinc, Titanium, Iron etc.) clays and different organic solvents. During the process large volumes of wastewater is generated. Figure 9.1 shows the wastewater sources during the general manufacturing process of Ayurvedic (Herbal) Medicine.

The global market for herbal drugs is growing rapidly (ISM&H, 2001). The current annual herbal global market was estimated as US $ 62.0 billion (Rs. 250,000 crores) and is projected to reach US $ 310 billion (Rs. 1250,000 crores) by the year 2010, *i.e.* five times as of today's market. To estimate, there are over 7000 licensed manufacturers of Ayurvedic medicines alone in India. Some 1200 more companies (700 Unani and 600-Siddha and the rest folklore) are operating in other areas of other Indian system of medicine-based drugs. The following could be a general break-up for Ayurvedic pharmaceutical industry in India (Brindavanam, 2002).

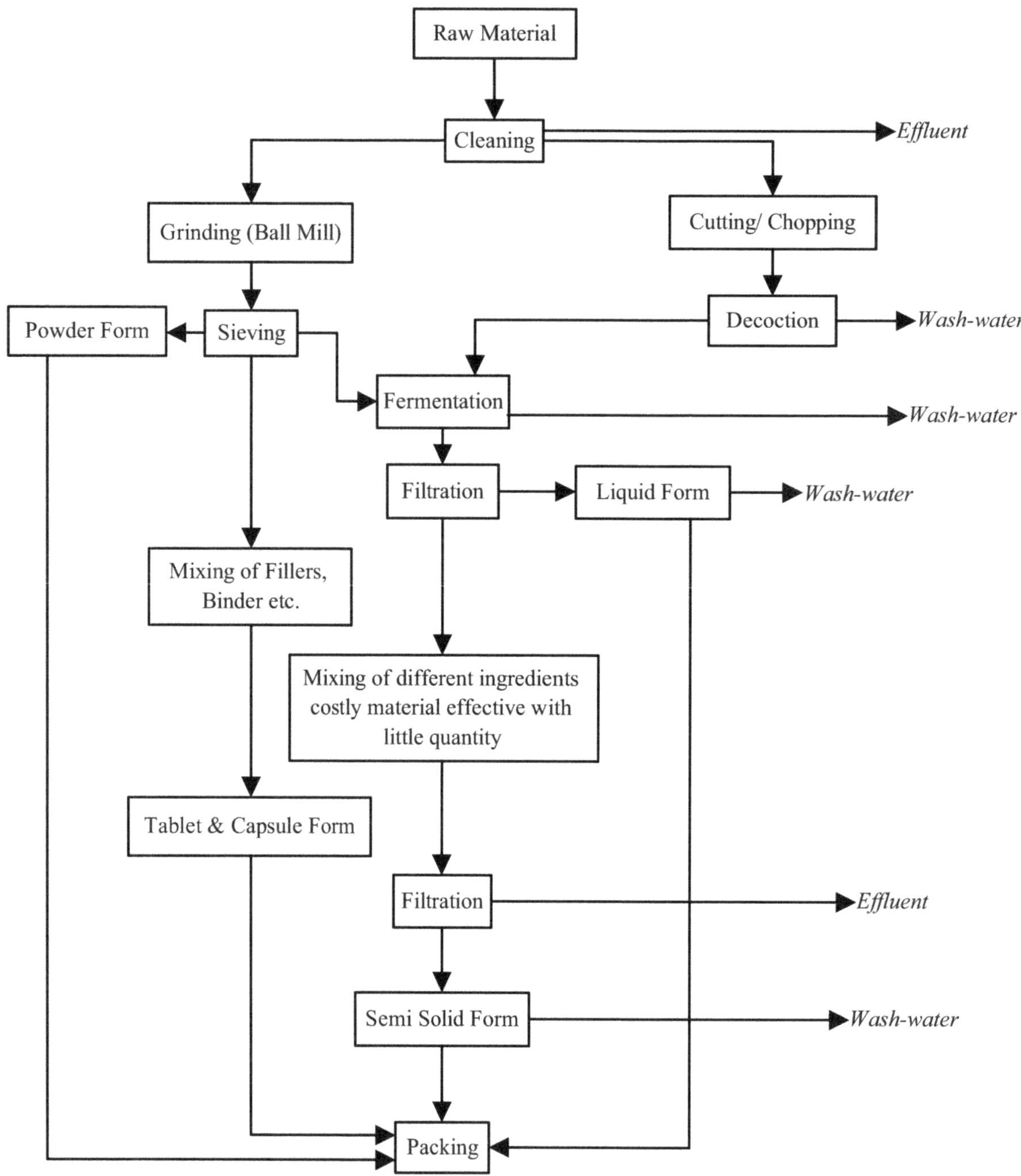

Figure 9.1: General Manufacturing Process of Ayurvedic (Herbal) Medicine

1. Large Scale Operators (750 Cr. Annual turnover): 10
2. Medium Scale Operators (5-50 Cr.): 25
3. Small Scale Operators (1-5 Cr.): 965
4. Very Small Operators (1 Cr. & less): 6000

Wastewater Generation and Treatability Studies (A case study)

Pharmaceutical wastewater distinguishes itself its wide range of pollution characteristics and poses great problems in its disposal due to variation in the types of medicines produced, raw materials used and based on market demands, the wastewater flow. Hence it is important to treat the wastewater prior to its disposal, as it will lead to environmental degradation. Literature on the treatment of herbal pharmaceutical wastewater is scanty. Few papers on the treatment of herbal pharmaceutical wastewater by anaerobic fixed film, fixed bed have been reported (Brindavanam, 2002; Nandy *et al.*, 1998; Debroy *et al.*, 1993) but no details are available as on today on economically viable and effective treatment scheme hence an attempt has been made to study in detail the treatment of the herbal wastewater. As a case study, typical unit producing herbal medicine was selected for the studies, which consumes 17 to 20 m^3/day of water for different processes and almost 95 per cent is generated as a wastewater.

The industries manufacturing theses medicines being small scale, based on low budget, a cost effective treatment is the need of the hour. An attempt has been made to develop a low cost treatment system using inexpensive coagulants coupled with commercially available synthetic polyelectrolytes in combination with biological treatment followed by tertiary polishing process This treatment is less susceptible to the frequent variation in the waste characteristics.

One of the pre-treatment technologies, which seems viable for the pollution problems is the physico-chemical treatment using conventional coagulants and has been successfully applied to various wastewaters. This technology has been tried for the herbal pharmaceutical wastewater to reduce the pollution load. No details are available on herbal pharmaceutical wastewater treatment using combination of coagulant and synthetic Polyelctrolyts. Recently a report on herbal pharmaceutical wastewater treatment by conventional coagulants has been reported by J. Garima and Atul P. Vanerkar (Jain *et al.*, 2000; Vanerkar *et al.*, 2005).

Experimental Procedure

For the characterization of herbal pharmaceutical effluent composite samples were collected and characterized. The samples were analyzed according to the procedures given in standard method (APHA, AWWA and WPCF, 1998). Detailed treatability studies are given below.

Primary Treatment

Physico-chemical treatment is used as a primary treatment (using jar test apparatus) to reduce the organic load [for further secondary (biological process),]. Experiments were carried out using different conventional coagulants (in the range of 50–600 mg/L) individually and in combination with different synthetic polyelectrolytes (in the range of 0.1–0.6 mg/L).

For detail studies stock solution of coagulants were prepared and used for the complete set of each test. While using jar test apparatus, initially the samples were flash mixed for one minute at high rpm of 100 and later at 40 rpm for half an hour. After half an hour of mixing the samples were settled for half an hour. Supernatant liquid was drawn and subjected to various parameters and sludge

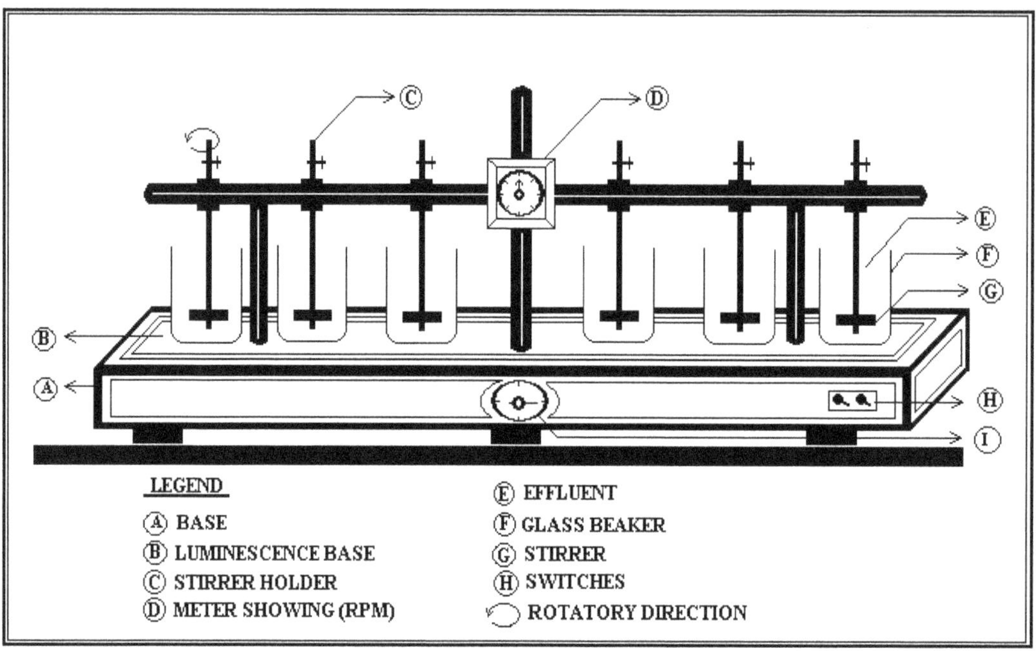

Figure 9.2: Schematic Diagram of Jar Test Apparatus

volume settled was noted. The laboratory unit shown in Figure 9.2.

Secondary Biological Treatment

Bench scale secondary biological treatment (Activated Sludge Process) were carried out to determine the treatability of the wastewaters by completely mixed activated sludge process. The laboratory unit shown in Figure 9.3 was operated with activated sludge acclimated to the wastewater. The required nutrients normally, Nitrogen and Phosphorus were provided in the form of Diammonium hydrogen orthophosphate for maintaining the nutrient to microorganisms in the proportion of BOD: N : P = 100 : 5 : 1.

In order to evaluate the performance of the unit at different F/M ratios (in the range of 0.2 to 1.2), the unit was fed with physicochemically treated wastewater having different BOD (obtained by using different dilutions) at constant HRT using peristaltic pump. Different HRT in the range of 6 to 48 hours were tried for each F/M ratios and for two different MLSS concentration of 3000 mg/L and 4000 mg/L. Oxygen level was maintained (in the form of compressed air) at the rate of 0.5 mg/L in the system. Once the system stabilized as indicated by the reduction in COD, BOD and SS, samples were withdrawn and analyzed for various parameters including MLSS, and MLVSS

Advance Oxidation Process

New technologies of wastewater purification leading to complete Mineralization of organic pollutants in water and wastewater by generating highly reactive Hydroxyl radical (OH^{*}), which is remarkably unstable but provided with very high oxidative potential of 2.8 E_0 (V) as compare to any other oxidant except Fluorine. In the advanced oxidation process like photocatalytic oxidation using

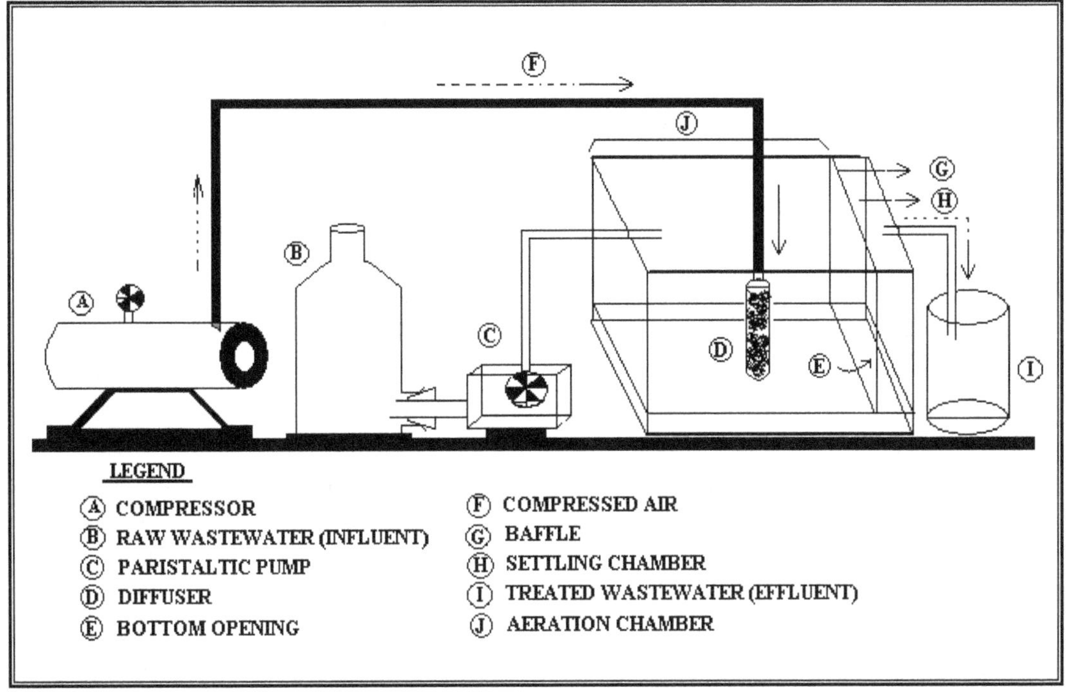

Figure 9.3: Reactor for Bench Scale Secondary Biological (Activated Sludge) Process

heterogeneous photocatalyst in combination of 1. Artificial-UV/TiO_2; and Fenton's oxidation were tried.

Fenton's oxidation of herbal pharmaceutical wastewater was carried out with chemically generated Fenton's reagent using jar test apparatus, shown in Figure 9.3. The oxidation condition was optimized by varying different parameters *viz.* pH, Contact Reaction Time (CRT) amount of $FeSO_4$ and amount of H_2O_2.

Results and Discussion

The herbal pharmaceutical wastewater shows large variation in COD, BOD, total solids, suspended solids, total nitrogen, total phosphate, sodium and some trace metals. This is due to the different additives used for different drug preparation and also depending on the nature of the products *i.e.* liquid or solid. COD, BOD, and SS values of the wastewater vary widely in the range of 21960-26000, 11200-15660, and 5460-7370 mg/L respectively. Physio-chemical characteristics of the wastewater used in the experiment are shown in Table 9.1.

Primary Physico-chemical Treatment

The Physio-chemical studies were of two-stage treatment, *i.e.* initial treatment by Lime to raise the pH of wastewater from initial pH of 4.2 to 7.0 pH, and then the second stage use of different coagulants were studied. Conventional coagulants used were Lime, Alum, ferric chloride, and ferrous sulphate on neutralized wastewater. Initially Lime being the cheapest and cost effective chemical was tried. During first stage of treatment the Lime required to neutralize the wastewater was 4.2 gm/L. Later neutralized wastewater was used for the experiments and individual coagulants were tried. Detail studies on the effectiveness of different anionic, nonionic and cationic polyelectrolyte on the conventional coagulants in organic removals were carried out.

Table 9.1: Characterization of Herbal Pharmaceutical Wastewater

Sl.No.	*Parameters*	*Raw Wastewater*	*Neutralized wastewater*
1.	pH	3.9–4.0	6.8–7.4
2.	Colour (Visual)	Dark yellow	Black
3.	Total Acidity	3000	–
4.	Total Suspended Solids	5460–7370	3320–4864
5.	Total Dissolved Solids	2564–3660	2680–4474
6.	Total Solids	8024–11030	6230–9340
7.	Chemical Oxygen Demand (COD)	21960–26000	17400–20000
8.	Biological Oxygen Demand (BOD, 5day–20°C)	11200–15660	9370–11910
9.	Sulfide as (S^{-2})	42–54	38–42
10.	Sulphates as (SO_4^{-2})	82–88	68–72
11.	Total Phosphates as (PO_4^{-2})	260–280	220–242
12.	Total Nitrogen as N	389–498	336–378
13.	Oil and Grease	140–182	75–98
14.	Sodium as Na^+	155–266	144–250
15.	Potassium as K^+	128–140	118–132
16.	**Heavy Metals**		
16a.	– Iron	65.6–65.7	26.2–41.4
16b.	– Copper	0.649–1.67	0.081–1.42
16c.	– Manganese	6.41–8.47	3.22–5.87
16d.	– Nickel	0.892–2.35	0.107–1.49
16e.	– Zinc	0.583–0.608	0.215–0.414
16f.	– Chromium	0.057–1.11	0.051–0.456
16g.	– Lead	0.559–6.53	0.460–4.08
16h.	– Cadmium	0.036–0.484	0.032–0.339
16i.	– Selenium	0.428–0.666	0.121–0.397
16j.	– Arsenic	0.0049–0.0076	0.0023–0.0040

* All values are expressed in mg/L except pH and colour.

Cationic polyelectrolytes (Oxyfloc-FL-11) dose of 0.25 mg/L with optimum Alum dose of 300 mg/L resulted in COD, BOD, and SS removals of 64.00 per cent, 69.40 per cent and 80.82 per cent, which shows the highest removals in organics by physicochemical treatability study this is due to the fact that herbal pharmaceutical wastewater comprises natural organic matters, which are anionic in nature, and hence its behaviour with cationic polyelectrolytes addition resulted in best removals in organics. This trend has also been reported in literature (Bolto and Braun, 1999; Chow and Leeuwen, 1999; Ratnaweera *et al.*, 1999; Heijman and Paassen, 1999; PCL, 1995).

Details studies on the effectiveness of different conventional coagulants individually and with different synthetic polyelectrolytes were carried out and results are shown in Table 9.2.

Table 9.2: Percent Removals of COD, BOD and SS at Optimum Doses of Various Conventional Coagulants (CC) Individually and in Combination With Synthetic Polyelectrolytes (SP) Doses

Sl.No.	*Conventional Coagulants + Synthetic Polyelectrolyte Doses*	*Optimum Doses of CC and SP (mg/L)*	*Removal Efficiency*					
			COD (mg/L)	*% R in COD*	*BOD (mg/L)*	*% R in BOD*	*SS (mg/L)*	*% R in SS*
1.	Lime	300	10684	38.60	5227	44.22	1494	55.00
2.	Alum	300	12389	28.80	6390	31.80	1882	43.30
3.	Lime + Alum	(300+150)	9814	43.60	4948	47.20	1706	48.60
4.	Ferrous sulphate	100	11310	35.00	5471	41.61	1555	53.30
5.	Ferric chloride	150	11971	31.20	5788	38.23	1743	47.50
6.	Lime + Magnafloc–E-207	(300+0.20)	8839	49.20	4198	55.20	1136	65.78
7.	Lime +Magnafloc–1011	(300 +0.20)	8770	49.60	4104	56.20	1138	65.71
8.	Lime + Zetag –7563	(300 +0.25)	7586	56.40	3729	60.20	1036	68.77
9.	Lime + Zetag –7650	(300 + 0.25)	7517	56.80	3636	61.20	1146	65.48
10.	Lime + Oxyfloc–FL-11	(300 + 0.20)	7378	57.60	3270	65.10	864	74.00
11.	Alum + Magnafloc–E-207	(300 + 0.25)	7656	56.00	3636	61.20	1018	69.33
12.	Alum + Magnafloc–1011	(300 +0.20)	7795	55.20	3729	60.20	1022	69.21
13.	Alum + Zetag–7563	(300 + 0.25)	7238	58.40	3448	63.20	940	71.67
14.	Alum + Zetag–7650	(300 +0.25)	6890	60.40	3373	64.00	892	73.13
15.	Alum + Oxyfloc-FL-11	(300 + 0.25)*	6266	64.00	2867	69.40	637	80.82

* Optimum dose of Conventional Coagulants and Synthetic cationic Polyelectrolyte in mg/L (*i.e.* Alum + Oxyfloc–FL-11).

In general the physico-chemical studies carried out indicate that the pharmaceutical wastewater is amenable to this treatment. It may be applied as a primary treatment before subjecting the wastewater to secondary biological treatment. This will help in efficient working of biological system, by reducing system, by reducing the organic load.

Secondary Biological Treatment

Under this study physicochemically treated herbal pharmaceutical wastewater was subjected to bench scale aerobic-biological- activated sludge treatment on different F/M ratios (0.1 to 0.7), different HRTs (24 to 54 hours) and two different MLSS concentrations of 3000 mg/L and 4000 mg/L.

The results shown in Table 9.3 indicated that COD, BOD, and SS removals on 4000 mg/L MLSS concentrations were found to be in the range of 896-944 mg/L (84.93 per cent to 85.63 per cent), 156 – 174 mg/L (93.93 per cent to 94.51 per cent) and 66-74 mg/L (88.40 per cent to 89.49 per cent) respectively on HRT of 42.00 hours with 0.18 F/M ratio.

It can be inferred from the results that an optimum HRT of 42 hours and 0.18 F/M ratio is a feasible combination where absolute concentrations of COD, BOD, and SS of 896-944 mg/L, 156-174 mg/L and 66-74 mg/L respectively were achieved. The values indicated that the treated effluent is not suitable for discharge or reuse, as it does not comply with stipulated standards for discharge of environmental pollutants (liquid effluents) prescribed by regulatory body *i.e.* Central Pollution Control Board, New Delhi under Schedule VI of Environment Protection Third Amendment Rules, 1993 [17]. This concentration is suitable for advance oxidation studies.

Table 9.3: Performance of Bench–Scale Aerobic Activated Sludge System Using Wastewater from Herbal Pharmaceutical Industry (F/M = 0.18; MLSS = 4000 mg/L; Influent Flow = 1.250 L/Day)

Set	Sample	HRT in Hrs.	pH	COD mg/L	% R in COD	BOD	% R in BOD	SS	% R in SS	MLSS
I	Influent	—	6.7	6266	—	2865	—	638	—	—
	Effluent	24	6.8	1944	68.98	663	76.896	162	74.61	3900
		30	6.8	1580	74.78	456	84.08	128	79.94	3930
		36	6.9	1276	79.64	300	89.53	102	84.01	3950
		42	7.0	944	84.93	174	93.93	74	88.40	4000
		48	7.0	752	87.99	123	95.71	50	92.16	4040
		54	7.1	572	90.87	87	96.96	30	95.30	4080
II	Influent	—	6.7	6236	—	2844	—	628	—	—
	Effluent	24	6.8	1896	69.60	642	77.43	150	76.11	3890
		30	6.9	1544	75.24	429	84.92	112	82.17	3920
		36	6.9	1248	79.99	282	90.08	94	85.03	3960
		42	7.0	896	85.63	156	94.51	66	89.49	4010
		48	7.1	716	88.52	114	95.99	46	92.68	4050
		54	7.1	560	91.02	78	97.26	28	95.54	4100

Tertiary–Advanced Oxidation Treatment

Only two types of advanced oxidation process were tried to treat the biologically treated herbal pharmaceutical wastewater *i.e.* Fenton's oxidation, and UV-TiO_2 photoassisted (heterogeneous photocatalytic) oxidation.

Fenton's oxidation process was carried out at ambient temperature and pressure, as the literature reveals that (Meyers, 1998; Herber and Weiss, 1934), thermodynamically this process is exothermic in nature hence, high temperature have negative impact on H_2O_2 stability, therefore the process has been carried out at ambient temperature, which favoured the maximum removal of organics. This process was applied to the herbal pharmaceutical wastewater, obtained from bench scale activated sludge process (with 0.18 F/M ratio, 42.00 hours of HRT and 4000 mg/L of MLSS concentration) having COD, BOD, SS, and TOC concentrations of 932 mg/L, 168 mg/L, 70 mg/L, and 450 mg/L respectively. This effluent was selected, based on the optimum removals achieved.

Thus after achieving all the optimal conditions *viz.* pH, CRT, dose of $FeSO_4$ and H_2O_2, herbal pharmaceutical wastewater was further subjected to Fenton's oxidation as a tertiary treatment. The results on all the optimized conditions are given in Table 9.4.

The results show that, under the optimum conditions *i.e.* pH of 3.5, CRT of 60 minutes, dose of

$FeSO_4$ of 200 mg/L and dose of H_2O_2 of 6.0 ml/L, the average optimum removals of organics in terms of COD, BOD, SS, and TOC were found to be 138 mg/L (85.19 per cent), 20 mg/L (88.10 per cent), 21 mg/L (70.00 per cent), and 98 mg/L (78.22 per cent) respectively, which is well below the standards prescribed by regulatory body [Central Pollution Control Board (CPCB), New Delhi, India] for discharging the wastewater in to inland water or surface water.

Table 9.4: Performance on Optimized Conditions Using Wastewater from Herbal Pharmaceutical Industry

Sl.No.	*Optimized Conditions*	*COD 932 mg/L*		*BOD 168 mg/L*		*SS 70 mg/L*		*TOC 450 mg/L*	
		mg/L	*%R*	*mg/L*	*%R*	*mg/L*	*%R*	*mg/L*	*%R*
1.	pH of Wastewater: 3.5; Contact Reaction Time: 60 min; Dose of $FeSO_4$: 200 mg/L; and Dose of H_2O_2: 6.0 ml/L	128	86.27	18	89.29	18	74.29	94	79.11
2.		156	83.26	24	85.79	26	62.86	104	76.89
3.		152	83.69	21	87.50	26	62.86	104	76.89
4.		132	85.84	18	89.29	20	71.43	96	78.67
5.		128	86.27	18	89.29	18	74.29	92	79.56
6.		140	84.98	21	87.50	22	68.57	98	78.22
7.		132	85.84	18	89.29	18	74.29	96	78.67
8.		136	85.40	24	85.79	20	71.43	98	78.22
Avg.		138	85.19	20	88.10	21	70.00	98	78.22

Conclusions

1. The characterization of herbal pharmaceutical wastewater shows that the wastewater from this type of industries is of moderate strength and should not be discharge without treatment in order to avoid the deterioration of aquatic environment.
2. Overall physico-chemical studies was indicated that the herbal pharmaceutical wastewater needs to be treated by physico-chemical treatment as a primary process to reduce the organic load and increase the performance efficiency of the secondary biological treatment process.
3. Secondary treatment concludes that for the process is efficient for removal of organics, where final concentration of COD, BOD, and SS were achieved in the range of 896-944 mg/L, 156-174 mg/L and 66-74 mg/L respectively. Though these removals are the maximum removals obtained but are not at par with the standards promulgated by regulatory bodies to use for other purposes. Hence it is inferred that tertiary (advance) treatment is essential for this wastewater.
4. The tertiary treatment of wastewater from herbal pharmaceutical industry shows that the wastewater is amenable to Fenton's oxidation process as well as heterogeneous photocatalytic oxidation process and thus achieves the optimum removal of organics as promulgated by regulatory agency for discharging the same to the inland water.

References

APHA, AWWA and WPCF, 1998. *Standard Method of the Examination of Water and Wastewater*, 20th edition, APHA, AWWA and WPCF.

Bolto, Brian and Gudrum Abbt Braun, 1999. Experimental evaluation of cationic polyelectrolyte for removing natural organic matter from water. *Water Sci. and Technol.*, 40(9): 71–79.

Brindavanam, N.B., 2002. Herbal Drugs: India's potential vis-à-vis global scenario. In: " *Proceeding of the National Conference on Utilization of Bioresources*, NATCUB–2002, RRL–Bhubaneswar ALLIED Publishers Pvt. Ltd.

Chow, C.W.K. and J.A. Van Leeuwen, 1999. The impact of the character of natural organic matter in conventional treatment with alum. *Water Sci. and Technol.*, 40: 97–104.

CSIR, 1987. National Science Day: A Commemorative Volume. Council of Scientific and Industrial Research, New Delhi (India), pp. 18–25.

Debroy, Mukti, Nandy, T. and S.N. Kaul, 1993. Anaerobic up flow bioreactor for pharmaceutical waste treatment. *Indian Journal of Environmental Protection*, 13(1): 41–46.

Heijman, S.G.J. and A.M. Van Paassen, 1999. Adsorptive removal of NOM during drinking water treatment. *Water Sci. and Technol.*, 40 (9): 183–190.

Herber, F. and J. Weiss, 1934. The catalytic decomposition of hydrogen peroxide by iron salts. In: *Proc. of Royal. Soc. London*, 147(A): 332–351.

ISM&H, 2001. *Indian System of Medicine and Homoeopathy, Export Opportunities, Technology Export Development Organization Report*, New Delhi, pp. 1–253.

Jaggi, O.P., 1984. *Impact of Science and Technology in Modern India: History of Science and Technology and Medicine in India*. Second; Atma Ram and Sons, Delhi and Lucknow (India), pp. 65–83, 1984.

Jain, Garima, Shanta Satyanarayan, P. Nawghare and S.N. Kaul, 2000. Treatment of pharmaceutical wastewater (herbal) by a coagulation/flocculation process. *International Journal of Environmental Studies*, U.K., Nov. 2000.

Meyers, Robert A., 1998. *Wiley Encyclopaedia Series in Environmental Sciences: Environmental Analysis and Remediation*, 3: 1661–1674.

Mukerji, S. K. and Subbarayappa, B.V. *Science in India: A Changing Profile*. Indian National Science Academy, New Delhi (India), pp. 4–16.

Nandy, T., S.N. Kaul and L. Szpyrowicz, 1998. Treatment of herbal pharmaceutical wastewater with recovery. *International Journal of Environmental Studies*, 54: 83–105.

PCL, 1995. Pollution Control Law Series: PCL/4/1995–96, "Standerds for Liquid Effluents, Gaseous Emissions, Automobile Exhaust, Noise and Ambient Air Quality". CPCB, June– 1995.

Pushpangadan, P., Ravi, K. and Santosh, V., 1997. *Conservation and Economic Evaluation of Biodiversity*, First. Oxford and IBH Publishing Co. Pvt. Ltd. : New Delhi and Calcutta (India), pp. 144–147.

Rajashekharan, P.E., 2002. *World of Science: Herbal Medicine*. Employment News, Sept 21–27; Ministry of Information and Broadcasting, India, New Delhi, pp. 3–4.

Ratnaweera, Harsh, EgilGjessing and Elvind Oug, 1999. Influence of physical-chemical characteristics of natural organic matter (NOM) on coagulation properties: An analysis of eight Norwegian water Ssources. *Water Sci. and Technol.*, 40(9): 89–95.

Vanerkar, Atul P., S. Satyanarayan and D.M. Dharmadhikari, 2005. Enhancement of organic removals in high strength herbal pharmaceutical wastewater. *J. Environmental Technology*, 26(4): 389–396.

Sustainable Environmental Management *Pages 75–88*
Edited by: **Dr. L.V. Gangawane & Dr. V.C. Khilare**
Published by: **DAYA PUBLISHING HOUSE**

Chapter 10

Rhizosphere and Rhizoplane Mycoflora Under The Influence of Deproteinized Leaf Juice of Lucerne

D.A. Doiphode[1] and A.M. Mungikar[2]

[1]Department of Botany, Vasantrao Naik Mahavidyalaya, Aurangabad– 431 003
[2]Department of Botany, Dr. Babasaheb Ambedkar Marathwada University, Aurangabad – 431 004

ABSTRACT

Investigations on rhizosphere and rhizoplane mycoflora associated with 9 crop plants were undertaken. Application of deproteinized leaf juice (DPJ), left after leaf protein (LP) extraction from lucerne (*Medicago sativa* L.), to the soil as a source of fertilizer affected soil mycoflora qualitatively as well as quantitatively. Growth of many pathogenic fungi was suppressed due to the application of DPJ to soil, while that of useful fungi emerged. The overall effect of DPJ on rhizosphere and rhizoplane mycoflora was found to be favorable to the crop plants under investigation. It is believed that the DPJ was analogous to the root exudates and hence its applications to the soil improved microbial status on and around the roots of crop plants.

Keywords: *Deproteinized leaf juice, Rhizosphere, Rhizoplane, Soil microbiology, Leaf protein.*

Introduction

Exploitation of green leaves for extraction of proteins and preparation of leaf protein concentrate (LPC), for use in human nutrition as protein-vitamin-minearl rich food, has been advocated (Pirie, 1942, 1971, 1978, 1987; Telek and Graham, 1983). For the preparation of LPC the process of green crop fractionation (GCF) is employed which, generates along with LPC, the deproteinised leaf juice (DPJ) in the form of brown 'Whey' or 'liquor' as a by-product. Random disposal of DPJ may cause environmental bio-pollution and therefore, its use in a proper way is essential. The DPJ is rich in soluble plant metabolites. Its use in microbial technology as a medium for cultivation of microbes has been recommended by Deshpande and Joshi (1971), Ghewande and Deshpande (1975) and Salve and Gangawane (1984). Ajay Kumar and Mungikar (1990b, 1990c) pointed out that *Aspergillus niger* and *Penicillium notatum* grow well on DPJ. Jadhav and Mungikar (2000) advocated the use of DPJ for commercial production of enzymes by cultivating fungi. Gogle et. al. (2001), Mungikar (2001), Jadhav and Mungikar (2001), Sayyed and Mungikar (2001, 2003) and Mungikar and Gogle (2001) suggested the use of DPJ for producing microbial biomass and metabolites.

In addition to the use of DPJ as a medium for growing microbes, studies undertaken at various places indicated that the DPJ can be used as a source of manure. Davys (1973) suggested the use of DPJ for irrigation as a nutrient source and soil conditioner. When DPJ is added to the soil, growth of gas forming bacteria increases with an improvement in the structure of intractable soil (Arkcoll, 1973). The quantities of P and exchangeable K in soil increases with the application of DPJ (Ream et. al., 1983). Studies in this laboratory by earlier workers have shown beneficial effects of adding DPJ to the soil. Salve and Gangawane (1984) added DPJ of lucerne to the soil and observed increased nodulation and growth of groundnut plants. They reported increase in the population of *Rhizobium, Actinomycetes* and other bacteria in the soil due to the irrigation with deproteinized leaf juice. The DPJ of lucerne is a good source of fertilizer to *Sorghum* (Dakore and Mungikar, 1986), maize (Ajaykumar and Mungikar, 1990a), bajra (Kawathekar, 2001),wheat, rice (Salve, 2002) and cowpea (Jadhav and Mungikar, 1998b). Jadhav (1997) however, reported phytotoxic effect of DPJ when applied at higher concentration to pea plants.

Apart from soil, composed of organic and inorganic substances, the root system of higher plants is associated with vast population of microorganisms. As the microflora (flora of microbes) around the living roots of higher plants is different from soil microflora, the plant thus create special sub-terranean habitat for soil microbes around their roots. The performance of plant is affected by the population of microbes around its root, as the root zone is the site from which mineral nutrients are obtained. This unique environment under the influence of plant roots affecting soil fertility and plant growth is called rhizosphere.

The rhizosphere may be either 'inner' on the very root surface, or 'outer' embracing the immediately adjacent soil. In the inner zone biochemical reactions between microorganisms and roots are more pronounced and it is termed as rhizoplane. Obviously the outer environment is called rhizosphere. Relationship between plant roots with rhisoplane and rhizosphere, include symbiotic (legumes and rhizobia), mycorohizal association, pathogenic (bacteria, fungi and nematodes) commensalisms and protocoopertion (Alexander, 1967).

The term rhizosphere was introduced by the German scientist Hiltner (1904), who defined it as the zone of soil under the influence of roots. The micro-population associated with rhizosphere is now known to extend a decisive influence on the physiological activities of plants (Subba Rao, 1977). The plants supply them nutrients in the form of residue and excretions which affect rhizoplane as well as rhizosphere microflora. Rhizosphere is characterised by greater microbial activity than the soil away from plant roots. It is now clearly established that greater number of bacteria, fungi and actinomycetes are present in rhizosphere soil than in non-rhizosphere soil (Lakshmi Kumari, 1961). The root surface microhabitat, called rhizoplane, is also equally important.

Work on fungi associated with rhizoplane and rhizosphere *i.e.* rhizoplane mycoflora and rhizosphere mycoflora of different crop plants received considerable attention. In India several workers like Agnihotrudu (1955, 1961), Bhuvaneshwary (1960), Rangaswamy and Vasantharajan (1962), Rao (1962), Das (1963), Mishra (1967), Gangawane (1972, 1985), and Subrahmanyam (1975) have made many significant contributions in this field. Lists of fungal species from rhizosphere of various crop plants have been compiled by many workers (Agnihotrudu, 1961; Naim, 1967, Peterson, 1958; Sunar and Chohan, 1971; Shrinivasan, 1958; Rao, 1962; Rangaswami and Venkatasan, 1964; Gangawane, 1972; Subrahmanyam, 1975). It was observed that number of species isolated from the rhizosphere of various plants differ considerably; rhizosphere of different plants harbour different genera. Most of the workers reported *Aspergillus, Penicillium, Rhizophus, Curvularia, Cladosporium, Fusarium, Alternaria* and *Trichoderma* as prevalent in the rhizosphere (Rao, 1962; Roy and Gujarati, 1967; Gangawane and Deshpande, 1972). Comparatively, less number of genera have been reported from rhizosplane which included *Fusarium, Cylindrocarpon, Rhizoctonia, Gliocladium, Mortierella, Rhizopus, Trichoderma, Aspergillus, Penicillium* and sterile mycelia (Gangawane, 1985).

The rhizosphere and rhizoplane mycoflora quantitatively and qualitatively change with season (Gangawane and Deshpande, 1977), ecological nitches (Padma and Manohara Chary, 1994), organic amendments, chemical fertilizers, bacterization (Gangawane and Deshpande, 1974) crop rotation (Gangawane and Deshpande, 1975), partial sterilization of soil (Gangawane and Deshpande, 1978; Kulkarni and Gangawane, 1982; Saler and Gangawane, 1994), foliar sprays (Gangawane, 1979) and viral infection (Gangawane and Deshpande, 1973). Application of farm yard manure (FYM), green manure, organic soil amendments, sludge etc. were found to alter rhizosphere and mycoflora (Singh, 1937; Gangawane and Deshpande, 1976; Gangawane and Kulkarni, 1985). Addition of chemical fertilizers to soil also gives somewhat similar results (Saksena, 1955). Gangawane and Deshpande (1977) showed favourable effect of chemical fertilizers on fungal population in rhizosphere; urea had highest effect followed by P_2O_5, K_2O and NPK mixture.

Different plant species often establish somewhat different subterranean floras due to variations in rooting habits, tissue composition, excretion products of plants in the form of root exudates (Alexander, 1967; Subba Rao, 1977). Legumes engender more pronounced rhizobial effect than grasses and grain crops. The age of the plant also alters the underground flora, while the stage of maturity controls the magnitude by rhizosphere effect.

An earlier investigation from this laboratory has well established that the deproteinised leaf juice (DPJ), which is a by-product of green crop fractionation (GCF) system can be employed as a source of manure. When applied to the soil, it improves the growth of *Sorghum* (Dakore and Mungikar, 1986), maize (Ajaykumar and Mungikar, 1990a), Bajra (Kawathekar, 2001), cowpea (Jadhav and Mungikar, 1998) and Groundnut (Salve and Gangawane, 1984). However, its application to leguminous crop plants at a higher level showed phytotoxic effect (Jadhav, 1997). Furthermore its micotoxic nature and inhibitory effect on seed germination has also been shown (Mainderkar and Mungikar 1994; Jadhav and Mungikar, 1998a). Very few attempts were, however, made to study the effect of application of DPJ to agricultural land, as a source of manure, on soil, rhizosphere and rhizoplane microflora. Salve and Gangawane (1984), however, recorded decrease in fungal population and number of species in rhizosphere of groundnut due to the application of lucerne DPJ to the soil. Attempts were, therefore made to evaluate the effect of lucerne DPJ on growth of 9 plants along with studies on rhizosphere and rhizoplane mycoflora. The results obtained were compared with those obtained due to the irrigation with either water or 2 per cent urea solution.

Preparation of DPJ

Fresh green foliage of lucerne (*Medicago sativa* L.) was pulped on IBP pulper (Davys and Pirie, 1969). The pulp was then pressed using laboratory scale press (Davys *et. al.*, 1969). The pressed crop residue (PCR) was kept aside for animal nutrition while the leaf juice released due to the pressing of the pulp was collected for the preparation of leaf protein concentrate (LPC). In a clean stainless steel container, about 150 ml water was taken and boiled. To the boiling water leaf juice was slowly added with stirring and heated to 90 ± 5°C. At this temperature leaf protein concentrate (LPC) appeared as a curd due to precipitation and coagulation of proteins in juice. The LPC was isolated from heated juice by filtration through cotton cloth and employed for use as a food grade product. The filtrate or whey, known as deproteinised leaf juice (DPJ) was collected in suitable container.

Cultivation of Crop Plants

Nine crop plants viz Bengal Gram (*Cicer arietinum* L.), Soybean (*Glycine max* Merr.), Groundnut (*Arachis hypogaea* L.), moth bean (*Phaseolus aconitifolius* Jacq.), Pea (*Pisum sativum* L.), Pigeonpea (*Cajanus cajan* (L.) Millsp.), Green gram (*Phaseolus aureus* Roxb.), Wheat (*Triticum aestivum* L. cv. Kalyan Sona) and Sorghum (*Sorghum bicolour* Moench) were selected for present study. The plants were raised in the pots (45 cm i.d.) filled with soil collected from the University Botanical Garden. The soil was sandy loam, alkaline in reaction (pH 7.8), having 101, 10.1 and 223 kg/ha of available N, P and K respectively. The organic carbon (O.C.) content of the soil was 0.48 per cent.

The sowing was done in the month of June 2000. after the emergence, extra seedlings were removed to maintain uniform plant population of 10 plants per pot. There were in all 3 treatments: Control (C): wherein the pots neither received urea nor DPJ; Urea treated (U): wherein the pots received 250 ml 2 per cent aqueous urea solution 15, 30 and 45 days after sowing; and DPJ treated (D): wherein the pots received 250 ml DPJ from lucerne 15, 30 and 45 days after sowing. The potted plants were raised under irrigation. At the age of 60 days the plants were separated from soil without damaging the root system.

Rhizosphere and Rhizoplane Mycoflora

Fungi associated with rhizosphere soil (around root zone) and rhizoplane (on root surface) were isolated following standard plate count (SPC) method which is also called as dilution plate method as described by Johnson *et al.* (1959). The rhizosphere soil was separated and soil suspension was obtained by shaking roots of individual plants in separate aliquots of sterile water in conical flasks. The technique was so adopted that 1 g soil was suspended in 100 ml sterile distilled water. One millilitre (ml) of the soil suspension was transferred into sterile petridish. To it, 15–20 ml melted rose bengal agar medium having not more than 46°C temperature, was added. The petridish was swirled gently to mix the soil extract and the medium. After solidification of the medium, the plates were inverted and incubated at 37 ± 0.5°C in incubator for 5–7 days.

For sampling of rhizoplane or root surface mycoflora, the roots, left behind after removing the soil adhering to it, were washed with sterile water several times until the clear root surface was exposed. The roots were then cut down into pieces of 1 cm length. Ten pieces of washed roots were plated on solidified rose bengal agar medium and incubated for 5 to 7 days.

The composition of rose bengal agar medium was: 5 g peptone, 1 g KH_2PO_4, 0.5 g $MgSO_4$, 10 g glucose, 20 g agar and 0.001 of Rose Bengal dissolved in 1000 ml distilled water. Rose Bengal was used as a bacteriostatic agent (Smith and Dawson, 1944) to prevent formation of bacterial colonies.

Observations for the number and types of rhizosphere and rhizoplane fungi present on the plates (in duplicate) were undertaken from 7th day onwards. Identification of fungal species was done on the basis of colony characters and using monographs and manuals referred by Padma and Manoharachary (1994). The colonies of fungi were isolated on normal potato dextrose agar (PDA) medium (Shinde, 1982) for identification of species. The numbers of fungal species observed on the plate along with number of colonies for each species were recorded. The studies on rhizosphere and rhizoplane mycoflora were undertaken with plants irrigated with either water (control), lucerne DPJ or urea.

Results and Discussion

In agriculture, alteration of rhizosphere mycoflora may take place due to the soil amendments with organic manures and inorganic fertilizers. Although qualitative differences have been pointed out in respect to the fungi from rhizosphere, particularly under the influence of soil amendments, it is difficult to interpret the exact significance of such observations.

One of the most important factors responsible for rhizosphere effect is the great variety of organic substances available at the root region by way of root exudates, which influence quality and quantity of microorganisms in the root region. The substances exuded by plant root include amino acids, sugars, organic acids, vitamins, nucleotides and many other unidentified substances. Studies undertaken by Chanda *et al.* (1984) on composition of DPJ from 5 plants also indicated the presence of minerals, glucose, lipids, nitrogen (N), phosphorus (P), calcium (Ca), potassium (K), iron (Fe); and among vitamins riboflavin, niacin and pantothemic acid. When microbial load of DPJ was studied by solidifying DPJ with agar and exposing it for 6 hours, she observed a maximum of 66, 40 and 20 colonies of bacteria, fungi and actinomycetes respectively per 25 ml DPJ. The common fungal isolates were *Cunninghamella, Mucor, Rhizophus, Aspergillus, Penicillium, Fusarium, Trichoderma* and *Phoma.* In view of the chemical constitution of DPJ, which was assumed as analogous to the root exudates, and microbial load, which indicated its ability to support growth of microbes, it was thought worthwhile to undertake studies on the effect of application of DPJ to the soil on rhizosphere and rhizoplane mycoflora.

Tables 1 to 9 gives an account on rhizosphere and rhizoplane mycoflora of 9 crop plants cultivated under the influence of either water (control) or 2 per cent solutions of either urea or DPJ of lucerne. Qualitative and quantitative changes in mycoflora were evident due to the application of either DPJ or urea. Due to the application of DPJ, the number of colonies decreased with very little change in number of fungal species associated with rhizosphere. However, an increase in colony count along with fungal species was evident when rhizoplane mycoflora was studied. The results thus gave an indication that the DPJ affected mycoflora on the root surface, rather than that around it. The effect of DPJ on rhizosphere and rhizoplane mycoflora of various crop plants may be summarised as under.

Pigeon Pea (*Cajanus cajan* L. Millsp)

The fungal population was increased due to the treatment with 2 per cent urea and DPJ when compared with control. Though the number of fungal types were same there was variation in the fungal species. In control rhizosphere percentage frequency was higher in case of *Fusarium oxysporum. Phoma* sp. appeared due to the treatment of urea, while *Aspergillus niger* due to DPJ. Only two fungi, *Fusarium oxysporum* and *Rhizopus stolonifer* were recorded on rhizoplane under all of the treatments. There was no variation in the percentage frequency of their occurrence even in the DPJ treated rhizoplane (Table 10.1).

Bengal Gram (*Cicer aritinum* L.)

The results given in Table 10.2 indicate that the total number of colonies were higher in the rhizosphere of water treated plants which were highly reduced due to the treatment with urea and

DPJ. A total of 8 fungal species were recorded when the plant received water while only 4 species were recorded with urea and DPJ treated plants. *Aspergillus niger* showed highest percentage frequency in control, while *Fusarium oxysporum* and *Aspergillus kanegawensis* showed maximum frequency in urea and DPJ treated plants. On rhizoplane, fungal species increased due to treatment of urea and DPJ. *Fusarium oxysporum* and *Fusarium semitectum* were found on control plants, whereas *Aspergillus niger* was an addition in the rhizoplane of urea treated plants. In DPJ treated rhizoplane *Penicillium funiculosum* appeared in addition to *Aspergillus niger* and *Fusarium oxysporum*.

Table 10.1: Effect of DPJ on Rhizosphere and Rhizoplane Mycoflora (per cent frequency/gm) of *Cajanus cajan* (Linn.)

Control		*Urea Treated*		*DPJ Treated*	
Fungal Species	*% Frequency*	*Fungal Species*	*% Frequency*	*Fungal Species*	*% Frequency*
Rhizosphere mycoflora					
Fusarium semitectum	17.3	*Fusarium oxysporum*	82.3	*Aspergillus niger*	6.8
Fusarium oxysporum	60.8	*Phoma* sp.	5.8	*Fusarium oxysporum*	86.2
Rhizopus stolonifer	21.7	*Rhizopus stolonifer*	11.7	*Rhizopus stolonifer*	6.8
Total colonies	**23.0**	**Total colonies**	**34.0**	**Total colonies**	**29.0**
Rhizoplane mycoflora					
Fusarium oxysporum	66.6	*Fusarium oxysporum*	66.6	*Fusarium oxysporum*	66.6
Rhizopus stolonifer	33.3	*Rhizopus stolonifer*	33.3	*Rhizopus stolonifer*	33.3
Total colonies	**15.0**	**Total colonies**	**15.0**	**Total colonies**	**15.0**

Table 10.2: Effect of DPJ on Rhizosphere and Rhizoplane Mycoflora (per cent frequency/gm) of *Cicer aritinum* (Linn.)

Control		*Urea Treated*		*DPJ Treated*	
Fungal Species	*% Frequency*	*Fungal Species*	*% Frequency*	*Fungal Species*	*% Frequency*
Rhizosphere mycoflora					
Aspergillus niger	35.2	*Aspergillus carbonarieus*	7.1	*A. kanagavaenisis*	96.6
A. terreus	9.8	*Fusarium oxysporum*	78.5	*Fusarium semitectum*	20.0
A. carbonarieus	9.8	*F. solanae*	10.7	*F. oxysporum*	10.0
A. fumigatus	7.9	Myrothecium roridum	3.5	*Penicillium* sp.	6.60
F. semitectum	10	*F. moniIforme*	23.5	*F. oxysporum*	5.8
Myrothecium roridum	3.9				
Total Colonies	**51.0**	**Total colonies**	**28.0**	**Total colonies**	**30.0**
Rhizoplane mycoflora					
Fusarium oxysporum	90	*Aspergillus niger*	60	*Aspergillus niger*	22.2
Fusarium semitectum	10	*Fusarium semitectum*	10	*Fusarium oxysporum*	22.2
Fusarium oxysporum	30	*Penicillium funiculosum*	55.5		
Total colonies	**10**	**Total colonies**	**10**	**Total colonies**	**09.0**

Pea (*Pisum sativum* L.)

The results obtained with pea are depicted in Table 10.3. It was interesting to note that quantitatively numbers of fungi were highly reduced due to the treatment of urea and DPJ. Six species were recorded when plants received water while only 4 or 6 species were recorded in the rhizosphere of plants grown under the influence of urea and DPJ treated soils respectively. *Fusarium oxysporum* appeared to be a dominant species in all of the cases while the appearance of *Phoma* sp. was suppressed due to urea. *Helminthosporium tetramera* on the other hand appeared due to the treatment of DPJ to the plant.

Table 10.3: Effect of DPJ on Rhizosphere and Rhizoplane Mycoflora (per cent frequency/gm) of *Pisum sativum* (Linn.)

Control		*Urea Treated*		*DPJ Treated*	
Fungal Species	*% Frequency*	*Fungal Species*	*% Frequency*	*Fungal Species*	*% Frequency*
Rhizosphere mycoflora					
Aspergillus carbonarieus	2.5	*Aspergillus niger*	33.3	*Aspergillus niger*	5.2
Aspergillus niger	23.5	*Fusarium semitectum*	8.3	Aspergillus nidulans	5.2
Fusarium semitectum	11.7	*Fusarium oxysporum*	37.5	*Aspergillus terreus*	26.3
Fusarium oxysporum	35.2	Rhizopus stolonifer	20.8	*Fusarium oxysporum*	52.6
Phoma sp.	2.9			*H. tetramera*	5.2
Rhizopus stolonifer	5.8			Rhizopus stolonifer	5.2
Total Colonies	**34.0**	**Total colonies**	**24.0**	**Total colonies**	**19.0**
Rhizoplane mycoflora					
Aspergillus flavus	18.1	*Fusarium oxysporum*	47.1	*Aspergillus niger*	20
Aspergillus niger	54.5	*Rhizopus stolonifer*	52.9	*Fusarium oxysporum*	50
Fusarium oxysporum	27.2			Rhizopus stolonifer	30
Total colonies	**11.0**	**Total colonies**	**17.0**	**Total colonies**	**10**

Rhizoplane mycoflora indicated quantitative increase of fungal colonies due to urea. The occurrence of *Aspergillus niger* was suppressed due to the treatment of urea while showed highest frequency with the control. The frequency of *Rhizopus stolonifer* and *Fusarium oxysporum* highly increased due to the application of either urea or DPJ to the soil.

Moth Bean (*Phaseolus aconitifolius* Jacq.)

With moth bean it was observed that fungal population increased in the rhizosphere due to the application of urea and DPJ. The fungal species increased qualitatively in urea and DPJ treated rhizosphere. The frequency of *Fusarium semitectum* was found to be higher irrespective of treatment. *Aspergillus terreus* and *Penicillium varians* appeared in urea treated rhizosphere, while *Aspergillus niger* and *A. fumigatus* appeared due to the treatment with urea and DPJ respectively (Table 10.4).

On rhizoplane, numbers of fungal species were less on urea treated plants. The occurrence of potential pathogen *Rhizoctonia bataticola* on rhizoplane was suppressed due to DPJ, whereas *Penicillium fumiculosum* appeared on root surface which received DPJ as a source of manure.

Table 10.4: Effect of DPJ on Rhizosphere and Rhizoplane Mycoflora (per cent frequency/gm) of *Phaseolus aconitifolius* (Jacq.)

Control		*Urea Treated*		*DPJ Treated*	
Fungal Species	*% Frequency*	*Fungal Species*	*% Frequency*	*Fungal Species*	*% Frequency*
Rhizosphere mycoflora					
Fusarium semitehtum	64.2	*Aspergillus terreus*	15.3	*Aspergillus niger*	9.0
Fusarium oxysporum	28.5	*Fusarium semitectum*	23.0	*Aspergillus fumigatus*	22.7
Rhizopus stolonifer	7.1	*Fusarium oxysporum*	23.0	*Fusarium semitectum*	40.9
		Penicillium varians	38.4	*Fusarium oxysporum*	27.2
Total colonies	**14.0**	**Total colonies**	**26.0**	**Total colonies**	**22.0**
Rhizoplane mycoflora					
Fusarium oxysporum	66.6	*Fusarium oxysporum*	90	*Aspergillus flavus*	11.1
Rhizoctonia bataticola	6.6	*Rhizopus stolonifer*	10	Fusarium oxysporum	66.6
Rhizopus stolonifer	26.6			*Rhizoctonia bataticola*	22.2
Total colonies	**15.0**	**Total colonies**	**10**	**Total colonies**	**9.0**

Green Gram (*Phaseolus aureus* Roxb.)

The number of total colonies increased due to the treatment with DPJ in comparison with control and urea treated rhizosphere (Table 10.5). The fungal species varied according to the treatment received. *Aspergillus flavus* was absent in both the urea and DPJ treated soils in the rhizosphere of green gram. On the other hand, the existence of *Penicillium fumiculosum* was suppressed in the rhizosphere from urea and DPJ treated soils. DPJ treated soil showed the presence of *Phoma* sp. On the rhizoplane, *Fusarium oxysporum* and *Rhizopus stolonifer* were recorded in all the treatments but *Rhizoctonia bataticola* was observed on DPJ treated rhizoplane. In general the fungal population on rhizoplane was slightly increased due to the treatment with DPJ of lucerne.

Groundnut (*Arachis hypogaea* L.)

Results are given in Table 10.6. With groundnut it was observed that the fungal population reduced in the rhizosphere due to the treatment of urea and DPJ. With the DPJ only four species of fungi were recorded against eight fungi recorded, when the soil received either water alone or urea. The presence of *Aspergillus terreus, Phoma* sp., *Penicillium fumigatus* and *Rhizopus stolonifer* was suppressed in rhizosphere due to the treatment with DPJ. In rhizoplane mycoflora a total of three species were recorded. The fungal population on root surface reduced due to manuring of soil with either urea or DPJ. *Aspergillus niger* did not appeared on the rhizoplane due to the irrigation of urea and DPJ to the soil.

Soybean (*Glycine max* Linn.)

The population of fungi with soybean, reduced in the rhizosphere due to the irrigation with urea as well as DPJ. The DPJ appeared to be more harmful to some of the fungi and the presence of only two species were recorded, in the rhizosphere of this crop. In the DPJ treated rhizosphere the growth of *Phoma herbarium, Rhizopus stolonifer,* and *Fusarium semitectum* was suppressed (Table 10.7).

Table 10.5: Effect of DPJ on Rhizosphere and Rhizoplane Mycoflora (per cent frequency/gm) of *Phaseolus aureus* (Roxb.)

Control		*Urea Treated*		*DPJ Treated*	
Fungal Species	*% Frequency*	*Fungal Species*	*% Frequency*	*Fungal Species*	*% Frequency*
Rhizosphere mycoflora					
Aspergillus niger	18.7	*Aspergillus terreus*	35.2	*Aspergillus niger*	11.1
Aspergillus flavus	6.25	*Fusarium oxysporum*	47.2	*Aspergillus terreus*	15.5
Fusarium oxysporum	37.5	Rhizopus stolonifer	17.6	*Fusarium oxysporum*	66.6
Penicillium funiculosum	37.5			*Phoma* sp.	6.6
Total colonies	**16.0**	**Total colonies**	**17.0**	**Total colonies**	**45.0**
Rhizoplane mycoflora					
Fusarium oxysporum	66.6	*Fusarium oxysporum*	62.5	*Fusarium oxysporum*	45.4
Rhizopus stolonifer	33.3	*Rhizopus stolonifer*	37.5	*Rhizoctonia bataticola*	36.3
				Rhizopus stolonifer	18.18
Total colonies	**15.0**	**Total colonies**	**16**	**Total colonies**	**22.00**

Table 10.6: Effect of DPJ on Rhizosphere and Rhizoplane Mycoflora (per cent frequency/gm) of *Arachis hypogae* (Linn.)

Control		*Urea Treated*		*DPJ Treated*	
Fungal Species	*% Frequency*	*Fungal Species*	*% Frequency*	*Fungal Species*	*% Frequency*
Rhizosphere mycoflora					
Aspergillus carbonarieus	8.8	*Aspergillus niger*	38	*A. carbonarieus*	23.8
Aspergillus niger	26.4	*Aspergillus terreus*	8	Fusarium semitectum	23.8
Fusarium semitectum	17.6	*Fusarium semitectum*	10	*Fusarium oxysporum*	38.0
Fusarium oxysporum	20.5	*Fusarium oxysporum*	24	Rhizopus stolonifer	14.2
Myrothecium roridum	5.8	Phoma *sp.*	2		
Penicillium sp.	14.7	Penicillium fumigatus	10		
Rhizopus stolonifer	5.8	Rhizopus stolonifer	8		
Total Colonies	**34.0**	**Total colonies**	**50**	**Total colonies**	**21.0**
Rhizoplane mycoflora					
Aspergillus niger	47.6	*Fusarium oxysporum*	18.1	*Fusarium oxysporum*	50.0
Fusarium oxysporum	4.7	*Rhizopus stolonifer*	81.8	Rhizopus stolonifer	50.0
Rhizopus sp.	47.6				
Total colonies	**21.0**	**Total colonies**	**11.0**	**Total colonies**	**20.0**

On rhizoplane the fungal population increased due to the treatment with urea and DPJ. Due to the treatment with urea there was increase in number of fungal species. It was interesting to know that

Rhizoctonia bataticola a potential pathogen appeared on rhizoplane due to the treatment of urea and DPJ.

Table 10.7: Effect of DPJ on Rhizosphere and Rhizoplane Mycoflora (per cent frequency/gm) of *Glycine max* (Linn.)

Control		*Urea Treated*		*DPJ Treated*	
Fungal Species	*% Frequency*	*Fungal Species*	*% Frequency*	*Fungal Species*	*% Frequency*
Rhizosphere mycoflora					
Aspergillus niger	4.1	*Aspergillus niger*	21.8	*Aspergillus niger*	14.0
Fusarium oxysporum	14.2	*Fusarium semitectum*	18.7	*Fusarium oxysporum*	85.7
Phoma herbarum	81.6	*Fusarium oxysporum*	50.0		
		Rhizopus stolonifer	9.3		
Total colonies	**49**	**Total colonies**	**32.0**	**Total colonies**	**7.0**
Rhizoplane mycoflora					
Fusarium oxysporum	53.8	*Aspergillus niger*	7.1	*Fusarium oxysporum*	47.1
Rhizopus stolonifer	46.1	Fusarium oxysporum	14.2	*Rhizoctonia bataticola*	52.9
		Rhizoctonia bataticola	7.1		
		Rhizopus stolonifer	71.4		
Total colonies	**13.0**	**Total colonies**	**14.0**	**Total colonies**	**17.0**

Sorghum (*Sorghum bicolour* Moench)

The result presented in Table 8 indicate that the population of fungi in rhizosphere around sorghum increased due to the presence of urea. On the other hand, DPJ reduced the population of fungal species. The number of fungal species, however, reduced due to treatment of urea while increased due to DPJ in the rhizosphere. *Fusarium semitectum* and *Fusarium oxysporum* were recorded in rhizosphere due to the presence of urea and DPJ. On the rhizoplane, the fungal population was slightly reduced due to DPJ. Though *Rhizoctonia bataticola* was present on the rhizoplane the presence of *Fusarium oxysporum* was suppressed due to the DPJ treatment (Table 10.8).

Wheat (c.v. Kalyan Sona) (*Triticum aestivum* L.)

The wheat plants were treated with either urea or DPJ wherein it was noticed that total number of fungal colonies increased in the rhizosphere grown in the DPJ treated soil. Treatment of urea reduced the number of fungal species in the rhizosphere. However, the *Rhizoctonia bataticola* appeared in the DPJ treated rhizosphere. On rhizoplane total number of colonies increased due to the treatment of DPJ, though *Rhizoctonia bataticola* was universally present irrespectively with various treatments (Table 10.9).

Application of DPJ, as well as urea during present study prevented the growth of some fungi while promoted that of others, depending on the species around the root zone of which the studies were undertaken. The numbers of fungal species recorded during present study were less in both rhizosphere and rhizoplane then those reported by Gangawane (1985), probably due to variation in strain, season and experimental methods. Further investigations are necessary to interpret the exact significance of such observations, especially on nutritional grouping of fungi in the rhizosphere.

Table 10.8: Effect of DPJ on Rhizosphere and Rhizoplane Mycoflora (per cent frequency/gm) of *Sorghum valgare* (Pers.)

Control		*Urea Treated*		*DPJ Treated*	
Fungal Species	*% Frequency*	*Fungal Species*	*% Frequency*	*Fungal Species*	*% Frequency*
Rhizosphere mycoflora					
Aspergillus niger	12.5	*Aspergillus niger*	9.0	*Aspergillus niger*	13.7
Aspergillus terieus	43.7	*Aspergillus flavus*	9.0	Aspergillus terieus	10.3
Penicillium sp.	21.8	*Fusarium semitectum*	63.6	*Aspergillus ustus*	27.5
Phoma sp.	12.5	*Fusarium funiculosum*	18.1	*Fusarium semitectum*	13.7
Rhizopus stolonifer	9.3			*Fusarium oxysporum*	17.2
				H. tetramera	17.2
Total Colonies	**32.0**	**Total colonies**	**55.0**	**Total colonies**	**29.0**
Rhizoplane mycoflora					
Fusarium oxysporum	38.4	*Fusarium oxysporum*	37.5	*Fusarium roseum*	40.9
Rhizoctonia bataticola	23.0	*Phoma* sp.	37.5	Rhizoctonia bataticola	40.9
Rhizopus stolonifer	38.4	*Rhizoctonia bataticola*	25.0	Rhizopus stolonifer	18.1
Total colonies	**26.0**	**Total colonies**	**24.0**	**Total colonies**	**22.0**

Table 10.9: Effect of DPJ on Rhizosphere and Rhizoplane Mycoflora (per cent frequency/gm) of *Triticum aestivum* (Linn.)

Control		*Urea Treated*		*DPJ Treated*	
Fungal Species	*% Frequency*	*Fungal Species*	*% Frequency*	*Fungal Species*	*% Frequency*
Rhizosphere mycoflora					
Aspergillus terieus	40.0	*Fusarium oxysporum*	23.2	*Aspergillus flavus*	5.7
Fusarium oxysporum	40.0	*Rhizoctonia bataticola*	53.8	*Aspergillus terieus*	34.2
Penicillium sp.	10.0	*Rhizopus stolonifer*	23.0	*Aspergillus niger*	11.4
Rhizoctonia sp.	10.0			Canninghamella *sp.*	2.8
				Fusarium oxysporum	45.7
Total colonies	**20.0**	**Total colonies**	**13.0**	**Total colonies**	**35.0**
Rhizoplane mycoflora					
Fusarium oxysporum	35.2	*Fusarium oxysporum*	28.5	*Fusarium oxysporum*	36.8
Rhizoctonia bataticola	41.1	Rhizoctonia bataticola	50.0	*Rhizoctonia bataticola*	21.0
Rhizopus stolonifer	23.5	*Rhizopus stolonifer*	21.4	*Rhizopus solanae*	26.3
				Rhizopus stolonifer	15.7
Total colonies	**17.0**	**Total colonies**	**14.0**	**Total colonies**	**19.0**

Acknowledgement

The authors thank Prof. L.V. Gangawane for encouragement.

References

Agnihothrudu, V., 1955. *Naturwissenshaften*, 42: 515.

Agnihothrudu, V., 1961. *Soil. Sci.*, 91: 133.

Ajaykumar, K. and Mungikar, A.M., 1990a. *Sci. and Cult.*, 56: 342.

Ajaykumar, K. and Mungikar, A.M., 1990b. *Geobios.*, 17: 188.

Ajaykumar, K. and Mungikar, A.M., 1990c. *Mendel*, 7(3–4): 389.

Alexander, M., 1967. "*Introduction to soil Microbiology*" IV Edn. John Wiley and Sons, Inc. New York.

Arkcoll, D.B., 1973. *J. Sci. Fd Agric.*, 24: 437.

Bhuvaneshwari, K., 1960. *Mem. Ind. Bot. Soc.*, 3: 152.

Chanda, S. Chakraborti, S. and S. Matai, 1984. In: *Current Trends in Life Sciences Vol. XI*. Progress in Leaf Protein Research (N. Singh Ed.) Today and Tomorrow's Printers and Publisher's New Delhi, pp. 377–389.

Dakore, H.G. and Mungikar, A.M., 1986. *Pollution Res.*, 5: 123.

Das, A.C., 1963. *Trans. Brit. Mycol. Soc.*, 46: 431.

Davys, M.N.G., 1973. Read to Ann. Conf. Brit. Ass. Of Grass Crop Driers, Nov., 1973.

Davys, M.N.G. and Pirie, N.W., 1969. *Biotech. Bioengng.*, 11: 528.

Davys, M.N.G., Pirie, N.W. and Street, G., 1969. *Biotech. Bioengng.*, 11: 517.

Deshpande, K.S.and Joshi, R.N., 1971. *Mycopath. Mycol. Appl.*, 45: 151.

Gangawane, L.V., 1972. Taxonomy and physiology of rhizosphere mycoflora of groundnut. *Ph.D. Thesis*, Marathwada University, Aurangabad.

Gangawane, L.V., 1979. *Ind. Phytopath.*, 31: 528.

Gangawane, L.V., 1985. *Indian Bot. Reptr.*, 4(2): 160.

Gangawane, L.V. and Deshpande, K.B., 1972. *Marathwada University, J. Sci.*, 10: 23.

Gangawane, L.V. and Deshpande, K.B., 1973. *Ind. Phytopath.*, 26: 323.

Gangawane, L.V. and Deshpande, K.B., 1974. *Marathwada Vidnyan Mandir Patrika*, 9: 41.

Gangawane, L.V. and Deshpande, K.B., 1975. *Indian Nat. Sci. Acad.*, 41B: 383.

Gangawane, L.V. and Deshpande, K.B., 1976. *Kavaka*, 4: 2.

Gangawane, L.V. and Deshpande, K.B., 1977. *Acta Agronomica*, 26: 68.

Gangawane, L.V. and Deshpande, K.B., 1978. *Rev. D'Ecol. Biol. Sol.*, 15: 311.

Gangawane, L.V. and Kulkarni, L., 1985. *Ind. Phytopath.*, 37: 418

Ghewande, M.P. and Deshpande, K.B., 1975. *Indian J. Microbiol.*, 15: 33.

Ghewande, M.P. and Deshpande, K.B., 1977. *Indian Phytopath.*, 30: 114.

Gogle, D.P., Jadhav, R.K. and A.M. Mungikar, 2001. In *Frontiers in Fugal Biotechnology and Plant Pathogen Relations*, (Ed.) C. Manohoarachary. Allied Publications, Ltd., Hyderabad, pp. 272–274.

Hiltner, L., 1904. *Arb Dtsch. Landw. Ges.*, 98: 59.

Jadhav, R.K., 1997. Studies on the preparation of protein concentrates from green leaves. *Ph.D. Thesis*, Dr. Babasaheb Ambedkar Marathwada University, Aurangabad.

Jadhav, R.K. and A.M. Mungikar, 1998a. *Int. J. Mendel.*, 15 (1 and 2): 21–22.

Jadhav, R.K. and Mungikar, A.M., 1998b. *J. Aqua. Biol.*, 13: 144.

Jadhav, R.K. and Mungikar, A.M., 2000. In: *Environmental Issues and Sustainable Development*, (Ed.) Wagh, S.B.. Vinit Publications, Aurangabad.

Jadhav, R.K. and A.M. Mungikar, 2001a. *Geobios.*, 28: 271.

Jadhav, R.K. and Mungikar, A.M., 2001b. In: *Plant Disease Management*, (Ed.) J. Deshpande. Kailash Publications, Aurangabad, pp. 142–144.

Johnson, L.F., Curl, E.A., Bond, J.H. and Fribourg, H.A., 1959. *Methods for Studying Soil Microflora: Plant Disease Relationship*. Burgess Publishing Co., Minnesota, Minn.

Kawathekar, S.U., 2001. Studies on the fractionation of green foliages. *Ph.D. Thesis*, Dr. Babasaheb Ambedkar Marathwada University, Aurangabad.

Kulkarni, L. and Gangawane, L.V., 1982. *Rev. Ecol. Biol. Sol.*, 19: 525.

Lakshmi Kumari, M., 1961. Rhizosphere microflora and host-parasite relationships. *Ph.D. Thesis*, University of Madras.

Maindarkar, K.G. and Mungikar, A.M., 1994. *Marathwada University J. Sci.*, 27: 34.

Mishra, R.R., 1967. *Plant and Soil*, 27: 162.

Mungikar, A.M., 2001. *Dr.B.A.M.U. J. Sci.*, 31(8): 59.

Mungikar, A.M. and Gogle, D.P., 2001. In: *Advances in Mycology and Plant Pathology*. Dr. K.S. Deshpande Commemoration volume Ed. L.V. Gangawane. Dr. Kalidas Deshpande Commemoration Committee, Aurangabad, pp. 126–138.

Naim, M.S., 1967. *Mycopath. Et. Mycol. Appl.*, 31: 296.

Padma, B. and Manohara Chary, G., 1994. *Indian Bot. Reptr.*, 16(1+2): 66.

Peterson, E.A., 1958. *Can. J. Microbiol.*, 4: 247.

Pirie, N.W., 1942. *Nature, Lond.*, 149: 251.

Pirie, N.W., 1971. Ed."*Leaf protein: its agronomy, preparation, quality and use*" (Pirie, N.W. Ed.), IBP Handbook No.20, Blackwell Scientific Publications, Oxford and Edinburgh.

Pirie, N.W., 1978. *Leaf Protein and Other Aspects of Fodder Fractionation*. Cambridge University Press, London.

Pirie, N.W., 1987. *Leaf Protein and its By-products in Human and Animal Nutrition*. Cambridge University Press, London.

Rangaswami, G. and Vasantharaj, V.N., 1962. *Can. J. Microbiol.*, 8: 473.

Rangaswami, G. and Venkatesan, R., 1964. *Curr. Sci.*, 33: 181.

Rao, A.S., 1962. *Plant and Soil*, 17: 260.

Ream, H.W., Jorgensen, N.A., Koagel, R.G. and Bruhn, H.D., 1983. In: *Leaf Protein Concentrates*, (Eds.) Telek, L. and Graham, H.D. AVI Publishing Co. Inc., Westport, Connecticut, pp. 467.

Roy, R.Y. and Gujrati, S., 1967. *Proc. 54th Ind. Sci. Cong.*, 3: 294.

Saksena, S.B., 1955. *J. Ind. Bot. Soc.*, 34: 261.

Saler, R.S. and Gangawane, L.V., 1994. *Indian bot. Reptr.*, 13 (1+2): 41.

Salve, P.B. and Gangawane, L.V., 1984. In: *Current Trends in Life Sciences Vol. XI.* Progress in leaf protein research (Singh, N. Ed.) Today and Tomorrow's Printers and Publishers, New Delhi, pp. 391.

Salve, U.S., 2002. Studies on mechanic fractionation of green foliage to produce high quality food and feed products. *Ph.D. Thesis*, Dr. Babasaheb Ambedkar Marathwada University, Aurangabad.

Sayyed, I.U. and Mungikar, A.M., 2001. In: *Plant Disease Management*, (Ed.) J. Deshpande. Kailash Publications, Aurangabad, pp. 138–141.

Sayyed, I.U. and Mungikar, A.M., 2003. *Int. J. Mendel*, 20 (3–4): 79.

Shinde, S.R., 1982. In: *Methods in Experimental Plant Pathology*, (Eds.) Mukadam, D.S. and Gangawane, L.V. COSIP–ULP, Botany Dept., Marathwada Univ., pp. 74.

Shinde, S.R., 1983. Studies on Physiology of Phoma herbarum with special reference to cellulolytic enzymes. *Ph.D. Thesis*, Marathwada University, Aurangabad.

Shrinivasan, K.V., 1958. *J. Ind. Bot. Soc.*, 37: 334.

Singh, J., 1937. *Ann. Appl. Biol.*, 24: 154.

Smith, N.R. and Dawson, V.T., 1944. *Science*, 58: 467.

Subba Rao, N.S., 1977. *Soil Microorganisms and Plant Growth*. Oxford and IBH Publishing Co., New Delhi.

Subrahmanyam, P., 1975. *Ph.D. Thesis*, Sri Venkateswara University, Tirupati (A.P.) Quoted from Gangawane.

Sunar, M.S. and Chohan, J.S., 1971. *Ind. J. Agric. Sci.*, 41: 38.

Telek, L. and Graham, H.D., 1983. *Leaf Protein Concentrates*. AVI Publishing Co., Inc. Westport, Connectincut.

Sustainable Environmental Management *Pages 89–95*
Edited by: **Dr. L.V. Gangawane & Dr. V.C. Khilare**
Published by: **DAYA PUBLISHING HOUSE**

Chapter 11

Role of Nitrogen Fixing Microorganism in Bioremediation of Salt Affected Soils

L.V. Gangawane

Soil Microbiology and Pesticides Laboratory, Department of Botany, Dr. Babasaheb Ambedkar Marathwada University, Aurangabad – 431 004, MS, India

ABSTRACT

Nearly 7 million hectares of land is salt affected in India. Many scientific workers suggest microbial bioremediation. Many microbes have been detected in saline soils. However, application of nitrogen fixers have been suggested as best source for bioremediation of saline soils by many authors. These microbes include *Azotobacter, Azospirillum, Anabaena, Scytonema, Oscillatoria,* etc. *Rhizobium* have been also suggested for the saline tolerant legumes. This paper reviews the possible application of such nitrogen fixers for bioremediation of salt affected soils.

Keywords: *N–fixation, Saline soil, Azotobacter, Azospirillum, Rhizobium*

Introduction

In India salt affected soils are known as 'Usar soils' covers more than 7 million hectares of land. These soils are distributed in Uttar Pradesh, Haryana, Delhi, Punjab, Bihar and parts of Rajasthan; medium and deep black soil areas of Madhya Pradesh, Gujarat, Maharashtra and Andhra Pradesh, coastal and deltaic alluvium of West Bengal, Orissa, Andhra Pradesh, Tamil Nadu, Kerala, coastal arid areas of Gujarat, arid soils of Rajasthan and Gujarat (Kaushik and Ummat, 1992). These soils are becoming useless for crop production day by day. The crop either fails completely or gives very poor yields in sodic and saline sodic soils mainly because of poor soil physical conditions and nutritional disorders related to high exchangeable sodium content in soil. The conventional method of reclaiming these soils is application of gypsum and heavy dose of organic matter with proper drainage system.

However, bioremediation with the application of salt tolerant microorganisms, particularly N-fixers have been found to be more fruitful. Their biology and possible applications are reviewed as under.

Microorganisms in Saline Soils

Microbial aspects of salt affected soils have received very little attention. Bajpai and Gupta (1979) reported sulphur metabolism of chemolithotrophic bacteria and other bacteria from saline and alkaline soil in U.P. Interaction between microorganisms, chemical composition and environment in salt affected soils was studied by Douka *et al.* (1983). They determined the microbial population over a period of 19 months. Simultaneously elemental chemical composition measured by X-rays fluorescence technique was done in uncultivated salt affected areas of Greece at Lantza and Klidi. It was noted that in Lantza soil both the microbial population (fungi, bacteria and actinomycetes) and chemical composition of soil remained constant for a 19 month of study while at Klidi there was alteration in both the microflora and chemical composition. This was due to ecological variations.

Free living nitrogen fixing microorganisms have been reported from saline soils in different countries. *Xanthohacter lavas* and *Alcaligem paradoxus* from strongly saline Takyar soil in USSR showed nitrogenase activity when they were grown in media containing 0.5 per cent to 2.0 per cent NaCl. The predominant growth of these bacteria in the complex of diazotrophs from strongly Takyar soils and utilization of various substrates have been noted (Kravchenko, 1988). Zafar *et al.* (1987) indicated the presence of *Klehsielh pnemonae* and *Beijerinkia* sp. in the rhizosphere of salt tolerant grass, *Leptochha fusca* and could tolerate 3 per cent NaCl in the medium.

Nitrogen fixation by *Azotobacter and Azospirillum* in saline soils have been reported by few workers. In Egypt 156 representative soil samples, particularly from Nile valley, showed the presence of *Azotobacter croococcum* and *A vinelandii.* Development of these species in saline soil was improved by amending the soil with 1 per cent maize straw. Tilak and Krishnamurti (1981) found that *Azospirillum* survives in saline and alkali soils and seed inoculation with this organism in these soils helps in increasing yields of a salt tolerant crop like barley in India. *Azospirillum brasileme* was isolated from Egyptian soils (Nadia *et at,* 1985). Jena *et al* (1988) noted that nitrogen fixation in *Azospirillum* sp. from saline soils decreased with an increase in salinity level at Cuttack. *A. brasileme* and A *lipoferum* were isolated from the roots of wheat grown in sandy saline soils in Iraq (Al Maadhidi, 1989).

Growth of some locally isolated strains of blue green algae (BGA) or cyanobacteria on different types of saline and alkaline soils was studied by Ibrahim *et al.* (1979). They noted that *Anabaena otyzae* and *A. navicuhides* could grow slightly in these soils. Subhashini and Kaushik (1981) found that algal growth resulted significant reduction in pH, electrical conductivity, exchangeable Na, hydraulic conductivity and aggregation status of the soil due to algal growth. Similarly, Venkatraman and Venkatraman (1991) found that fresh water cyanobacterium *Anabaena variabilis* ARM (668) was found to respond to 0.2 M NaCl stress. *Anabaena variabilis, Gheocapsa limnetica, Oscillatoria formosa, Nodularia spumigno* and *Nostoc* sp. have also been reported from saline soils of Saudi Arabia (Ibrahim *et al,* 1979).

In India, Singh (1961) reported the occurrence of species of *Microcoleus, Scytonema* and *Porphyrosiphon* on the loose as well as compact salt efflorescence of sodium on salty soils. The succession of these genera was followed with species of *Complonema, Cylindrospermum and Nostoc.* Kaushik (1983a) reported the occurrence of species of *Lyngbya, Microcoleus, Oscillatoria* and *Aphanocapsa* on saline sodic soils of Haryana, Maharashtra, Orissa, Punjab, Rajasthan and Uttar Pradesh. Species *of Anabaena, Cahthrix, Haplosiphon, Westielhpsis, Schizothrix, Microcoleus, Oscillatoria* and *Plectonema* also constitute the prominent form of cyanobacterial flora of saline soils (Kaushik and Ummat, 1992). Application of cyanobacteria to the salt affected soils has led to the enrichment of soils with nitrogen content (Subhashini and Kaushik, 1981; Kaushik and Subhashini, 1985). The increase in soil nitrogen varied from 17.9 to 289.5 per cent over initial soil nitrogen (Kaushik and Subhashini, 1985).

Salt Tolerant Rhizobia and N-fixation in Legumes

Symbiotic N_2 fixation is commonly limited by soil salinity. Wilson (1931) indicated that *Rhizobium trifolii* and *R. leguminosarum* is sensitive to salinity especially at more than 0.4 per cent NaCl. He further noted that NaCl inhibited nodulation of soybean. Wilson and Norris (1970) demonstrated reduced nodule formation due to salt stress (140 mM) of *Neonotonia wightii.* Bernstein and Ogata (1966) also showed inhibition of nodulation in soybean due to salinity.

Singh and Sharma (1971) isolated 16 rhizobial cultures from different legumes and identified them as *Rhizobium japonicum, R. meliloti, R. trifolii* and *R. leguminosarum.* Salt tolerance and utilization of carbon source varied from strain to strain of each species. *R. japonicum* utilized peptone sugars better than hexose. Yadav and Vyas (1973) found that the salt effect appeared to be ion specific where chlorides were more toxic than sulphates of Na, K and Mg to *Rhizobium* strain from *Sesbania cannabina, Crotolaria juncea* and *Glycine max.* $MgSO_4$ at 1 per cent had a stimulatory effect on bacterial growth. Both salt sensitive and salt tolerant strains of rhizobia were recorded. Subba Rao *et al.* (1972) reported that lucerne nodulating strains of *R. melilotii* were capable of tolerating NaCl concentration ranging from 2.75 to 3.0 per cent in the growth medium. Further it was noted that in lucerne, salinity resulted in a root system devoid of root hairs, mucilagenous layer and infection thread formation inspite of the optimum growth of rhizobia under salt conditions.

Toxicity of sodium and chloride ions to *Rhizobium trifolii* and *R. melilotii* in peat culture was studied by Steinborn and Roughley (1975). Calcium chloride was more toxic than sodium chlorides in broth and peat culture. The toxicity of NaCl was ascribed to chloride ions. *R. melilotii* strains grew on 3.5 per cent NaCl after adaptation during a long period while rhizobia for soybean and cowpea grew at 0.5 per cent NaCl and those for clover and pea at 1.0 per cent NaCl. Pandher and Kahlon (1978) showed that very few isolates *of Rhizobium leguminosarum* from pea *(Pisum sativum)* could show their growth on NaCl. $MgSO_4$ was found to be stimulatory while $MgCl_2$ and sodium bicarbonate were inhibitory even at low concentrations.

Study of plasmids, biological properties and efficiency of N-fixation in *Rhizobium japonicum* strains from alkaline soils was done by Dennis *et al.* (1979). They isolated plasmids from strains of *R. japonicum* serogroup 135 and indicated that there are four groups on the basis of difference in plasmid number and size. *Rhizobium melilotii* Be 151, an effective strain for *Medicago sativa* was a facultative halotolerant bacterium. The strain tolerated salt up to 600 mM, although 40 per cent inhibition in growth was observed at 400 mM NaCl. But addition of glycine betaine helped this strain to overcome the diminution growth in the presence of NaCl. This proved that extent of N-methylation of glycine is important in producing the antistress effect (Sauvage *et al.,* 1983).

Accumulation of amino acids in *Rhizobium* sp. from wild legume plant *Prosopis* spp. in response to NaCl salinity is known. Hua *et al.* (1982) found that this *Rhizobium* is able to grow in the medium containing NaCl up to 500 mM approaching that of sea water. Intracellular free glutamate was found to increase rapidly in response to osmotic stress by NaCl. It accounted for 88 per cent amino acid pool when the bacterium was grown in 500 mM NaCl. The glutamate dehydrogenase activity differed according to the strain. In this connection Yap and Lim (1983) found that *Rhizobium* sp. UMKL 20 also responded to increased NaCl concentration in the medium by elevating the intracellular concentration of K^+ and glutamate. Increase in K^+ occurred in a time course by synchronous with glutamate. Addition of uncoupler, 2,4-dinitrophenol significantly reduced K^+ uptake but had little effect on glutamate accumulation. New protein synthesis did not appear to be required for the stimulation of K^+ uptake by NaCl. Assays of enzymes involved in glutamate synthesis showed that under salt stress condition increased activities of glutamine synthetase and glutamate synthetase were detected indicating that the GS/GOGAT pathway is the major pathway for increasing intracellular glutamate concentration.

Similarly Hua *et al.* (1986) found that accumulation of high levels of intracellular glutamate was a common response of a variety of bacteria to osmotic stress. They characterized the NADPH-dependent glutamate synthetase (GOGAT) in salt tolerant *Bradyrhizobium* sp. strain WR-1001. This enzyme had a pH optimum in the region of 8.3 as compared to 7.6 for most of the strains. This enzyme was not inhibited by the end product glutamate in the range of 50-200 mM.

Variation in the response of salt stressed *Rhizobium* strains to betiness is recorded by Bernard *et al.* (1986). They studied 15 rhizobial strains belonging to *R. melilotii, R. japonicum, R. trifolii, R. leguminosarum, Rhizobium* spp. (from *Sesbania rostrata* and *Hedysarum coronarium*) for their growth rate under salt stress and noted that in the presence of inhibitory concentrations of NaCl, the enhancement of growth resulting from added glycine betine in *R. melilotii* strains and *Rhizobium* spp. only. The concentration of glycine betine required for maximum growth stimulation was 1 mM.

Characterization of an effective salt tolerant fast growing strain of *Rhizobium japonicum* USD A 191 in China was made by Yelton *et al.* (1983). They showed that this strain nodulates American soybean very effectively. Further, this strain was compared withi?. *japonicum* USDA 110 from America. The strain *R. japonicum* 191 has a doubling time of 3.2 h in complex medium and grows in concentration of up to 0.4 M NaCl while strain USDA 110 had doubling time of 12 h and is inhibited at 0.1 M NaCl. Under salt stress conditions, intracellular levels of K^+ and glutamate were shown to be increased. A comparison based on carbohydrate metabolism, DNA homology and protein patterns on polysaccharide gel reveals that strain USDA 191 is more closely related to fast growing rhizobia than to *R. japonicum.* The strain also retains the capacity to nodulate American soybean and cowpea cultivers effectively.

In Greece, Douka *et al.* (1984) studied salinity tolerance of *Rhizohium melilotii* strain isolated from salt affected soils. They saw the growth of this strain in YEM broth containing 0-1.2 per cent NaCl. They also noted the growth at same generation time and simultaneous onset of stationary phase while the total viable number of cells was the same for three continuous generations. The nodulation, plant yield and elemental composition *oiMedicago sativa* plants grown on agar slopes containing 0-1.2 per cent NaCl responded identically to all inoculum. Singleton *et al.* (1992) studied effect of NaCl on nodule formation in soybean in Hawaii. They found that *Rhizobium japonicum* colonization on inoculated root surface was affected by the salt treatment up to 79.9 mM NaCl. Total nitrogenase activity decreased proportionally in relation to nodule number and dry weight. Specific nitrogenase activity, however, was less affected by salinity and was not depressed significantly until 79.9 mM NaCl.

Passage of *Rhizobium trifolii* on media containing increasing concentration of NH_4NO_3 resulted in an extension of the range of tolerance of *R. trifolii* to this salt whereas ability to tolerate high levels of $NH4NO_3$ resulted in an improvement in growth on lower level passage batch. On NH4NO3 free media caused a gradual reversion to the original range of tolerance ability to tolerate high concentration of NH4NO3 was accompanied by a change to nodulation pattern on white clover (Berraj, 1983). *Rhizobium leguminosarum* TAL 271 could tolerate NaCl upto 2 per cent while $CaCl_2$ was more toxic. $MgSO_4$ was beneficial at lower concentrations (Helmish and El-Gammal, 1987).

In UK, EL-Sheik and Wood (1989) studied response of chickpea and soybean rhizobia to salt. They reported that there is influence of carbon source, temperature, pH on salt tolerant rhizobia through the osmotic and specific ion effects of salts. Chickpea and soybean rhizobia showed patterns of utilization of carbon sources characteristic of fast growers in the absence of NaCl. Out of four chickpea rhizobia and four soybean and Bradyrhizobia only chickpea *Rhizobium* strain 2-ICAR-MOR-Ch-192, soybean *Rhizobium* strains USDA 191 and USDA 201 were able to utilize some carbon sources in the presence of NaCl. All rhizobia were able to grow at 25 and 37°C in YEM broth. Salt stress was more severe at alkaline pH and lower temperature. Hence they showed that tolerance to salt by rhizobia is more dependent upon pH, temperature and carbon source. In addition, chloride ions of Na, K and Mg were found to be more toxic than the corresponding sulphate ions. The osmotic effect using polyethylene glycol (PEG) glycerol and sorbitol to raise the osmotic pressure to –0.1 and –2.0 Mpa indicated that PEG was toxic even at low concentrations. Although, the sensitivity of strain to NaCl was increased at higher osmotic pressure, the harmful effect of salt on the growth of chickpea *Rhizobium* could be attributed to some specific ion effects rather than osmotic effect. This inhibition also showed increased growth when glutamate was added to 0.34 M NaCl. Salt sensitive isolate of *Bradyrhizobium japonicum* showed slight improvement in its growth due to glycine betine.

There are reports on the salt tolerance by rhizobia and nodulation and N-fixation in legume plants. Susheng and Jilun (1989) noted that fast growing soybean *Rhizobium* RT-19 was salt tolerant but less efficient in nitrogen fixation. But they developed two transformants rhizobia RTt-19 and RTt-50 which were salt tolerant and fixes nitrogen effectively. These strains were obtained by transforming the DNA from strain RT-19 into cells of strain USDA-110. Free glutamate was found to increase rapidly in strain Rt-10 and also in transformed strains. Recent studies by El-Saidi and Ali (1983) in Egypt are interesting in relation to transfer of salt tolerant gene in *Rhizobium* and *Azotobacter* from *Bacillus* sp. strain tolerating 30 per cent NaCl and 55°C temperature. The transformed *Rhizobium* and *Azotobacter* could tolerate 10 to 20 per cent NaCl and were also effective in N-fixation. Higher yields of sugar-beets, oilseed, grape and barely were noted. Similarly, Biswal and Mishra (1992) isolated 24 strains of *Rhizobium* belonging to cowpea group from twenty-five legume species belonging to 15 genera of the family Fabaceae. These plants were obtained from coastal sand dunes and saline swapy area. Their compatibility with two varieties of mung bean was tested using tube and sand culture. Only 8 strains were compatible and survived under alkaline conditions. Artificial induction of tolerance to pH changes in these strains could be achieved to prolonged exposure to pH variation, which provided a possible way to use them as biofertilizer under alkaline soil conditions. Abo-Alla (1992) inoculated *Rhizobium leguminosarum* RCR 1001 to Faba bean Giza 3 in pot and irrigated with saline water (50 to 125 mM NaCl). He noted that there was significant decrease in nodule number, nodule fresh weight and total nitrogenase activity. Recent studies by Gangawane and Salve (1993) showed that isolates Id-51 and Iga-136(b) from *Indigofora duthei* and/. *glandulosa* var. *sykesii* could tolerate 3 per cent NaCl on the agar medium. Some isolates Igg-69, Igs-20, Id-21, Id-50, Igg-62, Ie-101(b), Ie-108(a) and Igs-136(a) tolerated NaCl up to 2.8 per cent. However, *Rhizobium* Em-110 from *Eleiotis monophylla* was stimulated by 0.01 per cent NaCl. Nitrogen fixation in groundnut (peanut) by five salt tolerant isolates indicated that Ie-108(a) was more effective in bottle jar and was also effective on groundnut Cv.SB-11 under saline field conditions. There was 20.81 per cent increase in the pod yield over uninoculated control.

Acknowledgement

This paper is prepared under UGC-DRS IPM programme. Thanks are due to UGC, New Delhi for the financial assistance under SAP.

References

Abo-Alla, M.H., 1992. *Symbiosis*, 12: 311–319.

Al–Maadhidi, J.F., 1989. *J. Univ. Kuwait Sci.*, 6: 343–348.

Bajpai, P.D. and B.R. Gupta, 1979. *J. Ind. Soc. Soil Sci.*, 27: 197–198.

Bernard, T., Pocard, J.A, Perroud, B. and D.L. Rudulier, 1996. *Arch. Microbiol.*, 143: 359–364.

Bernstein, L. and G. Ogata, 1966. *Agronomy J.*, 58: 201.

Berraj, R.D., 1983. *Can. J. Microbiol.*, 29: 563–569.

Biswal, A.K. and A. Mishra, 1992. Effect of *Rhizobium* as biofertilizer for green gram cultivation under acid and alkaline conditions. In: *Biofertilizer Technology Transfer,* (Ed.) Gangawane L.V. Assoc. Publ. Co., New Delhi, pp. 121–176.

Dennis, G., Vidaver, AK. and R.V. Klucas, 1974. *J. Gen. Microbiol.*, 114: 257–266.

Douka, C.E., Xenboulis, AC. and T. Paradellis, 1984. *Folia Microbiol.*, 29: 316–324.

El-Saidi, M.T. and A.M.M. Ali, 1993. Growing of different field crops under high salinity levels and utilization of genetically engineered rhizobia and *Azotobacter* salt tolerant strain. In: *Towards the Rational Use of High Salinity Tolerant Plants, Vol. 2. Agriculture and Forestry Under Marginal Soil Water Conditions,* (Eds.) H. Lieth and A.A. AL-Masoom. Kluwer Academic Publ., London, pp. 59–66.

EL-Sheikh, E.A.E. and M. Wood, 1989. *Soil Biol. Biochem.*, 21: 883–887.

Gangawane, L.V. and P.B. Salve, 1993. Salt tolerant rhizobia from wild legumes and nitrogen fixation in groundnut in semi–arid tropics. In: *Towards the Rational Use of High Salinity Tolerant Plants*, Vol. 2, (Eds. H. Lieth and A.A. AL-Masoom). Kluwer Academic Publishers, London/Boston, Dordrecht, pp. 53–58.

Helmish, F.A. and S.M.A. EL-Gammal, 1987. *Zent. Microbiol.*, 142: 211–214.

Hua, ST., Lichens, G.M., Guirao, A and V.Y. Tsai, 1986. *F.E.M.S. Microbiol. Letters*, 37: 209–213.

Hua, S.T., Tsai, V.Y., Lichens, G.M. and AT. Nema, 1982. *Appl. and Environ. Microbiol.*, 7: 135–140.

Ibrahim, AN., Elayouty, E.Y., EL-Sherbent, M.A and M.S. Khadr, 1979. *Egypt. J. Physiol. Sci.*, 6: 149–154.

Jena, P.K., Adhya, T.K. and Rajaram V. Mohanrao, 1988. *Microbios.*, 54: 157–164.

Kaushik, B.D. and D. Subhashini, 1985. *Proc. Ind. Nat. Sci. Acad.*, B51: 386–389.

Kaushik, B.D. and J. Ummat, 1992. Reclamation of salt affected soils with blue green algae (cyanobacteria): A technology development. In: *Biofertilizer Technology Transfer,* (Ed. Gangawane, L.V.). Assoc. Publ. Co., New Delhi, pp. 158–164.

Kravchenko, J.K. and T.A Kaliminskaya, 1980. *U.S.S.R. Microbiologia*, 52: 279–283.

Nadia, F.E., Fayaz, H., Makboul, H.E. and R. El-Shanwy, 1985. *Egypt, Z. Pflan. Bod. Kd.*, 147: 210–217.

Pandher, M.S. and S.S. Kahlon, 1978. *Ind. J. Microbiol.*, 18: 81–84.

Sauvage, D., Hamelin, J. and F. Larher, 1983. *Plant Science Letters*, 31: 291–302.

Singh, A and P.B. Sharma, 1971. *Ind. J. Microbiol.*, 11: 33–36.

Singh, R.N., 1961. Role of blue green algae in nitrogen economy of Indian Agriculture. In: ICAR Publ. New Delhi, p. 61.

Singleton, P.W., EL-Swaify, S.A and B.B. Bohlool, 1992. *Appl. Environ. Microbiol.*, 44: 884–890.

Stinbom, J. and R.J. Roughley, 1975. *J. Appl. Bact.*, 39: 133–138.

Subba Rao, N.S., Lakshmikumari, M., Singh, C.S. and S.P. Magu, 1972. *Ind. J. Agric. Sci.*, 42: 384.

Subhashini, D. and B.D. Kaushik, 1981. *Aust. J. Soil Res.*, 19: 361–366.

Tilak, K.V.B.R. and G.S.R. Krishnamurti, 1981. *Zbl. Bakt. II. Abt.*, 136: 641–643.

Venkatraman, S. and G.S. Venkatraman, 1991. *Ind. J. Microbiol.*, 31: 131–138.

Wilson, J.K., 1931. *J. Agric. Res.*, 58: 261–266.

Wilson, J.R. and Norris, D.O., 1970. *Proc. 11th Int. Grasl.Cong.*, pp. 455–458.

Yadav, N.K. and S.R. Vyas, 1973. *Folia Microbiol.*, 18:242–247.

Yap, S.F. and S.T. Lim, 1983. *Arch. Microbiol.*, 135: 224–228.

Yelton, M.M., Yang, S.S., Edie, S.A. and S.T. Lim, 1983. *J. Gen. Microbiol.*, 129: 1537–1547.

Zafar, Y., Malik, K.A and E.G. Niemann, 1987. *J. Appl. Microbiol. Biotechnol.*, 3: 45–56.

Sustainable Environmental Management *Pages* ***96–104***
Edited by: **Dr. L.V. Gangawane & Dr. V.C. Khilare**
Published by: **DAYA PUBLISHING HOUSE**

Chapter 12

Sewage Irrigation and Soil Microorganisms

***P.N. Jadhav*[1] *and L.V. Gangawane*[2]**

[1]*Department of Microbiology, Deogiri College, Aurangabad*
[2]*Soil Microbiology and Pesticides Laboratory, Department of Botany, Dr. Babasaheb Ambedkar Marathwada University, Aurangabad – 431 004, MS, India*

ABSTRACT

Sewage is rich source of plant nutrients and can be utilized for irrigation (Stone, 1955; Seep, 1971). The possibilities of application of sewage and sludge in agriculture have also been demonstrated in Jordan (Gharaibeh, 1989). However, use of raw sewage continues to be health hazards. Hence most of the researches are carried out on human pathogens (Dingress, 1969) and comparatively very little is known regarding other beneficial soil micro-organisms like nitrogen fixing *Azotobacter* sp. (Gangawane, 1989), This review takes an account of soil microorganisms present in sewage irrigated soil.

Keywords: *Sewage irrigation, Soil microorganisms, Azotobacter, Rhizobium.*

Introduction

Efficient utilization of water resources is vital to agricultural production. Because water sources are limited and are rather insufficient to meet the long term requirement of agriculture and consequently affects the rural development. There is also a large gap between the net available water supply and the amount required for intensive cultivation of crop plants. Moreover, natural rainfall is seasonal and generally distributed throughout the country. The technology has brought rapid urbanization and with the multifold expansion of cities, sewage management has become a big task. The following discussions reviews the past and present trends of utilizing sewage for irrigation and the role of different microorgaisms including nitrogen fixing free living *Azotobacter* (Kanwar, 1970).

Sewage Farming

In each of the ancient civilization there is evidence of the attempts made for the utilization of sewage for irrigating land for crop production. In many European countries this practice has a long tradition. In 1890's a combined sewer system of 100 km length was built in Braunscheweiz. In about 1843, at Edinburgh, Ashburton and Devon first sewage fields associations were established for the purpose of manuring the soil with sludge. In Germany, land application of wastewater was about 11,444 hectares in 1955. The sewage utilization association (SUA) was founded at Braunschweing in 1954. About 350 farmers were members of this association. They made available for surface flooding, a total area of 4,200 ha including 3,000 hectares of irrigated crop land within Mexico city and at 20 per cent of the municipal waste water was already being utilized for watering parks (Jorge and Eloy, 1977).

A study by Henry *et al.* (1954) indicated the influence of sewage irrigation on mineral contents; organic contents, soil structure and other aspects of soil quality. In their studies, it was seen that the most important mineral problem was of boron since it has become an important additive to household soaps and detergents for bleaching purposes. Zinc and copper were also important metals additives to soil; but as galvanised pipe usage declines and zinc may become less important.

In India, there are about 145 cities and towns, where sewage farming is practised. These farms cover approximately 13,000 hectares of land using about 255 million gallons of sewage per day. The sewage is a rich source of plant nutrients and other organic wastes. Average content of N, K and P in raw sewage is shown to be 66.2, 42.1 and 22.6 mg/Lit respectively (Kanwar, 1970). A series of chemical analysis of sewage-irrigated soils were conducted by Dye (1958). The total soluble salts in the effluent were within the range of ordinary tap water. There was also increase in HCO_3, NaP_2O_5 and total N in sewage. Irrigation of soils with sewage adds 2.5–3.0 kg of N, 1.0 Kg of P, and 1.3–1.6 kg of K per capita per year.

The major constituents of organic waste are polysaccharides (cellulose, starch, pectin, hemicellulose), lignin lipids (esters of higher fatty acids with glycerol) and proteins. It also contains heavy metals like Cu, Zn and Cd that is readily available for transport in soil. Sidle and Kardos (1977) showed presence of less than 1 per cent of Zn and about 56.2 per cent of Cd in the sludge while Jorge and Eloy (1977) showed the increase in the boron concentration in Mexico city fields. Effect of sewage sludge on physical and chemical properties of soil is also known (Epstein *et al.*, 1976).

There are many references showing use of sewage for irrigation. In USA study of wastewater reclamation and utilization revealed beneficial effects of sewage or sludge on grasses, shrubs and field crops like rye, wheat, corn and alfalfa (Anonymous, 1967). Bachmann (1954) found 52 per cent increase of hay yields due to sewage spray. Similar observations were made by Bendixon *et al.* (1968) in case of grasses when trickling filter effluent was used.

Day *et al.* (1963) conducted experiments for two years to compare the grain yield, grain quality and malt quality of two barley varieties Atlas and Hauchan (1954) irrigated soil with sewage effluent and normal well water. He showed that kernel weight and kernel size were reduced on plots irrigated with sewage effluents. However, nitrogen content was increased in sewage effluent. Although the sewage effluent plots produced a higher yield of malt extract percentage from both varieties, the high nitrogen in sewage effluent tended to reduce barley and malt quality.

Kardos (1967) found that hay yield was increased by 139 per cent in corn grain by 78 per cent and oat grains by 70 per cent due to the use of sewage effluent while Odum *et al.* (1977) noted 3 times increase of growth of cypress trees. Day and Tucker (1959) obtained 11.14 tonnes winter pasture forage yields from barley irrigated with sewage effluent with no additional fertilizer. Similarly wheat and oat production was 263 per cent and 249 per cent higher respectively than those, which received only pump water. Barley was more sensitive to show detrimental effects of sewage effluent than wheat and oat. Ball (1977) cultivated eight perennial different legumes using municipal wastewater and found differential yields of these crops.

Application of municipal sewage to agricultural lands in Japan improved both soil fertility and increased yields of Italian rye grass, tulip and tomato (Yoshihisa *et al.*, 1974). Grape vines irrigated with sewage effluent increased weight of the fruits (McCarthy, 1981). Sewage irrigation also increased maize fodder yield due to increase of N uptake (Sharma and Kausal, 1984).

Effluents from sugar factory were high in various solids, COD, BOD, chlorides, sulphates and showed alkaline pH. Germination of kidney bean seeds was 100 per cent in water-irrigated soil while it ranged 91 per cent to 99 per cent in the concentration of effluent. The water irrigated soil and the soil irrigated with 25 per cent effluent were most suitable for germination (Mohammad and Khan, 1983). Sewage was utilized for short day's plants and for high yielding rabi paddy cultivation which resulted in better growth and crop yield. Soil fertility was also improved day by day with slow deposition of suspended solid matters (Naskar *et al.*, 1986). On the basis of the estimation of the population, large amounts of sewage sludge have revealed the possibility of its use in agriculture (Gharaibehn, 1989). Studies by Juwarkar *et al.* (1991) indicated that yield of vegetable crops, oil seed crops, fruit crops and other cash crops is increased due to application of untreated sewage and diluted sewage at Nagpur.

The increasing agricultural reuse of treated effluent serves goals such as promoting sustainable agriculture preserving scarce water resources and maintaining environmental quality. Also, irrigating with wastewater may reduce purification levels and fertilization costs, because soil and crops serve as bio-filters and wastewater contains nutrients (Harwy and Nava, 1997). In Egypt fruiting of Anna apples in response to application of sewage sludge was studied. The fate of heavy metals that is cadmium, selenium, lead and chromium in the leaves and fruit juice as well as the yield and fruit quality of Anna apples in response to the addition of sewage and sludge (250 gm N/tree) is combined with mineral N, source (500 gm N/tree) were investigated during 1996 and 1997 seasons in Egypt. Results showed that the leaves and fruit juice contained higher chromium, lead, selenium and cadmium. Although the concentration of heavy metal in the juice in fruits produced trees from treated with sludge was not toxic, more studies are needed for the safety of sludge for fruit crops (Ahmed *et al.*, 1998).

Occupational environmental health associated with both industrial and domestic sewage reuse for food production in Athi River town Kenya has been studied (Kingsley *et al.*, 1999). In New Zealand, stream water impact of sewage effluent irrigation onto steeply sloping land. In pilot study indicated that how irrigation of secondary sewage effluent onto steeply sloping land effected soil physical chemical and biochemical properties. Irrigation significantly improved total P status of the soils and greatly enhanced nitrification potential (Speir *et al.*, 1999).

Hong Kong produces over two millions tonnes of municipal wastewater each day. Until recently 50 per cent of wastewater volume entered water sources, rivers and coastal water without treatment. In 1986, a flexible framework of environmental management master plan was designed to be implemented over the following two decades. The master plan comprises (1) establishment of water control zones and pollution control legislation (2) upgrading of services and facilities for management of municipal sewage and chemical wastes (3) construction of the strategic sewage disposal scheme and 4) implementation of a "polluted pays policy" (Chua, 1999).

In India, the crop of sugarcane (*Saccharcum officinarum*) was grown at the agricultural farm of the mathur oil refinery in a simple randomised block design. The experimental plots were irrigated with ground water or treated wastewater, the plants gave better response to wastewater. The quantity of the required nutrients was comparatively more in wastewater (Ahmed and Inam, 2003).

Sewage Farming and Soil Microorganisms

Researches on other microorganisms in sewage or sewage-irrigated soils are actually very few. Effluents receive large number of fungal species in the treatment plants and also effluents discharged into the crop fields. Soil microorganisms play a very important role in soil fertility. Sewage irrigation supported the population of N fixers such as *Azotobacter*, *Rhizobium* and also fungal organisms supporting organic matter decomposition. As many as 61 fungal species were identified from the field growing tomato, cabbage, pumpkin, banana, brinjal, lucerne, maize and sugarcane. Potential pathogens like *Aspergillus niger*, *A. flavus*, *Alternaria* sp., *Colletotrichum* sp., *Helminthosporium* sp., *Phytophthora* sp., *Rhizoctonia bataticola*, *Fusarium oxysporum* etc. were noteworthy. On the plant surface (phyllosphere) many species occurred of which few were cattle and human pathogens. Sludge treatment, however reduced disease producing organisms (Kulkarni, 1981).

Sewage components like soaps, household pesticides (DDT, phenyl, Dettol) altered microbial population considerably. Cloth washing soaps and household pesticides affect microbes adversely. *Rhizobium*, however, responds differently to the sewage components. Phenyl, mustard oil, extracts of cumin, pepper, onion and chillies are harmful whereas sunlight soaps supports the growth of this organism (Kulkarni and Gangawane, 1980, 1982, 1985). Root rot diseases, however, are minimized due to sewage components such as sunlight, det, human bathing soaps, DDT, dettol and phenyl household pesticides. Chillies, garlic and onion extracts were inhibitory to these diseases. This is due to the inhibition of enzyme metabolism (pectolytic and cellulolytic) in the organism required for the process of infection.

Sewage supports the growth of algae very rapidly under the sunlight. Blue green algae also multiply in the untreated sewage and are the source of N fixation. Soil irrigated with sewage is enriched with such algal forms. A total of 56 blue green algal species are reported from the sewage-irrigated soils. Here *Anabaena anomols*, *Aulosira fertilissims*, *A. doliolum*, *Nostoc calcicola*, *N. puretiformae*, *Tolypothrise tenuis* were dominant (Khalil, 1983) all these are recommended as biofertilizers. Sewage is the best source for their multiplication and preparation of inoculum. From 9 x 40m^2 sewage pond, nearly 20–25 kg dry blue green algae can be obtained. This yield can be increased under suitable environmental conditions. Growing algae over the sewage also reduces the concentration of toxic heavy metals and consequently reduce the soil and water pollution. Disposal of sewage in an urban area therefore, may be arranged in such a way that algal biomass is produced for making biofertilizers or manure and the sewage partially or fully purified in this way may be disposed off in the agricultural fields for irrigation purpose.

Burns and Hardy (1975) reported that approximately 83 per cent of the N is fixed annually originates from biological N_2 fixation while only 14 per cent is from manufacture of fertilizers. Knowles (1978) suggested that free living N_2 fixing microorganisms are widely distributed and found in almost every ecological niche in soil and rhizosphere. Dixon and Wheeler (1986) also noted that in nature diazotrophy is restricted to some prokaryotes and Archaebacteria. About 87 species are known to show diazotrophy. Among the free living nitrogen fixer *Azotobacter* has occupied an important place over the years in many countries including India as it is useful biofertilizer for cereals, oil seeds, vegetables and economically important non leguminous plants (Mishustin and Shilnikava, 1969;

Subba Rao, 1979; Pandey and Kumar, 1989). Beijerinck (1901) first time discovered *Azotobacter*. The family Azotobacteraceae is a coherent group of aerobic, free living heterotrophic bacteria whose main characteristics is its ability to fix atmospheric nitrogen in a nitrogen free medium with organic carbon compound as energy source. Members of this group are known as non symbiotic nitrogen fixer consisting four genera like *Azotobacter, Azomonas, Beijernckia* and *Derxia* (Johnstone *et al.*, 1959).

The number of *Azotobacter* in soil in various part of the world is usually below 10^4gm^{-1}. An exception to this general situation is the occurrence of high number of *Azotobater i.e.* 10^7gm^{-1} as in Nile valley (Abdel–Malak, 1971). In Indian soils their number is generally low amounting to not more than few thousands. The maximum number per gram reported is 1.1 x10^4 in Haryana soils (Sindhu and Lakshiminarayan, 1986) and 8 x10^4 in forest soil of Karnataka (Channal *et al.*, 1989). *Azotobacter chroococcum* have been also reported to occur in parenchymatous cells of root cortex and in leaf sheath (Tippanawar and Reddy, 1989).

Effect of Agrochemicals

Bactericidal effect of heavy metals have been reported by Maslyukov *et al.* (1992) on *Aspergillus niger, Azotobacter, E. coli, Pseudomonas* and *Tetrahymena.* It was concluded that a complex of their physicochemical properties determines the bactericidal effect of heavy metal.

Macarty *et al.* (1993) studied influence of Mn^{2+} and Mg^{2+} on the activity of glutamine synthtase in various organisms and the effect of different concentration of Mn^{2+} and Mg^{2+} on assimilatory No_3 reduction and NH_4 assimilation in culture of *Pseudomonas fluorescens* ATCC 13525 and *Azotobacter chroococcum* ATCC 9043.

Page *et al.* (1996) reported that the growth yield of *Azotobacter salinestris,* a Na+ dependent, microacrophilic nitrogen fixing bacterium was inhibited by more than 60 per cent by 5 mg Zn++. This organism was much more sensitive to Zn^{2+} than the obligate acrobe *Azotobacter vinelandii.* Population of *Azotobacter* sp. and *Rhizobium* sp. were enhanced by more than that of 6 to 10 fold by the application of copper to soil, the increase was pronounced on 30th and 40th day respectively (Murugesan, 1996).

Mineev *et al.* (1998) studied the effect of phosphorous fertilizer as a component of PK and NPK fertilizer on the growth of soil bacteria *viz., Azotobacter*, aerobic, mycolytic and actinolytic bacteria. Bezbaruh and Saikia (1998) reported production of catechol and Azotobactin by *Azotobacter chroococcum* strain in medium containing Cu, Co, Mn, Zn, Pb, Fe, Mo and N; after 24 hours and 48 hours respectively. Fungicidal potency of purified siderophore was more in culture amended with Cu and Mn.

Osman *et al.* (1999) carried out lab experiments to study the effect of fungicide Amilsar on growth of some beneficial microorganisms and on three strains of *Phytophthora infestants.* Actually no risk to beneficial microorganisms has been demonstrated. There were *Pseudomonas putida, Klebsiella planticola, A. chroococcum* and *Clostridium acetobutilicum.*

Gonnzaley Lopez *et al.* (1999) studied the effect of some herbicides on the production of lysine by *Azotobacter chroococcum.* Production of lysine by *Azotobacter chroococcum* strain H23 was studied in chemically defined media amended with different concentration of alachor, metalochar 2, 4-D, 2, 4, 5-T and 2, 3, 6–T BA. The presence of 5, 10 and 50 mg/ml of alachor or 2, 3, 6 T BA significantly decrease quantitative production of lysine. However the presence of 2, 4-D or 2,4 5-T at concentration of 10 and 50 mg/ml/enhanced the production of lysine. Quantitative production of lysine was not affected as consequences of the addition of metalachor to the culture media showing that the release lysine to the culture media by *A. chroococcum* was not affected by that herbicide.

Onwurah (1999) noted that role of diazotrophic bacteria in the bioremediation of crude oil polluted soil. N_2 is one of the most limiting inorganic nutrients in the process of bioremediation of crude oil polluted environments. Enhanced remediation of crude oil polluted soil was achieved *in situ* by accelerating the biodegradation process through seeding with adapted *Azotobacter* which not only acted as supplier of mixed N_2 to the indigenous crude oil degrading bacteria, but also performed some co-metabolic activities. This work describes the capability of *Azotobacter* in providing activities that are useful in the bioremediation of crude oil polluted soil and biological nitrogen fixation when in association with indigenous oil degrading bacteria.

It was thought very essential to find out the fate of nitrogen fixing *Azotobacter* sp. in the sewage irrigated field soils and rhizosphere of crop plants. *Azotobacter* population was studied in the rhizosphere of 15 crop plants along with the soil samples during various seasons. It was noted that the number of *Azotobacter* varied among the rhizosphere and soil of different crops. This variation was also recorded during the different seasons. During rainy season highest population of *Azotobacter* was recorded in the rhizosphere and soil of sugarcane, during winter highest number was recorded in the rhizosphere and soil of ginger while in the summer highest number was recorded in the rhizosphere and soil of brinjal. Hence the R/S effect was also found to be different.

Presence of *Azotobacter* on the phyllosphere of eight crops irrigated with sewage was studied. Here also there was large variation in the population of *Azotobacter* on the phyllosphere of different plants. Seasonal variations were also recorded. The maximum number was seen on phyllosphere of tomato, groundnut and brinjal.

Morphological and biochemical characteristics of 20 *Azotobacter* isolates were studied. Colony size and margin and colour appeared to be variable. The bacterium was gram negative rods and capsulated. Among the biochemical characteristics catalase, Nitrate reduction, gelatinase activity, amylase activity, production of acid and gas etc. were studied. Among these 5 isolates gave amylase activity while on Rhamnose the activity was negative. On dextrose 14 isolates gave acid and gas, three isolates gave only acid production. On sucrose nine isolates produced acid and gas, on mannitol 10 isolates produced acid and gas and 8 produced only acid, three isolates gave negative reaction, on lactose 11 isolates gave acid and gas, on maltose 7 isolates produced acid and gas while 12 produced only acid. All the isolates give negative reaction for the production of H_2S gas.

The growth of *Azotobacter* isolates was luxuriant at pH 7, while 16 isolates showed the growth at pH 8 and some isolates could grow at pH 9. Very few isolates showed growth at pH 5 to 6. Temperature 30°C was most favourable for the growth of all isolates, and 0.2 per cent NaCl was most favourable for the growth of all 20 isolates.

On the basis of morphological, physiological and biochemical characteristics the isolates showed the presence of five species such as *Azotobacter chroococcum, A. paspali, A. insignis, A. vinelandii* and *A. beijerinikii*. The maximum belonged to *A. chroococcum*. This was done according to the Bergey's manual of Bacteriology.

Efficacy of *A. chroococcum* on N fixation in tomato was studied. It was observed that *A. chroococcum* was effective on tomato plants. The shoot and root length and dry matter and N content was highly increased with *A. chroococcum* isolates. There was also increase in N_2 fixation due to sewage irrigation.

Effect of sewage and sewage component on the growth of *A. chroococcum* was studied. The sewage components consisted of detergents, soaps, antibiotics, spices, oils and fats, toothpastes, household pesticides, heavy metals and also vegetable extracts. There was a large variation in the growth of *A. chroococcum* in the different component of the sewage. Detergents were found to be inhibitory.

In addition soaps were also inhibitory. Among the antibiotics oxytetracycline was highly inhibitory. Moreover Streptomycin, Amoxycillin and Gentamycin were also found to be inhibitory. Among the spices the Dalchini was highly inhibitory than other spices. Oils and fats showed variation in the growth of *A. chroococcum.* Oils like groundnut, sesamum were non inhibitory whereas, cotton, mustard, coconut and Dalda were inhibitory for the growth of *A. chroococcum.*

Tooth pastes such as close up, Babool, Colgate, Meshwak and Colgate herbal were inhibitory. However, Herbal Colgate was less inhibitory when compared with other toothpastes. Among the household pesticides phenyl was highly inhibitory. Moreover, heavy metals such as $MgSo_4$, $FeSo_4$, $CuSo_4$, $ZnSo_4$, $HgCl_2$ and Ag No_3 also showed reduction due to the toxicity. $AgNo_3$ was highly inhibitory when compared with other metals. Interestingly extracts of eight vegetables crops showed stimulatory effect on *A. chroococcum.*

Effect of sewage and sewage components on the occurrence of *A. chroococcum* in the rhizosphere of tomato was studied. Again the similar components such as soaps and detergents, antibiotics, spices, toothpastes, household pesticides, vegetable extracts and oils and fats were incorporated in the soil. It was observed that *A. chroococcum* was increased considerably due to all the soaps and detergents in rhizosphere and soils except Lifebuoy. Other bacteria were also increased but the fungal population was reduced in rhizosphere due to application of Vim bar, Cinthol, Chandrika, Lifebuoy and Surf excel. Antibiotics also reduced *A. chroococcum* in the rhizosphere and soil as well. Other bacsteria were also reduced but fungal population was increased due to the use of antibiotics. All the nine spices increased population of *A. chroococcum* of rhizosphere of the tomato. This was also seen in case of other bacteria and fungi in the rhizosphere and soil. Among the oils soybean and cotton oil considerably increased the population of *A. chroococcum* in the rhizosphere and mustard oil reduced the number of *A. chroococcum*. Among the tooth pastes Colgate reduced the number of *A. chroococcum* whereas the Meshwak was more favourable for this bacterium in the rhizosphere. Other bacteria were, however increased due to the application of these toothpastes. Fungal population was also slightly increased due to the toothpaste. Among the household pesticides phenyl and BHC reduced population of *A. chroococcum* in the rhizosphere and soil of tomato. Heavy metals such as $HgCl_2$ and $BaCl_2$ were completely inhibitory for these bacteria in the rhizosphere. Of the 8 vegetable extracts onion, garlic, ginger, spinach potato and methi reduced the number of *A. chroococcum* in rhizosphere and soil. Other bacteria were, however reduced due to the application of all vegetable extracts. Tomato extracts was, however, found to be stimulatory for *A. chroococcum*. The fungal numbers were also increased with the vegetable extracts.

The sewage components may also affect the N_2 fixation in tomato. Hence effect of different components such as detergents and soaps, antibiotics, spices, oils, tooth pastes, heavy metals and vegetable extracts as the growth and N_2 fixation was studied. Detergent soaps, antibiotics spices like black pepper, guava, clover, chilli, turmeric, mustard, oils such as soybean, cotton and mustard, tooth pastes such as Colgate, heavy metals like $BaCl_2$, $AgNo_3$, and vegetable extracts such as methi, coriander, ginger and onion indicated reduction in N_2 contents of tomato plants.

Interaction between *A. chroococcum* and rizosphere mycoflora of tomato was studied by agar well technique. It was seen that some all the fungi were inhibitory. Among fungi *Trichoderma viride, Fusarium semitectum* and *Alternaria solani* were more antagonistic.

In order to find out whether the purification of sewage is possible due to *A. chroococcum* isolates, the isolate A-1, A-7, A-12 and mutant were inoculated in the sewage. Uninoculated sewage served as control. The O.D. of inoculated sewage was found to be highly reduced due to inoculation of *A.*

chroococcum. To find out whether the purification of sewage is due to the gene on main chromosome or plasmid, the isolates were subjected to curing with acridine orange. Plasmid analysis show that only chromosomal DNA bands in both the cases of cured and non-cured isolates. On testing the sewage for O.D. both the isolates showed only a minor difference in purifying ability when compared with non-cured strains (Jadhav, 2004).

References

Abd-el-Malak, Y., 1971. *Plant and Soil Special Volume*, 423–442.

Ahmed, A. and Inam, A., 2003. *J. Envi. Bio.*, 242: 141–146.

Ahmed, F.F., Ragab, M.A., A.E.M. Monsour and Gohara, A.A., 1998. *Egyptian Journal of Horticulture*.

Anonymous, 1967. California, Pub. No. 18.

Bachmann, G., 1954. *Wass Tech.*, 4: 191–261, *J. Ref. 1955, Yik*, 157–344.

Ball, R.C. 1977. *Wastewater Renovation and Reuse*, (Ed.) Frank M.D Ifri, 7: 205–253.

Bendixon, T.W., Hill, R.D., W.A. Schwartz and G.G. Robeck, 1968. *Jour. Sanitary Div. ASCE*, 94: SAL: 147–157.

Bezbarwah, B. and Saikia, N., 1998. *Ind. J. Expt. Biol.*, 38: 680–687.

Burns, R.C. and Hardy, R.W.F., 1975. *Springer Verlag*, New York.

Channal, H.T., Alagawadi, A.R., Bharama Gowdert, T.D. and Udupde, S.G., 1989. *Curr. Sci.*, 58: 70–71.

Chua, H., 1999. *Water Sci. and Tech.*, 40: 91–96.

Day, A.D. and Tucker, T.C., 1959. *Agronomy Journal*, 51: 569–572.

Day, A.D., Dickinson, A.D. and Tucker, T.C., 1963. *Agronomy Journal*, 55: 317–318.

Dingress, W.R., 1969. Unpublished material by Division of Waste water Technology and Survillance, Texas State Dept. of Health.

Dixon, R.O.D. and Wheeler, C.T., 1986. Scientific Publishers, Jodhpur, India.

Dye, E.O., 1958. *Sewage and Industrial Wastes*. 30: 825–828.

Epstein, F., Tylor, J.M. and Chaney, R.L., 1976. *J. Environ. Qual.*, 5: 422–426.

Gangawane, L.V., 1989. Ashish Publishing House, New Delhi.

Gharaibeh, S., 1989. Z. KULTURTECH LANDENTWICKLUNA, 30: 196–199.

Gonnzaly-Lopez, J. and Martinez, M.V., 1999. Amino acids (Viena), 165–173.

Haruvy and Nava, 1997. *Agriculture Ecosystems and Envt.*, 66: 113–119.

Henry, C.D., Moldenhauer, R.E., Englebert, L.E. and Trong, E., 1954. *Ind. Wastes*, 26: 123–133.

Jadhav, P.N., 2004. *Ph.D. Thesis*, Dr. B.A. Marathwada University, Aurangabad, MS, India.

Johnstone, D.B., Pfeffer, M. and Blancherd, G.C., 1959. *Can. J. Microbiol.*, 5: 299–304.

Jorge, A.M. and Eloy, U.T., 1977. (Eds.) Frank, M.D. and Otri. Marcel Dekkar Inc. N.Y. pp. 1–33.

Juwarkar, A.S., Asha Juwarkar, P.B., Seshbhratar and A.S. Bal, 1991. RAPA Report Bangkok (7): 178–201.

Kanwar, T.S., 1970. BARC, Trombay, Bombay, 85: 356–364.

Kingsley, Demas and A. Oteino, 1999. *Science and Technology*, 39: 343–346.

Knowles, R., 1978. Plenum Press, New York, pp. 25–40.

Kulkarni Leela and Gangawane, L.V., 1982. Rev. *Du. Ecol. Biol. Ser.*, 19: 523–533.

Kulkarni Leela and Gangawane, L.V., 1985. *Ind. Phytopath.*, 38: 756–757.

Macarty, G.W. and Bremner, J.M., 1993. *Proc. Natl. Acad. Sci.*, USA, 90: 9403–9407.

Maslyukov, A.P., Rakhmanin, Y.A., Matyushin, G.A. and Dyumaev, K.M., 1992. *Doklakad Nauk*, 323: 1180–1185.

McCarthy, M.C., 1981. *Am. J. Ecol.*, VI TI (3213): 189–196.

Mineev, V.G., Gomonova, N.F., Skvorstova, I.N. and Dickson, J., 1998. *Agrokhimiya*, 12: 5–9.

Mishustin, E.N. and Shilnikova, V.K., 1969. In: *Soil Biology, Reviews of Research*, pp. 72–124 (UNESCO Publication).

Mohammad Ajmal and Khan Ahsanullah, 1983. *Env. Pollut. Sev. Eco. Bio.*, 30: 135–142.

Murugesan, K., 1996. Jodhpur, 23: 46–49.

Naskar, K., Sha, S.K., Chakraborty, H.M., 1986. In: *Rural Dev.* Oct. 3–5, 1986, Salem, p. 7.

Odum, H.T., Ewel, K.C., Mitsch, W.J. Ordway, J.W., 1977. In: *Wastewater Renovation and Reuse,* (Eds.) Frank, M.D. Itri, pp. 33–64.

Onwurah, I.N.E., 1999. *J. Chem. Biotech.*, 74: 957–964.

Osman, A.D., Emstev, V.T. and Kalinin, V.A., 1999. Izvestiya Timiryazevskoi sel Kokhozyaistvenrio Akademmi, 1: 139–145.

Page, W.J., Manchak, J. and Yohemas, M., 1996. *J. Microbiol.*, 42: 655–661.

Seep, E., 1971. Bureau of sanitary Engineering Dept. Hlth. State of California.

Sharma, V.K. and Kausal, B.D., 1984. *Ind. J. Ecol.*, 11: 77–81.

Sidle, R.C. and Kardos, L.T., 1977. Longman, London, New York, pp.160–175.

Sindhu, S.S.and Lakshminarayan, K. 1986. *Environ. Ecol.* 4: 536–540.

Speir, T.W., Van Schaik, A.P., 1999. *Journal of Environmental Quality* 28(4): July– Aug, 1105–1114.

Stone, 1955. *American Inst. of Chem. Eng.*, p. 668.

Subba Rao, N.S., 1979. In: *Recent Advances in Biological Nitrogen Fixation*, pp. 406–420.

Tippanenawar, C.M. and Ramachandra Reddy, T.K. 1989. *Curr. Sci.*, 58: 1342–1343.

Yoshishisa, M., Hirsohi, N., Toshiko,K., Toshiro, T., Shigeto, Y. and Akira, F., 1974. *Buoll. Fac. Agri. Shimane Univ.*, pp. 116–133.

Sustainable Environmental Management *Pages* ***105–116***
Edited by: **Dr. L.V. Gangawane & Dr. V.C. Khilare**
Published by: **DAYA PUBLISHING HOUSE**

Chapter 13

Biodegradation of Lignin Sub-structure Model Compounds by *Penicillium*

S.R. Joshi* and S.A. Dutta

National Environmental Engineering Research Institute, [NEERI], Nehru Marg, Nagpur – 440 020, MS, India

ABSTRACT

Due to stringent environmental laws and public awareness, it has become customary to treat industrial wastes waters. Pulp and paper mill wastewater is very difficult to treat biologically as it contain not only high BOD and COD but also is associated with colour, which is due to plant originated lignin molecule. Lignin polymer is a highly complexed chemical with aromatic rings, methyl groups, and irregular carbon-to-carbon and ether linkages, which are difficult to degrade by microbes. However it is known that some fungal cultures are capable of degrading highly polymerized substances such as lignin cellulose, hemicelluloses, chitin etc., Lignin molecules contain about 16 different low molecular weight compound that contain chemical structure known to occur in lignin. A fungal culture identified as *Penicillium* was isolated and experiments were carried out to examine its efficacy in degrading different lignin model compounds. Results obtained with such few compounds are presented in this article.

Keywords: *Lignin, Model compounds,* Penicillium, *Phenol Veratrole, Vanillic acid, Syringic acid, Veratraldehyde, Homovanillic acid, Cinamic acid, Coniferyl alcohol, Ferulic acid, Guaicylglyceryl ether.*

Introduction

It is an extremely difficult task to study the biochemical mechanisms whereby microorganisms degrade the plant polymer, lignin. A primary reason for this difficulty is pronounced structural

* Ex-Scientist.

complexity of the lignin molecule (Sarakaren and Ludwig 1971). Unlike other biopolymers, 1ignin contains no readily hydrolyzable linkages ocurring at regular intervals along a linear backbone. Instead, lignin is a three-deminsiona1 amorphous polymer containing many different stable carbon and ether linkages between phenylpropanoid monomeric units. Theoretically, one way to circumvent this problem of chemical complexity is to study the microbial degradation of simple lignin model compounds of known chemical structure. In this experimental approach, one uses low molecular weight compound that contains chemical structures known to occur in lignin (Crawford and Crawford 1980). Moreover, lignin model compounds are found as intermediates during lignin degradation.

Microbial degradation of lignin substructure model compounds has been studied by many workers and has been reviewed by Crawford (1981). The purpose of our study regarding degradation of lignin substructure model compounds is also due to the fact that these compounds are present in pulp and paper mill effluent and its removal is necessary during wastewater treatment process. Vanillic acid,Syringic acid and p-Hydroxybenzoic acid has been most consistently reported single ring products of lignin biodegradation (Chang et al 1980). Several single ring, phenylpropanoid products have also been reported as trace products of lignin degradation by ligninolytic fungus (Ishikawa *et al.*, 1963 and Subba Rao *et al.*, 1971). In our study lignin structure model compounds were selected for the study of their degradation by *Penicillium* culture.

Materials and Methods

Lignin Substructure Model Compounds

Phenol, Veratrole, (C_6 compounds); Vanillic acid, Syringic acid, Veratraldehyde, ($C_{6-}C_1$ compounds), Homovanillic acid, (C_6–C_2 compounds), Ferulic acid, Coniferyl alcohol, Cinnamyl alcohol, (C_6–C_3, compounds) and Guaicyl glyceryl ether (B aryl ether compounds) were selected and all were purchased from Sigma chemical Co. USA.

Medium for Degradation Studies

Basal Mineral Medium

It contains $NaNO_3$–0.5 g, K_2HPO_A l-g $MgSO_4$–0.5 g, KC1–0.5 g, yeast extract–100 mg. Trace "element–5 ml, Distilled water–1000 ml, pH–5.5. The whole medium was autoclaved at 15 lbs pressure for 15 min. before adding the membrane filter sterilized trace element solution. Trace element composition: Ferric citrate 12 mg, $MnSO_4.H_2O$-4 mg, $ZnSO_4.H_2O$–7 mg, $CaCl_2.6H_2O$–l mg, $CuS0_4$-5 H_20–1 mg, Distilled water–Two ml.

Carbon Source

The model compounds were dissolved separately in minimum amount of ethanol and added to the sterilized basal mineral medium. The whole medium was then filter sterlized through 0.45-micron millipore membrane filter aseptically under laminar flow. The 50 ml portion of the sterilized medium was distributed aseptically into 250 ml erlenmeyer flask. The final concentration of the model compound was 200 mg L

Inoculum

One week old culture slants of *Penicilliurn* maintained on potato dextrose agar were used as inoculum. The spores were suspended in phosphate buffer containing 0.1 *per cent* sodium deoxycholate for uniform suspension and 0.1 ml of the suspension containing 7 x 10^5 spores was inoculated into 50 ml of the medium. Flasks were kept on a rotary shaker for 48 hrs. The 48 hours grown culture was used as an inoculum into the corresponding flask. One ml of the suspension containing about 10 mg of mycelium (dry wt on GFC filter paper) was centrifuged at 800 rpm for 15 minutes. The supernatant was discarded and the mycelium was washed thoroughly with sterile phosphate buffer and then added into the flask. Three replicates for each compound were kept on the rotary shaker and the degradation of model compound was monitored. In control flask the inoculum was heat killed vegetative mycelium instead of live ones. Heat killing was accomplished by autoclaving cultures at 116°C for 5 minutes.

Monitoring of Model Compounds

A small portion of about 3 ml was pipetted out aseptically after every 24 hours from experimenta1 and control flask and filtered throgh GFC. The filtrate was analyzed for the presence of model compound in UV spectrophotometer by spectrophotometeric method (Kawakami, 1980) after suitable dilution of 1:10 at the corresponding wavelength maximum which was determined earlier with double distilled water. The pH was adjusted to 7.0. The percentage degradation for each compound was measured for the reduction in absorbance at wave length maximum relative to the absorbance of the compound incubated with an identical heat killed mycelium. For degradation studies flasks were kept in triplicates.

Results and Discussion

Biodegradation of C_1 Compounds

Compounds used for biodegradation studies were Phenol, and veratrole.

Phenol

Figure 13.1 shows the biodegradation of Phenol by *Penicillium* culture. In control there was no decrease in absorbance of the filtrate while in the experimental flasks, there was decrease in absorbance of the culture filtrate indicating the biodegradation of these compounds. The decrease in absorbance due to mere binding of the compound on mycelia can be ruled out due to the fact that heat killed vegetative mycelia control flask did not have any effect on absorbancy. The maximum degradation (81 per cent) was observed after 72 hours. Initial absorbance in experimental flask was 0.452 which after 72 hours was decreased to 0.087. The absorbance in control flask remained unchanged throughout the period. The *Penicillium* culture could easily degrade Phenol as it contains enzymes to break down the aromatic nuclei. Many workers have reported the degradation of simple aromatic like Phenol (Crawford *et al.*, 1973). Veratrole is the compound with two Methoxyl groups at 3 and 4 position of the aromatic ring. The Veratrole degradation by *Penicillium* culture is shown Figure 13.2. It can be seen from the figure that Veratrole degradation was comparatively very slow. Even after 6 days of incubation period the absorbance decreased from 0.366 to 0.327 only. As Veratrole has two methoxyl groups at 3 and 4 position of the aromatic ring, ring cleavage becomes difficult.

Biodegradation of C_6-C_1 Compounds

In this category (C_6-C_1) different compounds were used for testing its biodegradability. They were Vanillic acid, Syringic acid and Veratraldehyde

Vanillic Acid

Vanillic acid is an important intermediate formed during lignin biodegradation. It could be observed that 93 *per cent* of Vanillic acid was degraded in 4 days (Figure 13.3). However, most of the degradation (81 *per cent*) was occurred in 2 days. In control there was not much change in absorbance indicating that mycelial adsorption was not taking place. Vanillic acid is easily degradable substrate and is usually demethylated by microorganisms yielding protocatechuic acid which then under goes

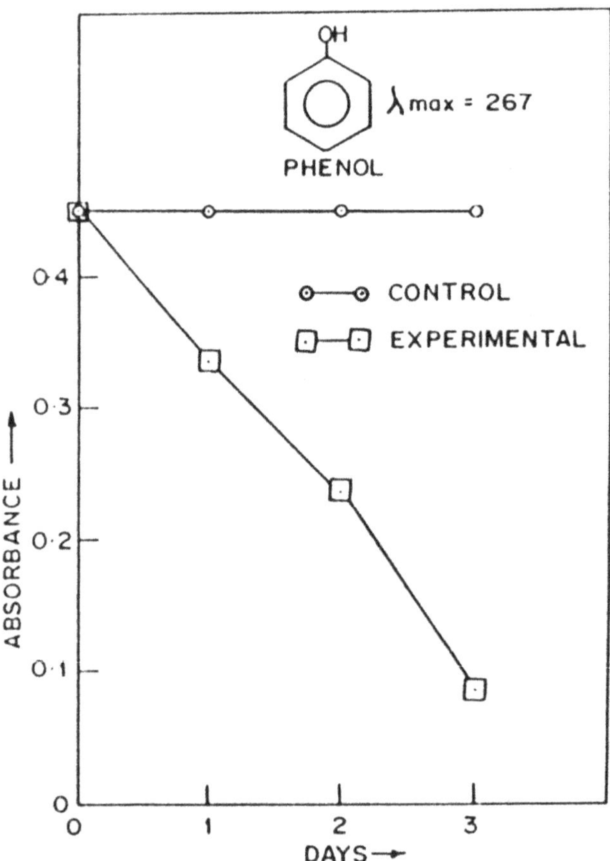

Figure 13.1: Biodegradation of Phenol by *Penicillium* sp.

dioxygenase catalysed ring fission (Crawford *et al.*, 1973, Cartwright and Smith 1967, Buswell *et al.*, 1977). Vanillic acid is oxidatively decarboxylated by *P.chrysosporium* and subsequently may be demethy lated to form hydroxy quinol which is metabolized further (Yajima *et al.*, 1979). As Vanillic acid and related compounds are prominent intermediates of lignin degradation, enzymes involved in Vanillic acid degradation are of importance in lignin degradation.

Results on degradation of Syringic acid showed that after one day incubation period there was no degradation (Figure 13.4). After 2 days, degradation of the Syringic acid was observed and after 4 days, 91 percent of the compound was degraded. The absorbance from 1.340 was reduced to 0.130 after 4 days. The increase in absorbance in control with the formation of co1our in the medium might be due to formation of quinone like compounds by auto-oxidation. Kawakami (1980) and Leatham *et al.* (1983) observed 80 percent degradation of Syringic acid in 3 days by *Pseudomonous* culture and *Phanerochaete* culture respectively, while during this study the *Penicillium* culture degraded 66 percent in 3 days. The concentrations used by Kawakami and Leatham *et al.* were 150 mg L and 100 mg L respectively while the concentrations used during this study was 200 mg L In Syringic acid molecule

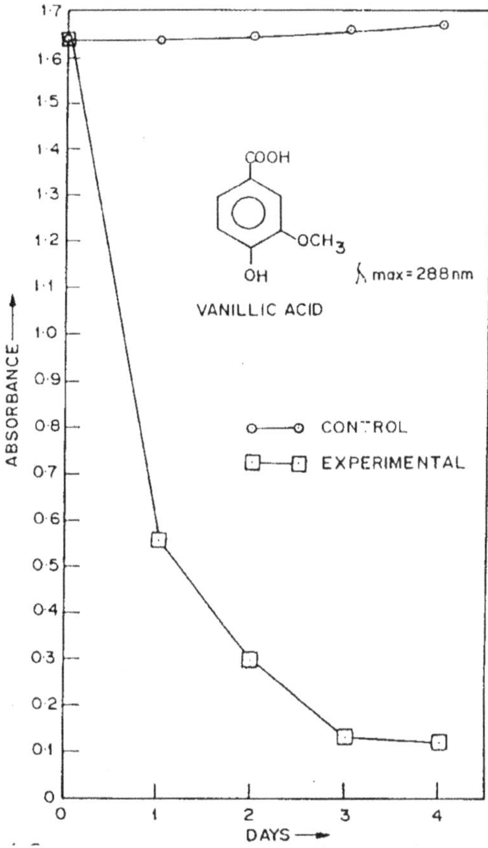

Figure 13.2: Biodegradation of Veratrole by *Penicillium* sp.

there are two methoxyl groups at position 3 and 5 of the phenolic ring where as in Vanillic acid only one methoxyl group is present on the phenolic group. The addition of methoxyl therefore did not have any affect on the degradibility of the compound. Veratraldehyde unlike veratryl alcohol was degraded easily (Figure 13.5). The *Penicillium* culture degraded 94 percent of the Veratraldehyde in 4 days. In control flask no degradation was observed. Kawakami (1980) and Leatham *et al.* (1983) observed 80 percent degradation of Veratraldehyde in 4 days while in same period 74 percent of degradation was observed during this study.

Biodegradation of C_6-C_2 Compounds

Homovanillic acid was the C_6-C_2 compounds used for degradation studies. This compound have 2 carbon chain attached to aromatic ring at 1-position. Experimental results (Figure 13.6) on degradation of Homovanillic acid have shown that like Vanillic acid, Homovanillic acid was found to be easily degraded by the *Penicillium* culture. About 90 percent of the Homovanillic acid was found to be degraded in 4 days. After 2 days of incubation period 60 percent of the degradation was observed. Kawakami (1980) observed 90 percent degradation of Homovanillic acid by *Pseudomonous* sp in 7 days whereas the *Penicillium* degraded to the same extent in 4 days.

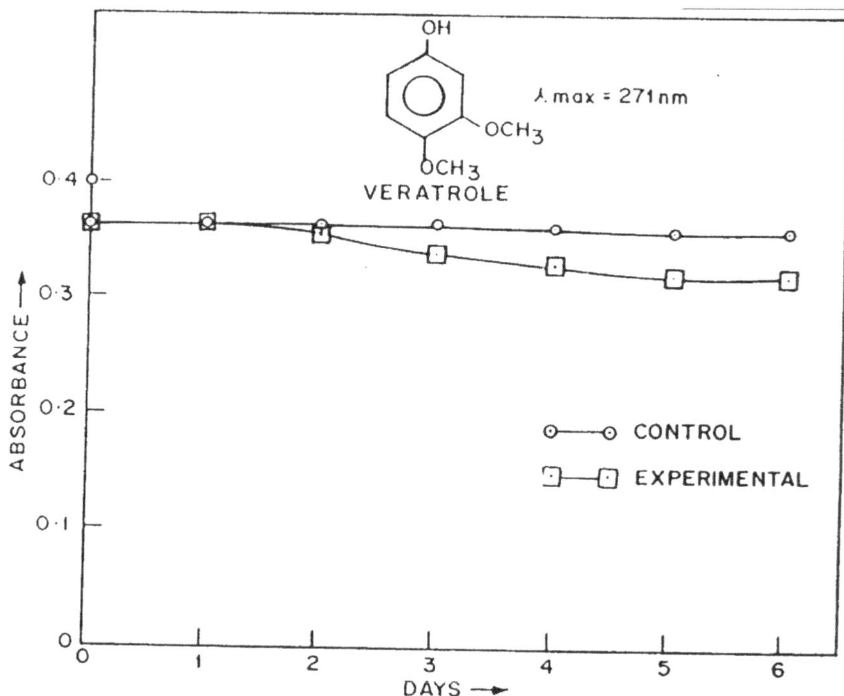

Figure 13.3: Biodegradation of Vanillic Acid by *Penicillium* sp.

Biodegradation of C_6-C_3 (Phenol Propanoids) Compounds

Some of the important C_6-C_3 compounds were selected for degradation studies. These compounds were Ferulic acid, Coniferyl alcohol, Cinnamyl Alcohol. The C_6-C_3 compounds have 3 carbon chains (propane) attached to 1 position of the aromatic ring (Phenol).

Ferulic Acid

It can be seen from Figure 13.7 that 94 percent of the Ferulic acid was degraded in 3 days. Most of the degradation was observed after 1 day of incubation period. This indicate that Ferulic acid was degraded by the *Penicillium* culture. Kawakami (1980) observed that only 2 days were required for complete degradation of 150 mg of Ferulic acid by *Pseudomanas* culture. Leatham *et al.* (1983) reported complete degradation of 100 mg of Ferulic acid in 3 days by *Phanerochaete* culture where as the *Penicillium* culture degraded 94 percent of 200 mg of Ferulic acid in 3 days. Ferulic acid, like Vanillic acid and Homovanillic acid has one methoxyl group and one OH group at 3 and 4 position of the aromatic ring respectively. In case of Ferulic acid 3 carbon chain is present while in Homovanillic acid and Vanillic acid 2 and 1 carbon chain are present respectively. The increase in carbon chain did not have significant effect on its degradation pattern.

Coniferyl Alcohol

The degradation of Coniferyl alcohol by *Penicillium* culture is given in Figure 13..8. Results have indicated that it required 5 days incubation period for this fungus to degrade 90 percent of the Coniferyl alcohol. However, most of the compound was degraded (80 percent) in 3 days. Kawakami (1980)

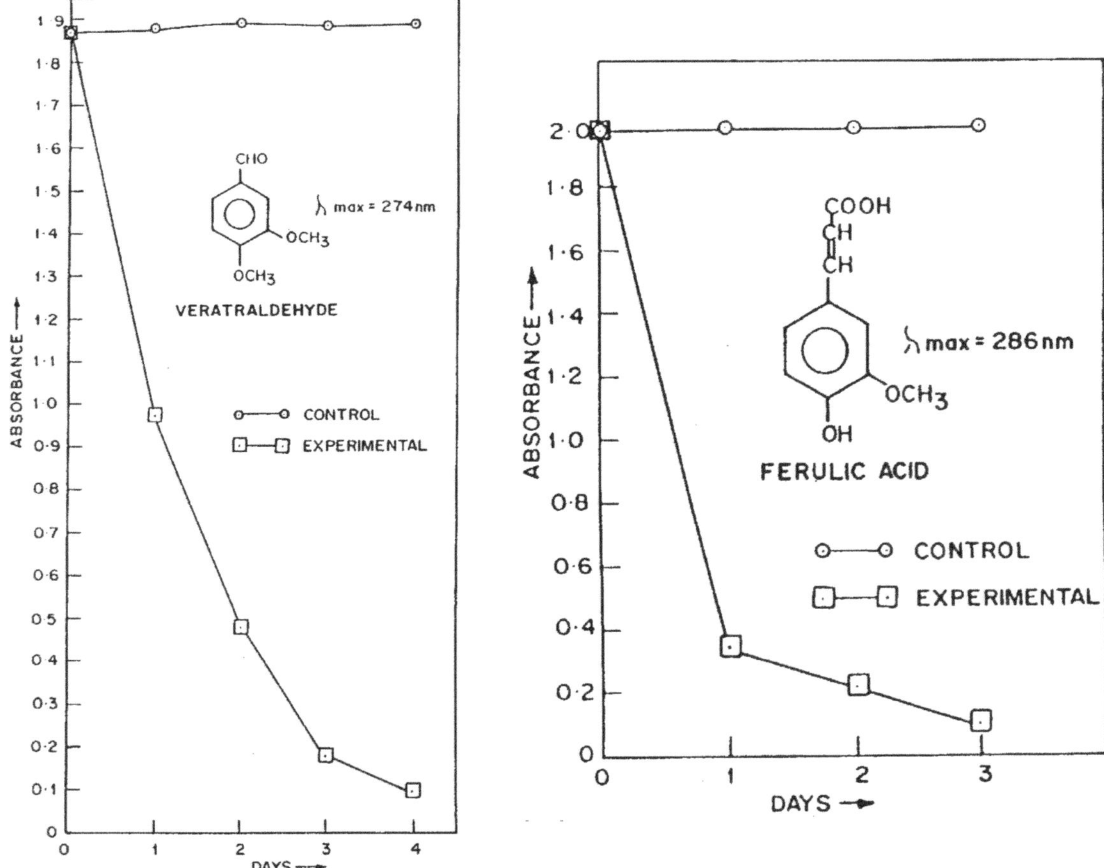

Figure 13.4: Biodegradation of Syringic Acid by *Penicillium* sp.

Figure 13.5: Biodegradation of Veratraldehyde by *Penicillium* sp.

observed about 90 percent of degradation of coniferyl alcohol by *Pseudomonas* in 3 days. When the degradation rates of Coniferyl alcohol and Ferulic acid were compared, Ferulic acid was found to degrade earlier than Coniferyl alcohol. The alcohol group of Coniferyl alcohol takes longer time to degrade than a aliphatic carboxyl group of Ferulic acid (Crawford 1981). The results on the biodegradation of Cinnamyl alcohol compound are shown fin Figure 13.9. Cinnamyl alcohol was found to degrade upto 46 percent after one day and further significant degradation could not be observed even after 5 days. As there is no functional group on the benzene ring in Cinnamyl alcohol, the degradation of the compound beacame difficult (Kawakami 1984).

Dimeric B Aryl Ether Compound

Only one compound Guaicyl glyceryl ether (GGE) was used for testing the degradation of dimeric B aryl ether compound. Figure 13.10 shows the degradation pattern of Guaicyl glyceryl ether by *Penicillium* culture. It could be seen from the figure that GGE was *very* resistant to degradation. Only 22 per cent of the compound was degraded in 2 days after which there was an increase in absorbance of the culture filtrate indicating that some intermediate might be getting accumulated in the medium

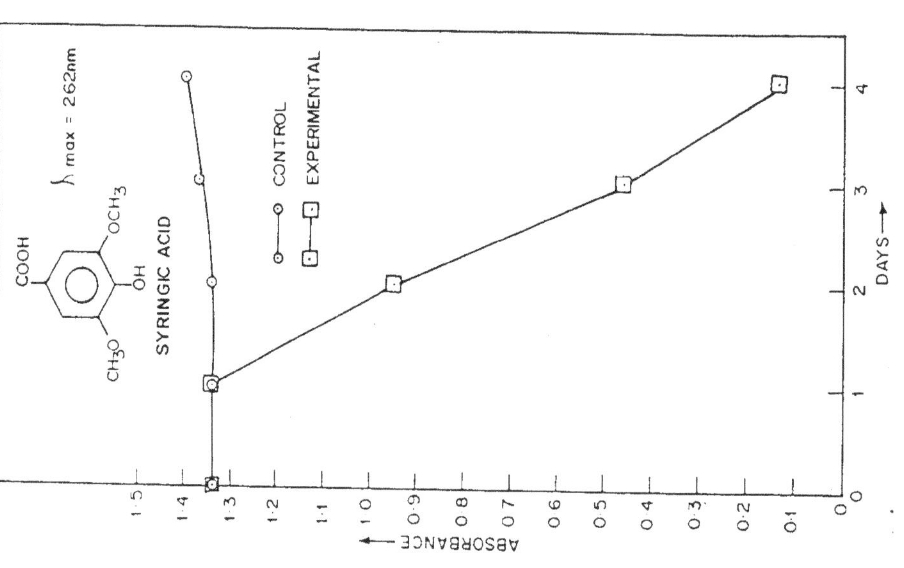

Figure 13.6: Biodegradation of Homo Vanillic Acid by *Penicillium* sp.

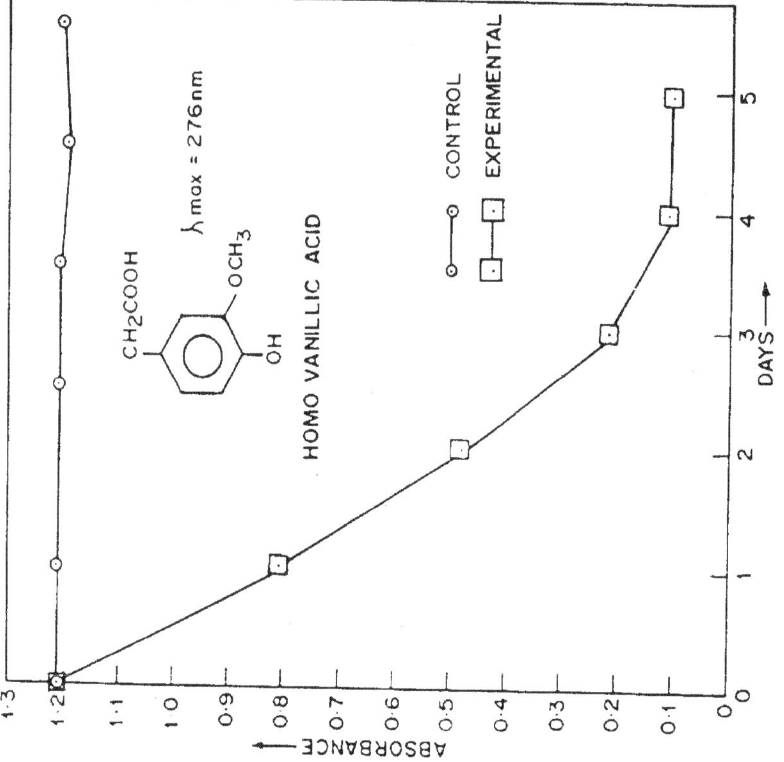

Figure 13.7: Biodegradation of Ferulic Acid by *Penicillium* sp.

which is ultimately *very* difficult to degrade. Kawakami (1980) also observed the resistant nature of this compound. The *Pseudomonous ovalis* culture degraded 50 percent of the compound in 5 days after which no degradation was observed. It has been reported that aryl-alkyl ether bond, an important unit in the lignin structure is cleaved by certain basidiomycetes (Fukuzumi *et al.*, 1969) and by bacteria of genera *Agrobacterium* (Trojanowksi *et al.*, 1970) and *Pseduomonas* (Crawford 1981). The *Penicillium* culture was unable to degrade the compound. However, the presence of additional carbon source might have enhanced the fungus in de-gradating the compound. Lignin related aromatic model compounds were tested for degradation by *Penicillium* culture. *Penicillium* culture thus appears to be one of the most versatile and non specific de-graders of aromatic molecules examined. Based on the disappearance of UV absorbance using 10 mg of compound per culture C-1 compounds were degraded by more than 80 percent The degradation assay used here was based on the difference in soluble phenolic compound concentration between live and heat killed control which effectively rules out artifacts caused by phenolic compound binding to mycelium. The degradation assay probably underestimates the extent of degradation, because UV absorbing degradation products are not distinguishable from starting material especially because some of the probable intermediates eg. muconic acids have higher extinction coefficient than do the starting material. Furthermore, at high concentration tested, toxicity of the aromatic compound might cause an underestimation of

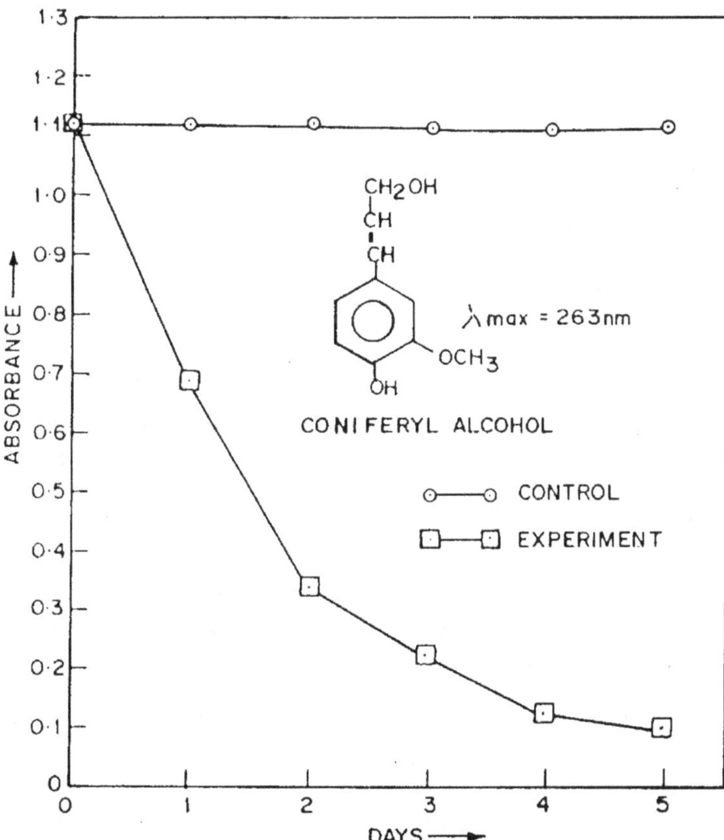

Figure 13.8: Biodegradation of Coniferyl Alcohol by *Penicillium* sp.

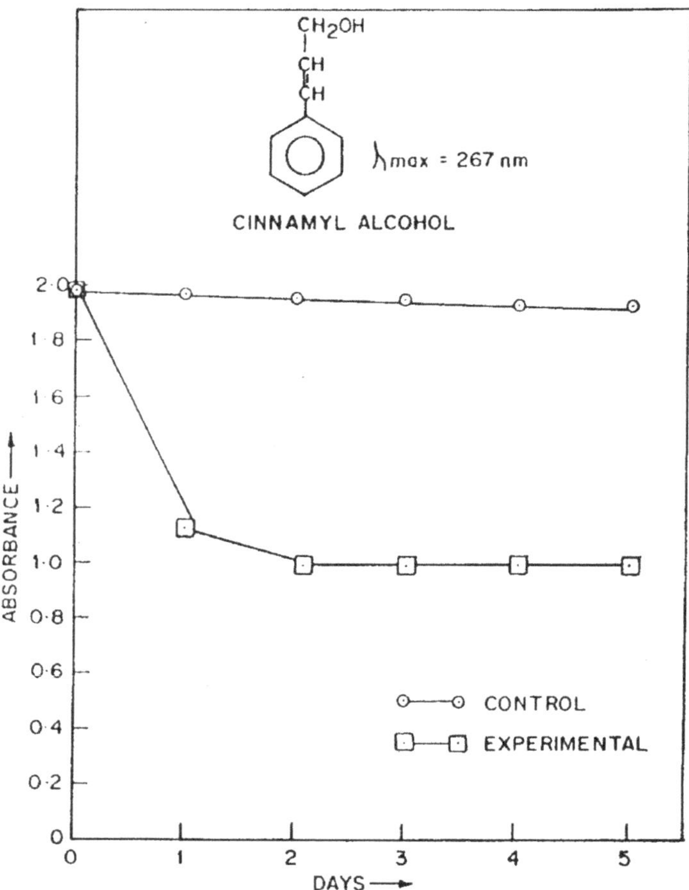

Figure 13.9: Biodegradation of Cinnamyl Alcohol by *Penicillium* sp.

degradability as compared with degradation at lower concentration. It has not been established with certanity that general use of lignin model compound as substrate for study of polymeric lignin biodegradation is a valid experimental approach. It is clear that microorganism isolated for their ability to degrade lignin often do not degrade the lignin macromolecule (Crawford *et al.*, 1973). However, known lignin de-graders generally are able to decompose lignin model compounds.

The major criteria of the use of model compound to study lignin biodegradation are that the model compounds are usually simple low molecular weight water soluble compounds while lignin is a complex macromolicule water insoluble substance. Enzymes used by microorganisms to degrade model compounds are typically intracellular and highly substrate specific and are unlikely to attack an extracellular complex polymer such as lignin. Enzymes that attack the lignin macro molecule thus may be different from that attack models. However, both group of enzymes may {perform the same types of chemical transformation (Kirk and Chang 1975, Chang *et al.*, 1980). In these studies absorbance was monitored for degradation. The mycelial growth was so less that it could not be monitored though fungus was utilizing these compounds as sole source of carbon.

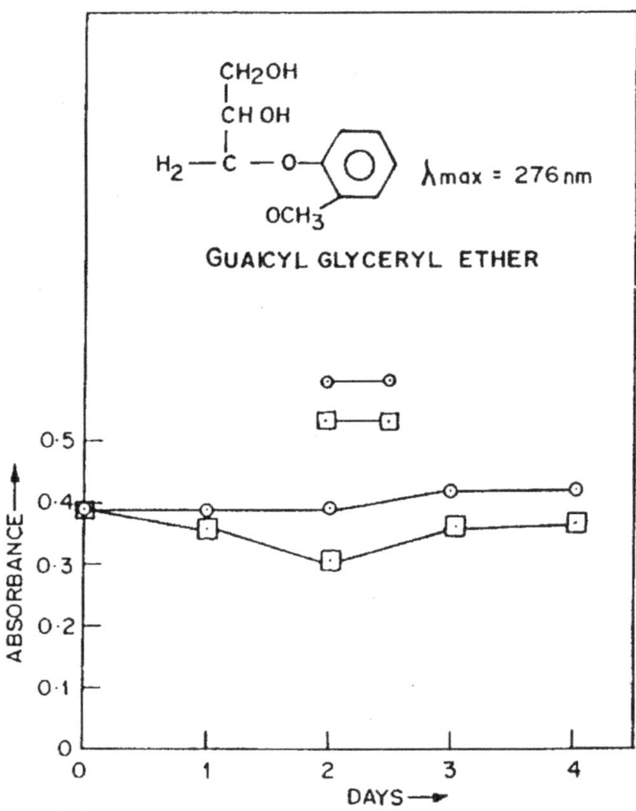

Figure 13.10: Biodegradation of Guaicyl Glyceryl Ether by *Penicillium* sp.

Conclusions

The use of lignin model compound as substrate for the study of polymeric lignin biodegradation and colour removal could lead to a better understanding of the mechanism involved during lignin degradation. The *Penicillium* culture was tested for degradation of the various substructure model compounds. During this study 10 lignin substructure model compound were selected. Amongst the C6 compounds Phenol was easily degraded whereas Veratrole was found resistant to degradation. Amongst the C_6-C_1 compounds Vanillic acid, Syringic acid and Veratraldehyde were found to degrade easily Amongst the C6-C2 compounds, Homovanillic acid was found to degrade easily Experiments with C6-C3 phenylpropanoid revealed that Ferulic acid Coniferyl alcohol, degraded easily as compared to Cinnamyl alcohol. Guaicyl glyceryl ether was found to be resistant to degradation.

References

Buswell, J.A., Ander, P., Peterson, B. and Eriksson, K.E., 1977. *FEBS Lett.*, 103 : 98.

Chang, Hm, Chen, C.L. and Kirk, T.K., 1980. Vol. I. CRC Press Inc., Boco Raton Fla., p. 215.

Crawford, R.L., Mcoy, E., Harkin, J. M., Kirk, T.K. and Obst, J.R., 1973. *App. Microbiol.*, 26: 176.

Crawford, D.L. and Crewford, R.L., 1980. *Enz. Microbiol Techno.*, 2 : 11.

Crawford, R.L., 1981. Wiley Interscience Publication.

Fukuzumi, T., Takatuka, H. and Meriami, K., 1969. *Arch Biochem Biophys.*, 129 : 396.

Gupta, J.K., Hamp, G., Buswell, J.A. and Eriksson, K.E., 1981. *Arch Microbiol.*, 128 : 349.

Ishikawa, H. Schubert, W. J. and Nord, FF 1963. *Arch. Biochem. Bio. Phys.*, 100 : 310.

Kawakami, H., 1960. Vol.1. CRC Press Inch. Boca raton Fla., pp. 105.

Leatham, C.F., Crawford, R.L. and Kirk, T.K., 1983. *Appl. Env. Microbiol.*, 46 : 191–197.

Sarakaren, K.W. and Ludwig, C.H., 1971. Wiley insterscience, N.Y.

Subba, Rao P.V., Nambudri, A.M.D. and Bhatt, J.V., 1971. *J. Scien. Ind. Res.*, 30 : 63.

Trojanowski, J., Wasclewska, M.L. and Wolska, B.J., 1970. *Acta. Microbial. Pol. Ser.*, B 2: 3.

Yajima, Y., Enoki, A., Mayfield, M.B. and Gold, M.H., 1979. *Arch Microbiol.*, 59: 302.

Sustainable Environmental Management *Pages* ***117–131***
Edited by: **Dr. L.V. Gangawane & Dr. V.C. Khilare**
Published by: **DAYA PUBLISHING HOUSE**

Chapter 14

Some Considerations on Genetically Modified Organisms and Aquatic Biodiversity

G.D. Khedkar[1] and A.S. Ninawe[2]

[1]*Aquaculture Research Laboratory, Department of Zoology, Dr. Babasaheb Ambedkar Marathwada University, Aurangabad*
E.mail: *gdkhedkar@gmail.com*
[2]*Director, Department of Biotechnology, Ministry of Science and Technology, GOI, New Delhi*

ABSTRACT

No one knows the scope of biodiversity, how many species of plants and animals share the planet with human beings. Whatever the actual number of species but the preservation of biodiversity itself is vital to humanity. Current patterns of resource exploitation by Humankind's do not bode well for the future of biodiversity. Ecologist Norman Myers recently estimated that some 600,000 species have been vanished since 1950. Today, two of every three species are estimated to be in decline. Aquatic biodiversity is extremely large and relatively poorly understood. Therefore the risk associated with use of GMOs or any genetically distinctive strain used in aquaculture is large. Aquaculture, beyond doubt, is the fastest growing food-producing sector in the world. The production of appropriate genetically modified organisms or (GMOs) offers considerably opportunities for more efficient and more effective aquaculture across a wide range of species. Although this potential is being realized in agricultural crop production globally. In aquaculture to a very negligible extent this potential is tested to commercial lines. In this review, the nature of GMOs, the range of aquatic species in which GMOs have been developed, the methods and target gene employed, benefits and effects on aquatic biodiversity, problems associated with the use of GMOs were discussed to length.

Keywords: *Genetically modified organisms, Biodiversity, Aquaculture, Fisheries, DNA, Transgenesis.*

Introduction

The production of appropriate genetically modified organisms or GMOs (in some cases combined with other forms of genetic improvements) can pave excellent opportunities for more efficient and more effective aquaculture practices across a wide range of species. It can be achieved by eliminating or reducing current constraints for better production technologies like enhanced growth, disease resistance, tolerance to environmental variables, better food conversions, better flesh test and colour, etc. The value of genetically modified crops in agriculture is already widely accepted, as the area shown to transgenic crop species worldwide exceeds 60 million hectares and this area is increasing rapidly year after year. Regarding the definition of GMOs there are lot of ambiguities prevailed. GMOs or living modified organism (LMOs) means any living organism that possesses a novel combination of genetic material obtained through the use of modern biotechnology. In this definition *"living organism"* means any biological entity capable of transferring or replicating genetic material, including sterile organisms, viruses and viroids. The term *"Modern biotechnology"* refers to the application of *in vitro* nucleic acid techniques, including recombinant deoxyribonucleic acid (DNA) and direct injection of nucleic acid into cells or organelles. It can also be defined as fusion of cells beyond the taxonomic family, that overcome natural physiological reproductive or recombination barriers and that techniques are not used in traditional breeding procedures.

The first transgenic animal produced was a mouse (Palmiter *et al.*, 1982). The first recorded instances of production of transgenics in aquatic species are those of Maclean and Talwar (1984) in rainbow trout and Zhu *et al.* (1985) in goldfish. Since then many species have been used to produce GMOs. However, both in terrestrial and aquatic animal species, we are not having any sustained evidence of commercial use of GMOs.

In this chapter, our purpose is to discuss, what are GMOs, nature of GMOs, benefits of GMOs, Health, biodiversity and Environment related aspects of GMOs.

Table 14.1: Teleost, Mollusc, Crustacean and Echinoderm Species Used in Transgenic Research

Common Name	*Latin Name*
TELEOST	
Tilapia	*Oreochromis niloticus, Cichlidae*
African catfsh	*Clarias gariepinus, Clariidae*
Channel catfsh	*Ictalurus punctatus, Ictaluridae*
Northern pike	*Esox lucius, Esocidae*
Medaka	*Oryzias latipes, Oryziatidae*
Zebrafish	*Danio rerio, Cyprinidae*
Loach	*Misgurnus anguillicaudatus, Cyprinidae*
Goldfsh	*Carassius auratus, Cyprinidae*
Red crucian carp	*Carassius carassius auratus, Cyprinidae*
Common carp	*Cyprinus carpio, Cyprinidae*
Brown trout	*Salmo trutta, Salmonidae*
Atlantic salmon	*Salmo salar, Salmonidae*
Rainbow trout	*Oncorhynchus mykiss, Salmonidae*
Cutthroat trout	*Oncorhynchus clarkii, Salmonidae*

Contd...

Table 14.1–Contd...

Common Name	Latin Name
Coho salmon	*Oncorhynchus kisutch, Salmonidae*
Chinook salmon	*Oncorhynchus tshawytscha, Salmonidae*
Black sea bream	*Acanthopagrus schlegeli, Sparidae*
MOLLUSC	
Abalone	Haliotis rufescens, Haliotis iris, Haliotidae
Surfclams	*Mulinia lateralis, Mactridae*
CRUSTACEAN	
Brine shrimp	*Artemia franciscana, Artemiidae*
ECHINODERM	
Sea urchin	*Arbacia lixula, Arbaciidae, Hemicentrotus pulcherrimus, Paracentrotus lividus, Echinidae*
	Strongylocentrotus purpuratus, Strongylocentrotidae
	Temnopleurus toreumaticus, Temnopleuridae

Genetic Modification Process

Genetically modified organism production is a multistage process, which includes following steps in general:

1. Identification of gene of interest;
2. Isolation of gene of interest;
3. Production of many copies of gene by amplifying it;
4. Associating the gene with promoter, poly A sequence and insertion into plasmids;
5. Multiplying the plasmid in bacteria and recovering the cloned construct for injection;
6. Transfer the construct into the recipient genome;
7. Integration of gene in recipient genome;
8. Expression of gene in recipient genome and
9. Inheritance of gene through further generations.

Selection of Gene of Interest

Most popularly used genes in aquatic species is growth hormone (GH) genes for better enhanced growth (Figure 14.1). GH has been widely used in terrestrial/mammalian species and as the gene sequence is highly conserved, the product is readily utilized across species boundaries. Another example is regarding extreme cold conditions where a low water temperature is a major problem in aquaculture in temperate climates when an unusually cold winter can severely damage both production and brood stocks of fish. Some marine teleosts have high levels of serum anti-freeze proteins (AFP) or glycoproteins (AFGP) which reduces the freezing temperature by preventing ice-crystal growth. Fletcher, *et al.* (2001) has demonstrated that there is one class of AFGP and four classes of AFP. Most are expressed primarily in the liver and some show clear seasonal changes (Melamed *et al.*, 2002). Production of AFP from the winter flounder (*Pleuronectes americanus*), the gene has been successfully introduced into the genome of Atlantic salmon, integrated into the germ line and passes on to F3 offspring where it was expressed in the liver. However, a number of Ala, Pro-specific endopeptidases are required for production of mature proteins and these are not present in Atlantic salmon. Furthermore, the AFP gene in winter flounder and possibly other Arctic species exists in many copies. Likewise techniques are having applied significance in order to develop effective antifreeze activity in Atlantic salmon (Hew *et al.*, 1999). Work on AFP has also been conducted in goldfish (Wang *et al.*,1995) and milkfish (Wu *et al.*,1998).

Genetic manipulation has also been undertaken in order to increase the resistance of fish to pathogens. This is currently being addressed by the use of DNA vaccines (encoding part of the pathogen genome) and antimicrobial agents such as lysozyme (Demers and Bayne, 1997). An example is the injection of Atlantic salmon with a DNA sequence encoding infectious hematopoeitic necrovirus (IHNV) glycoprotein under the control of the cytomegalovirus promoter (pCMV). Eight weeks later it was observed that a significant degree of resistance had been achieved. The fish were still resistant and were shown to have developed antibodies three months later (Traxler *et al.*, 1999). Similar studies have been undertaken for other fish diseases eg. Haemorrhagic septicaemia virus (VHS) (Lorenzen, *et al.*, 1999) and work of this kind appears to have great potential value for aquaculture (Melamed *et al.*, 2002). Evidence regarding integration of vaccine DNA into the recipient genome is not evident. In India Muzumdar and his group is working on DNA vaccine project.

Isolation of the Gene of Interest

Now a days many numbers of sequenced genoms are readily available on-line. Isolation and PCR amplification is required to develop more copies of it. If, however the gene is to be selected from a genome not previously investigated/sequenced, a more complex procedure will need to be followed. After this several million PCR amplified copies of the genes are needed for the generation of construct (Figure 14.2).

Cloning the Gene of Interest

After generation of many copies of the target gene, it can be placed in a "construct". Once the gene of interest has been ligated enzymatically into the construct, the whole complex is ligated into bacterial plasmids which act as production vectors and enable the gene to be replicated many times within the bacterial cells (Figure 14.3). The bacteria are then plated out. Specific colour (generally blue) of the colony confirms the presence of insert. Many times amplified DNA construct is then cut out of the plasmids using restriction enzymes. It is now ready to be used for insertion into egg of host species using various techniques like Microinjection into the nucleus or cytoplasm, Electroporation of the spermatozoa or eggs, Lipofection of the spermatozoa or eggs, Particle bombardment, Retroviral transfection, Embryonic stem cell derived, Nuclear localization signal mediated.

The Construct

A construct is a piece of DNA, which functions as the vehicle or vector carrying the target gene into the recipient organism. It has several different regions (Figure 14.4). There is a promoter region, which controls the activity of the target gene, a region where the target DNA is inserted, usually some type of reporter gene to enable one to ascertain whether the target has combined successfully with the construct and a termination sequence. The source of these several DNA sequences may be different species although promoter and target genes would ideally be derived from the same species. From early 1990s research focused on developing all fish constructs in preference to using mammalian promoters (Table 14.2). Use of all-fish constructs has dramatic effects on expression of transgenes, eg.

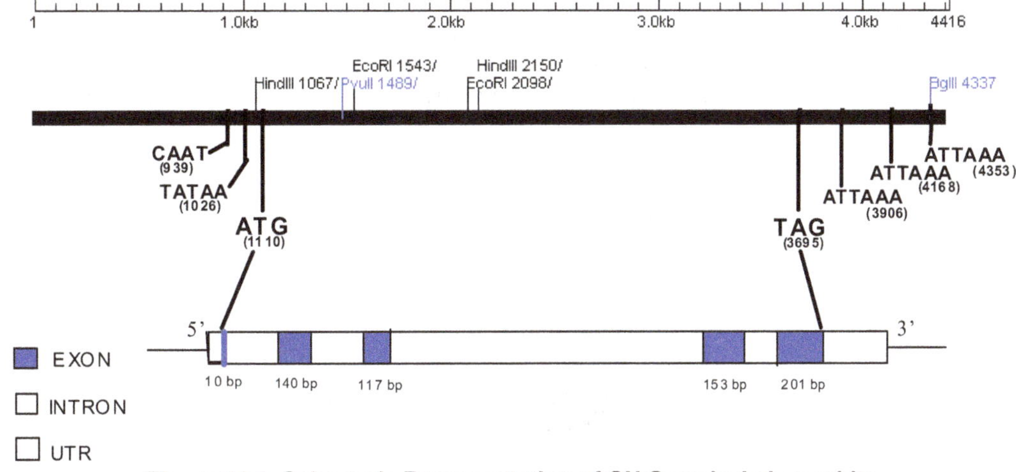

Figure 14.1: Schematic Representation of GH Gene in *Labeo rohita*

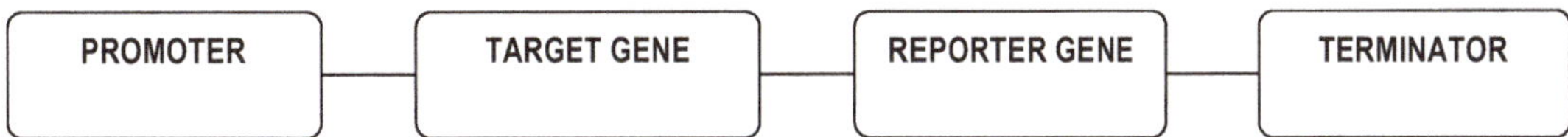

Figure 14.2: A Typical DNA Sequence of a Construct for Designing Transgenic Organism

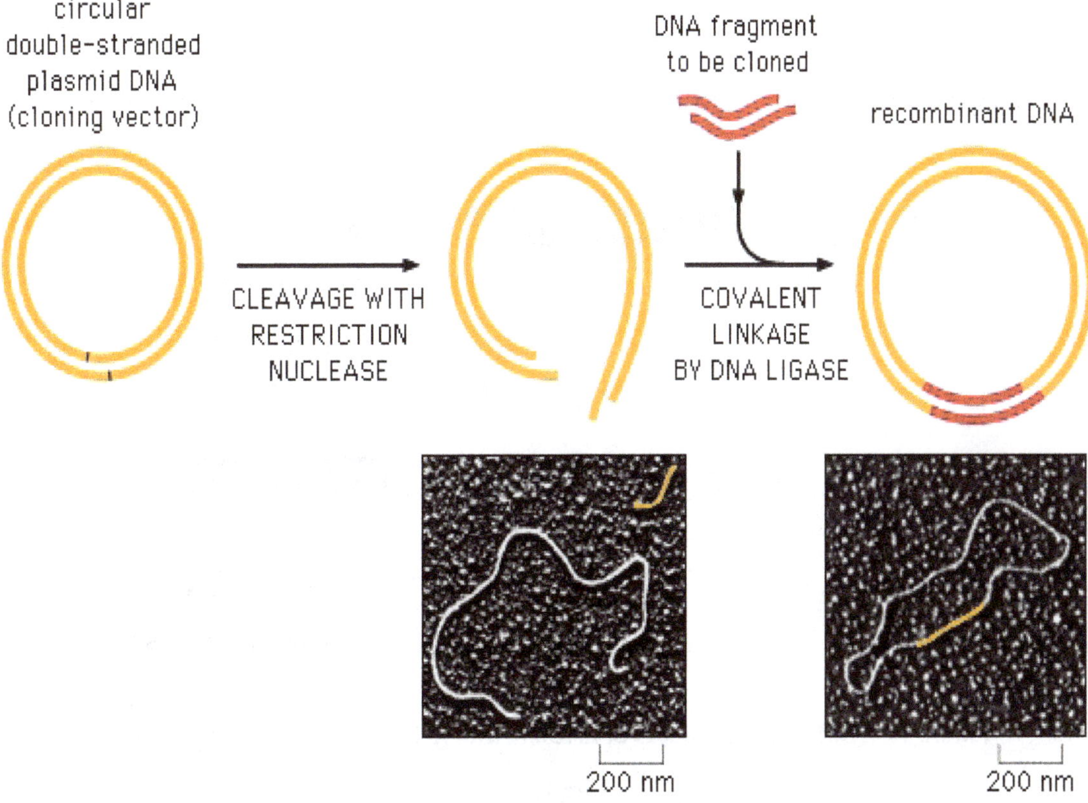

Figure 14.3: Steps Involved in Cloning

Devlin *et al.* (1994), developed an all salmon gene construct which accelerates the growth of transgenic salmonids by over 11 fold. In tilapia, Maclean (1994) found that using carp beta actin instead of rat beta actin promoter led to a ten fold increase in production of hormone in transgenic animals. Muzumdar and his team developed self-transgenic *L. rohita* using growth hormone having growth more than 10 fold.

Integration Site

The factors determining sites of integration are still poorly understood though research in this direction is increasing. It is particularly important to gain greater accuracy in controlled site of integration because of the unpredictable effects of uncontrolled integration on resident genes (Figure 14.5).

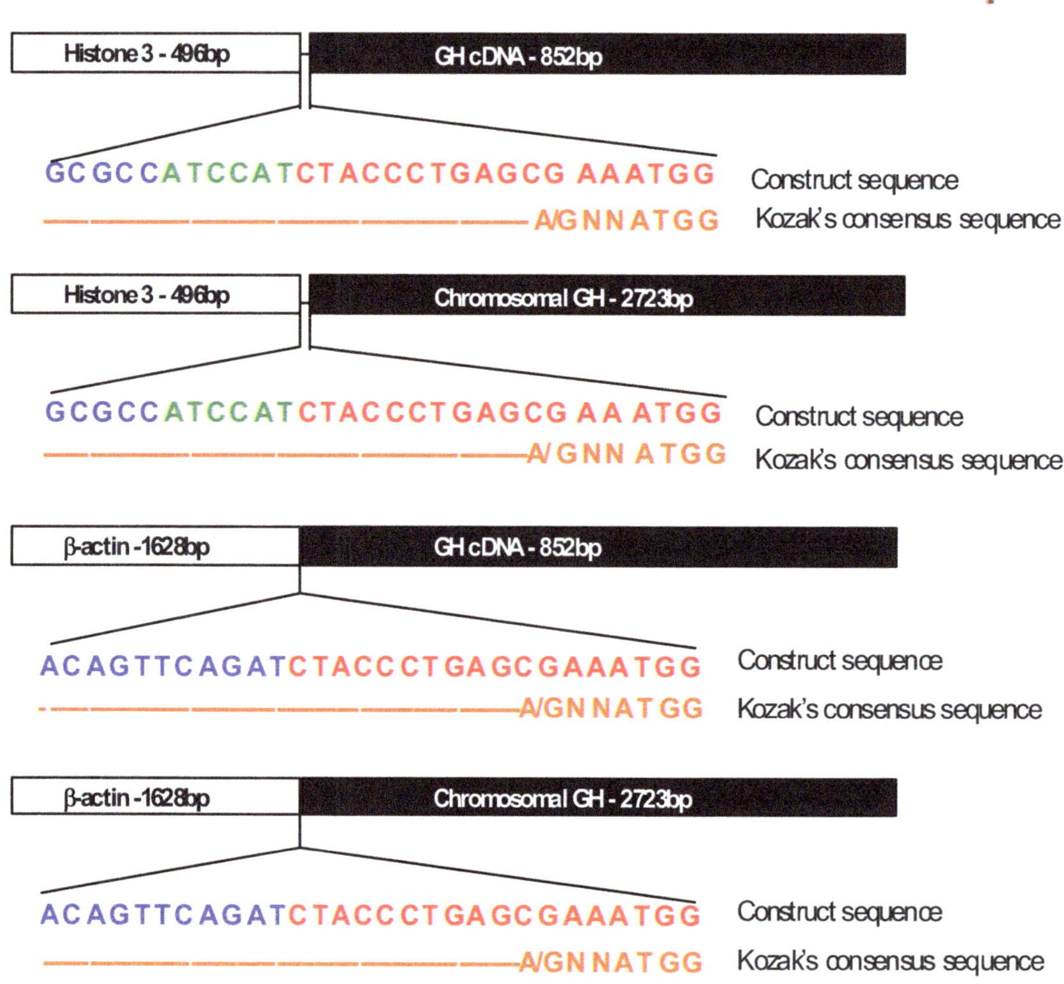

Figure 14.4: Different Constructs of *L. rohita* with Kozak's Consenses Sequence

Table 14.2: Summary of Major Research Effort in Inducing GMOs in Aquatic Species

Species	*Target gene*	*Typical construct*	*Induction method*	*Number of studies*
Salmon spp.	GH/AFP	Ocean pout AFP Linearized DNA	Microinjection	17/92
Rainbow trout	GH	Ocean pout AFP	–do–	14/92
Tilapia spp.	GH	Cytomegalovirus (CMV)	–do–	12/92
Carp	GH	Rous Sarcoma Virus Long Tandem Repeat	–do–	17/92
Zebrafish	Luciferase	pMTL plasmid	–do–	16/92
Medaka	CAT	AFP	–do–	11/92

Expression of gene

The uptake and integration of a transgene does not guarantee that the gene will express itself in the new genetic environment. Tests must be carried out to determine whether there is expression and if there is expression, at what level this takes place (Figure 14.6). Clearly, in commercial aquaculture only those transgenes expressing the target gene at a sufficiently high level will be of interest.

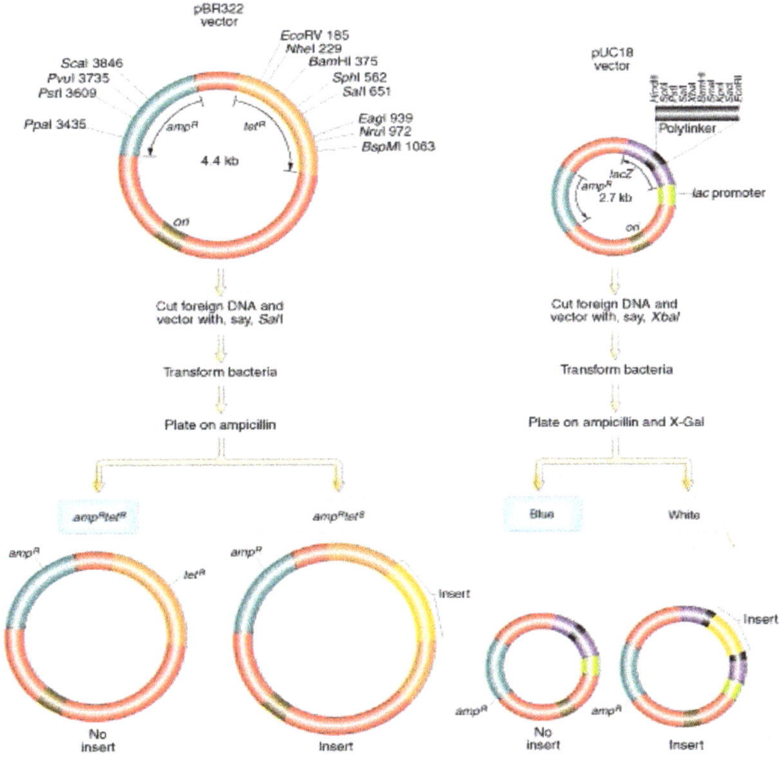

Figure 14.5: Integration Sites in Plasmid Vectors

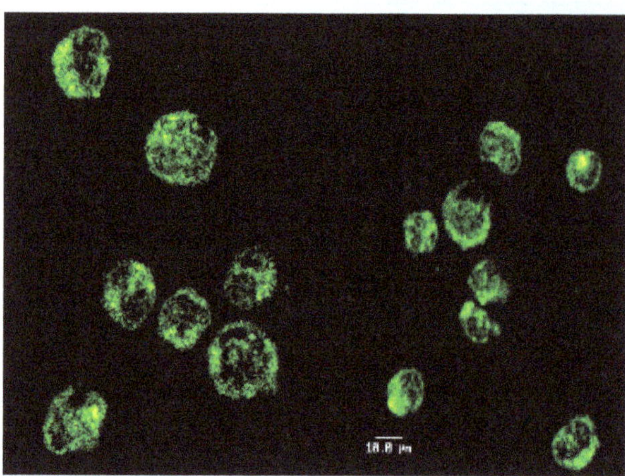

Figure 14.6: Expression of Green Fluroscent Protein Derived Histone Promoter in AK 5 Cells

Inheritance of Gene

A fish which expresses the target gene at an acceptable level may not be able to transmit the gene to progeny. This is because many transgenics are mosaic individuals and unless the gonads are included in the tissues possessing the transgene the transgenic animals will not breed true. Appropriate breeding tests must be carried out. The high proportion of mosaic individuals is one reason why the proportions of progenies of different genotypes resulting from parents that are putatively hemizygous for a transgene do not necessarily conform to mendelian expections. Another reason is the integration of two or more copies of the transgene at different sites in the recipient genome. Further breeding tests will be required in order to establish a pure breeding line of transgenic fish.

Genetic Considerations for Transgenesis

Most of the phenotypic characters of interest to aquaculture are quantitative rather than qualitative. It is therefore important to understand the genetic architecture of such characters. The polygenic theory of quantitative characters (Mather, 1943) envisaged a fairly large number of loci each with relatively small and equal effects acting in a largely additive way on a quantitative character. The theory made no assumptions about the heritability of the character so that two discrete characters might have similar numbers of loci concerned in their determination but very different values for genetic and environmental variance when these were partitioned from the total phenotypic variance. Over the years it has indeed been observed that relatively large number of loci may be involved but also that effects of dominance and epistasis are frequently involved and that the magnitude of effect produced by each locus can vary considerably (Mather, 1979).

Desired quantitative character will be contributed by:

1. A number of locus/loci which can reach several tenmers in number;
2. Genes acting in ways which may be additive, dominant, epistatic and interactive with environmental factors and,
3. Considerable variation in individual locus effect (including many small effects from genes whose primary effect is elsewhere on the phenotype through pleiotropic effects) *i.e.* a small number of loci account for a very large fraction of the variation in the character.

The success in achieving the desired phenotype in transgenic animals will depend on the nature of the genetic architecture of the character concerned.

Advantages Arising from Use of Genetically Modified Organisms

Aquaculture

Evidence of real benefit in terms of economically significant characters comes mainly from work on growth hormone (GH). The overall conclusion from the studies of several workers is that fish GH transgenesis is having markedly superior growth than those non transgenics. Studies have revealed enhancement of growth in transgenic fish from 3 to 20 folds as compared to non-transgenic (Normal) fish control.

The economic gains are much higher from the use of GMOs (Melamed *et al.*, 2002). According to Dunham *et al.* (1992) lines resulting from different transgenic events with the same construct in the same population may give different results which were confirmed in field trials. Transgenics can offer considerable potential in antifreeze protein genes, salinity tolerance, sterility, control of sexual phenotype, disease resistance, behavioural modifications. Also there is a possibility of modifying the genome of organism to allow greater production of nutrient rich characteristics of the flesh like, protein, vitamins, omega-3 fatty acids, etc. (Donaldson, 1997).

Other Uses of Transgenics in Aquatic Species

In addition to obtain enhanced production in aquaculture, there are several other potential uses like living pollution monitors can be achieved by incorporating a pollution sensitive promoter in transgenic animal. A typical example would be a green fluorescent protein structural gene (GFP) deriven by a metallothionin promoter. If the promoter is inactivated by heavy metal pollution the GFP is switched off and the colour change is visible. This can be applicable to any other environmental parameter to be monitored.

Another use in aquaculture is using fish as production system for valuable gene products which can be extracted in a comparable manner to similar production in mammalian species. Such products might include vitamins, proteins and substances related to some diseases like factor VII, one of the human blood clotting factor from fish like tilapia (Maclean, 2002). The use of AFP can be helpful in protection of membranes from cold and freezing damage (Rubinsky *et al.*, 1992, Arav *et al.*, 1993).

Economical Significance

Demand for fish is increasing dramatically every year and the fish production from capture fisheries is declining. Likewise, although aquaculture production is increasing the market for further expansion in aquaculture production is likely to be very good for many years to come.

The data available on GH transgenics suggest that the monetary benefits to be obtained from use of these fish will be large (Figure 14.7). For comparison, the use for the single step genetic change represented by monosex genetically male tilapia (GMT) in Nile tilapia increased the production almost by 30 percent and effectively doubled the net return of the farmers in Philippine (Mair and Abella, 1997). In India still we have not commercialized/emerged any aquatic organism, work so far is restricted to laboratory conditions/trials only.

Risk Factors Associated with GMOs

Most important areas of risks which need to be considered in the use of transgenics are:

Figure 14.7: Growth Variation in Transgenic Rohu for GH Gene and Normal Rohu

1. Human health
2. Biodiversity
3. Animal welfare

In each of these fields there exists a multiplicity of pathways by which effects could be brought about. Rational and responsible assessment of risk requires that the following properties are all to be considered:

1. The target gene DNA source;
2. The non target DNA segments source used for the construct;
3. Incorporation site(s) of the transgene within the recipient genome;
4. Product of the transgene;
5. Interaction of the transgenic product with other molecules in host, its environment and the consumer;
6. Possible molecular change(s) in transgene product during processing;
7. Pleiotropic effects of transgene;
8. Tissue specificity of transgenic expression and
9. Number of transgenic organisms capable of interacting with other organisms and natural system.

Human Health

Most of the dietary DNA get degraded by digestive enzymes relatively quickly (Royal Society, 2001) but use of viruses as vectors, must increase the risk factor significantly as these organisms are adapted to integrating into host genomes and can represent risk of induction of cancer. Zhixong Li *et al.* (2002) induced leukemia in mice using retroviral vector in making transgenics for a commonly used marker gene. Recently a child get induction of leukemia while undergoing gene therapy for x-SCID using retrovirus (Hawkes, 2002). The use of autotransgenics are having less risks than allotransgenics. The major risks from the production of the transgene will lie in the use of novel proteins or other bioactive molecules produced by the transgenic organisms in native form or in modifications in the human body, such molecules could be problematic to human health. Use of such substances should not be tested except the urgent necessity can be used under the expert supervision.

Other risks nay lie in incorporation of transgenic DNA into the genome of resident gut microflora (Probiotic use) or change in the pathogen spectrum of the transgenic fish leading it to hosting new pathogen which could be the pathogen to consumer.

Biodiversity

Aquatic biodiversity is extremely large and relatively poorly understood (Mair, 2002). Therefore the risk associated with use of GMOs or any genetically distinctive strain used in aquaculture is large. Also there is always danger of escape of genetically distinct farmed fish or GMOs to the open aquatic environment. The genetically improved forms including GMOs, are developed for a specific set of environmental conditions in which they enjoy an advantage conferred by human decisions. In nature, however such genetically distinct forms may legitimately be regarded as mutant forms of wild type. Therefore probability of survival of mutant forms is extremely low because they are disadvantaged in viability and fertility under natural conditions. If the GMOs raised for more generations there are chances to have adaptability and selection pressure on the organism genome to get acclaimed with changing environmental conditions. It can result in posing for success of inbreeding or breeding with natural population, which can threat the natural population or diversity. Some times due to sudden changes occurred can cause certain mutations, which lead to new speciation which is not expected. There are a lot of factors, which cannot be predicted without best containment measures are adapted. As a general rule and adapting a precautionary approach (OECD, 1995), it is however clear that each individual case needs careful study and appraisal.

The Cartagena Protocol on Biosafety, negotiated under the Convention on Biological Diversity, mandates the "advance informed agreement" of an importing country prior to the transboundary transfer of certain GMOs (called "living modified organisms" or LMOs). The objective is to allow an LMO receiving country to assess potential risks to biological diversity and human health that could be posed by such transfers. It is examined in detail the negotiating history of the Cartagena Protocol, including the manner in which the new and contested concept of "biosafety" has been framed in devising international obligations in this area (Gupta 1999, 2000a, 2000b). The demand for a biosafety protocol and for "informed consent" prior to GMO trade came originally from developing countries, led by Malaysia, during negotiation of the Convention on Biological Diversity (CBD) in the early 1990s. While green groups and Nordic countries like Denmark supported this developing country demand, it was opposed by agricultural GMO producer and exporter countries such as the United States and Australia, as well as by biotechnology industry groups. Developing countries called for a biosafety protocol out of concern that they might become the testing grounds for what they perceived to be novel substances, which they did not have the capacity to deal with. However, those opposed to a protocol argued that GMOs did not pose risks different from those associated with other techniques of genetic manipulation such as traditional breeding, and hence did not merit separate international regulation (Gupta 1999). Although the European Union initially offered only lukewarm support for a biosafety protocol under the CBD, negotiations of the protocol have unfolded over a four-year period of expanding public concern in Europe over ecological and food safety concerns relating to genetically modified organisms. This has been accompanied by an escalating trade conflict between the United States and the European Union. A *de facto* moratorium has been in effect against entry of transgenic crops/organisms into Europe over the last two years, as the European Community has debated amendments to its regional directives on contained use and deliberate release of GMOs and has halted new approvals until such amendments are in place. This has transformed what began largely as a developed versus developing country issue into a growing intra- OECD conflict, with repercussions for the scope and clarity of the protocol's obligations. The evolving nature of the alliances in this negotiation is reflected in the fact that towards the end of the failed Cartagena meeting in February 1999, five distinct negotiating groups had emerged. These included the Miami Group, consisting of six agricultural exporting countries (Argentina, Australia, Canada, Chile, Uruguay and the United States); the European Union; the Like-Minded Group (developing countries, excluding Argentina, Chile and Uruguay); Central and Eastern Europe; and the Compromise Group, consisting of OECD countries that are not agricultural exporters nor part of the European Union (Japan, Mexico, New Zealand, Norway, Singapore, South Korea and Switzerland). While the Miami Group was concerned about the impact of the protocol on the agricultural commodity trade, the European Union was responding to increased public concern about import into the Union of genetically modified foods. The Compromise Group with its eclectic membership represented a mix of such concerns, since it included leaders in biotechnology research such as Switzerland and major agricultural importing countries such as Japan.

Animal Welfare and Ethical Issues

The direct or indirect effects of transgenesis upon the GMOs fish welfare in aquaculture are very poorly understood. The notions regarding the cruel or unnatural treatments in mammalian species are not as it is applicable in many cases. But as life forms with highly developed nervous system and with a range of behavioural patterns and learning abilities, fish qualify for welfare considerations.

In transgenic coho salmon for anti freeze protein and growth hormone is reported for changes in colour, deformities of cranium with overgrowth of operculum and lower jaw deformities by Devlin *et al.* (1995b). After one year of growth, anatomical changes due to growth of cartilage in the cranial and opercular regions were more severe and reduced viability was also reported. Data on farmed species is available on this issue in large quantities, which shows dysfunctional development leading to acromegaly, lameness and infertility in some GH transgenics in pigs and sheep. But very scanty data is available on aquatic GMOs, probably because animal welfare is not sufficiently recognized as an issue regarding the use of GMOs in aquaculture.

Changes to the genetic structure of a population may occur through artificial selective breeding, genetic drift and gene mutation. Mutations occur in natural population at a low frequency. However, artificial selection may alter the genetic structure of a population more rapidly. The artificial introduction of a fusion gene to produce a transgenic fish, in theory, is not different from the natural processes, but it is a more rapid approach to transfer new genes into a fish. This, in effect, is putting a population through an artificial bottleneck (Devlin and Donaldson, 1992). Whether or not transgenic fish will have a significant impact on the environment is debatable and it is difficult to predict. However, one ethical issue is the method by which a genetic change is introduced. Further, the phenotypic changes, such as increased growth rate, are usually more prominent in the transgenic fish than those obtained by artificial selection or through efficient feeding regime. Individuals carrying a deleterious mutation are not likely to reach sexual maturity and are often eliminated from the population. However, transgenic fish designed for a chosen phenotype will remain in a population if it can survive in the natural environment. The mixing of the transgenic fish with the wild-type population may have important implications on the survival of the native species through competition (Jonsson *et al.*, 1996) and the conservation of the natural genetic diversity of populations. In view of such possible adverse effects on the environment and the genetic characteristics of the natural fish populations, the incorporation of risk assessment and risk management into public policies on genetically modified finfish and shellfish has been strongly advocated by the scientific community (Devlin and Donaldson, 1992; Hallerman and Kapuscinski, 1995). National policies on oversight of genetically modified aquatic organisms have been developed in a number of countries including Australia, Canada, New Zealand, Norway, the United Kingdom and United States. Countries that do not have any policies should be strongly encouraged to develop guidelines for transgenic fish research and the release of genetically modified aquatic organisms into the environment. Legislative power should be available for government agencies to oversee research using genetically modified fish and the release of such fish into the environment. For example, in New Zealand all transgenic organism research has to be approved by the Environmental Risk Management Authority (ERMA), which is under the Ministry for the Environment.

EMRA is responsible for assessing and approving all hazardous substances and new organisms, including the development and field-testing of genetically modified organisms. Development of biological containment of transgenic fish by induced sterility should be considered as a priority when considering the development of genetically modified fish for commercial exploitation. A tissue-specific gene promoter such as that for protamine can be used in the construction of a fusion gene coding for a toxic protein. Because protamine is a sperm-specific protein, a transgene driven by a protamine promoter will express the gene in testicular cells only. Using this approach, sterile male mice have been produced (Peschon, 1989; Braun *et al.*, 1990). In principle, such an experimental approach may also be applicable to fish (Maclean and Penman, 1990).

The Biosafety Regulatory Framework in India

Genetically modified organisms are regulated in India under the purview of the 1986 Indian Environment (Protection) Act (EP). The broad objective of the EP Act is the protection and improvement of the environment. To meet this objective, the Act calls for regulation of "environmental pollutants" which are defined as "any solid, liquid or gaseous substance present in such concentration as may be, or tend to be, injurious to the environment" (MOEF 1986). The Ministry of Environment used the broad definition of "environmental pollutant" and Forests in 1989 to issue rules to govern use of genetically engineered organisms under the EP Act. The 1989 "Rules for the Manufacture, Use, Import, Export and Storage of Hazardous Microorganisms, Genetically Engineered Organisms or Cells" (1989 Rules) constitute the legally binding regulatory framework for GMOs in India (Ghosh and Ramanaiah, 2000).

As evident from the title, GMOs are placed here in the same category as hazardous microorganisms and their regulation under the EP Act is justified by their alleged potential to be hazardous substances or environmental pollutants. Since the 1989 Rules call for guidelines to be developed to give them effect, biosafety guidelines were issued by the Department of Biotechnology under the Ministry of

Science and Technology in 1990, and have been consistently revised and expanded in the last decade. Development of this regulatory framework dating from the early 1990s has been in response to the need for regulatory oversight of the growing domestic community engaged in biotechnology research in both the agricultural and pharmaceutical sectors. It was also in response to debates within the European Union about safe use of genetically modified organisms, during negotiation of regional directives on contained use and deliberate release of GMOs. The possibility that India might be a future importer of GMOs from OECD countries thus also spurred development of a domestic regulatory framework.

References

Arav, A., Rubinsky, R., Fletcher, G. and Seren, E., 1993. *Molecular Reproduction and Development.* 36: 488–493.

Braun, R.E., Lo, D., Pinkert, C.A., Widera, G., Flavell, R.A., Palmiter, R.D. and Brinstein, A., 1990. *Biol. Reprod.*, 43: 684–693.

Demers, N.E. and Bayne, C.J., 1997. *Developmental and Comparative Immunology*, 21: 363–373.

Devlin, R.H. and Donaldson, E.M., 1992. Containment of genetically altered fish with emphasis on salmonids. In: *Transgenic Fish,* (Eds.) C.L. Hew and Fletcher, G.L. World Scientific Publication Co., Singapore, pp. 229–265.

Devlin, R.H., Yesaki, T.Y., Donaldson, E.M., Jun Du, S. and Saho J.D., 1995b. *Canadian Journal of Fisheries and Aquatic Science,* 5: 1376–1384.

Devlin, R.H.,Yesaki, T.Y., Biagi, C.A., Donaldson, E.W., Swanson, P and Whoon Clong Chan, 1994. Growth enhancement of salmonids through transgenesis using an "all salmon" gene construct, in High Performance Fish. In: *Proceedings of an International Fish Physiology Symposium at the University of British Columbia in Vancouver*, Canada, July 16–21, 1994, (Ed.) D.D. MacKinlay, pp. 343–345.

Donaldson, E.M., 1997. The role of biotechnology in sustainable aquaculture. In: *Sustainable Aquaculture,* (Ed.) J.E. Basdad. Wiley, pp.101–126.

Dunham, R.A., Duncan, P.L.Clien, T.T., Lin, C.M., and Powers, D.A., 1992. Expression and inheritance of salmonid growth hormone genes in channel catfish, *Ictalurus punctatus*, and effects on performance traits. "Aquaculture 92": Growing towards the 21st Century, pp. 83–84.

Fletcher, G.L., Hew, C.L. and Devies. P.L., 2001. *Annual Reviews in Physiology*, 63: 259–306.

Ghosh, P.K. and T.V. Ramanaiah, 2000. *Journal of Scientific and Industrial Research,* 59: 114–120.

Gupta, Aarti, 2000a. Governing trade in genetically modified organisms: The cartagena protocol on biosafety. *Environment*, 42(4): 23–33.

Gupta, Aarti, 2000b. *International Journal of Biotechnology*, 2(1/2/3): 205–230.

Gupta, Aarti, 1999. *Framing "Biosafety" in a Transnational Context: the Biosafety Protocol Negotiations under the Convention on Biological Diversity* (ENRP Discussion Paper E–99–10, Kennedy School of Government, Harvard University, Cambridge, MA) Available at <http://environment.harvard.edu/gea>.

Hallerman, E.M. and Kapuscinski, A.R., 1995. *Aquaculture,* 137: 9–17.

Hawkes, N., 2002. Cancer risk fails to halt "bubble boy" gene therapy. The Times, 4 October, p.7.

Hew, C.L., Poon, R., Xiong, F., Gauthier, S., Shears, M., King, M., Davies, P., and Fletcher, G., 1999. *Transgenic Research*, 8:405.

Jonsson, E., Johnsson, J.L. and Bjornsson, B.T., 1996. Growth hormone increases predation exposure of rainbow trout. *Proc. R. Soc. Lond.* 263B, 647–651.

Maclean, N. and Talwar, S., 1984. *J. Embryol and Exp. Morphol.*, 82(Supp): 187.

Maclean, N. and Penman, D., 1990. *Aquaculture*, 85: 1–20.

Maclean, N., 1994. *Animals with Novel Genes.* C.U.P.

Maclean, N., 2002. Personal communication.

Mair, G.C., 2002. Personal communication.

Mair, G.C. and Abella, T.A. (Eds.), 1997. Technoguide on the production of Genetically Male Tilapia. Freshwater Aquaculture Centre, Central Luzon State University, The Philippines.

Mather, K., 1943. Polygenic inheritance and natural selection. *Biol. Rev.*, 18: 32–64.

Mather, K., 1979. Historical overview; Quantitative Variation and Polygenic Systems. In: *Quantitative Genetic Variation*, (Eds.) J.N. Thompson and J.M. Thoday. Academic Press, pp. 5–34.

Melamed, P., Gong, Z.Y., Fletcher, G.L. and Hew, C.L., 2002. *Aquaculture*, 204: 255–269.

OECD, 1995. Proceedings of workshop on Environmental Impacts of Aquatic Biotechnology (1992 Trondheim Norway), OECD, Paris.

Palmiter, R.D., Brinster, R.L. and Hammer, R.E., 1982. *Nature*, 300: 611–615.

Peschon, J.J. 1989. Ann. NY Acad. Sci. 564, 186–197.

Royal Society, 2001. The use of genetically modified animals. Policy document 5/01. Royal Society, London.

Rubinsky, B., Mattioli, M., Arav, A., Barboni, B. and Fletcher, G.L., 1992. *American Journal of Physiology*, 262: R542–R545.

Traxler, G.S., Anderson, E., Laptra, S.E., Richard, J., Shewmaker, B., and Kurath, G., 1999. *Dis Aquat Org.*, 38: 183–190.

Wang, R., Zhang, P., Gong, Z., and Hew, C.L., 1995. *Molecular Marine Biology and Biotechnology*, 4: 20–26.

Wu, S.M., Hwang, P.P., Hew, C.L., and Wu, J.L., 1998. *Zoological Science*, 37: 39–44.

Zhixong, Li, Dullman, J., Schiedmeier, B., Schandt, M., von Kalle, Meyer, J., Forster, M., Stocking, C., Wahlers, A., Frank, O., Ostertage, W., Kuhlde, K., Eckert, H-G., Fehse, B. and Baum, C. 2002. *Science*, 296: 497p.

Zhu, Z.Y., Li, G., He, L. and Chen, S., 1985. *Journ. Appl. Ichthyol.*, 1: 31–34.

Sustainable Environmental Management *Pages* **132–151**
Edited by: **Dr. L.V. Gangawane & Dr. V.C. Khilare**
Published by: **DAYA PUBLISHING HOUSE**

Chapter 15

Environmental Impacts on Fish Germplasm

G.D. Khedkar

Aquaculture Research Laboratory, Department of Zoology,
Dr. Babasaheb Ambedkar Marathwada University, Aurangabad
E-mail: gdkhedkar@gmail.com

ABSTRACT

Fisheries resources of India are highly diversified in type and nature. Nearly 22,000 species of finfishes are known worldwide. About 10 per cent (2,200 species) of the global fish diversity is found in Indian waters. Out of these, about 400 species are commercially important which include cultured, cultivable and wild species, however, only few are being presently utilized in aquaculture. These fishes have acquired a wide variety of forms and habitats which have been reflected in their adaptations to survive in markedly varying biotypes, ranging from cold torrential mountain streams to the dark byssal depth of the seas. Too many factors are having their impacts on fish germplasm. Fish genetic resources, especially in the developing countries like India are under harsh environment attributable to the multipronged degradation, which may be due to over exploitation or water pollution or habitat shrinkage. As a result some of the prime fish germplasm are either depleting or gradually declining day by day. Factors responsible for habitat shrinkage are industrialization, urbanization and over harnessing of aquatic resources for commercial purposes. It includes impoundment of lotic eco-system, destruction of luxuriant flora in the catchments area, reclamation of wetland for agriculture and allied activities, layout of roads along the river courses, natural calamities like prolonged drought can destruct the aquatic diversity to a great extent. Introduction of genetically modified organisms or introduction of exotics is also threatening the aquatic biodiversity to a large extent. Overexploitation of natural resources is already imparting tremendous pressure on natural resources. In fisheries sector overexploitation and destructive fishing by using small sized nets, ghost fishing by using dynamite, poisons, etc. are eliminating fish germplasm rapidly. Increasing pressures on aquatic resources dictate that fish conservation can no longer be treated in isolation and an integrated approach to aquatic resource management is required.

Keywords: *Germplasm, Environment, Pollution, Resources, Diversity, Fish fauna, Exotics.*

Introduction

India is endowed with vast network of rivers, characterized by very rich and highly diverse fish fauna. The enormous and unmanageable magnitude of rivers and recognition of riverine fish stock as common property resources with open assess are the factors inhibiting sustainable exploitation of these natural renewable resources. We Indians are fortunate to possess vast and varied fish germplasm resources distributed widely in vivid aquatic ecosystems. The aquatic resources of country includes 2.02 million square km. area of Exclusive Economic Zone (EEZ) of surrounding seas, more than 29,000 km length of rivers, about 1,13,000 km of canals, about 2.90 million ha of existing water-spread in the form of reservoirs, about 2.254 million ha in the form of tanks and ponds and about 0.6 million ha of stagnant, derelict, swampy water-spread areas (Jhingran, 1991). The marine fish landings in India during 2003-2004 were estimated at 2.64 million tones. The current level of fish production from inland resources has been estimated to be about 2.1 million tones out of which 1.35 million tones is from aquaculture indicating more than 50 per cent contribution.

To meet the internal projected demand for inland fishes, two way strategies are being pursued. First is to conserve the stocks in open water systems especially rivers, estuaries, large reservoirs and wetlands. Second is to encourage both extensive and intensive aquaculture in manageable water particularly tanks and ponds. Traditionally the aquaculture practices in the country, both in freshwater and coastal saline waters were characteristically low-input and low-production systems depending on wild seed collection and stocking in natural ponds or impoundments without any further management measures. But during the last decade, aquaculture has been transformed into a business activity with high input and high yield. It resulted in increasing fish production at rate the of 5.21 per cent a year. This increase in growth is mainly due to increase in growth of inland fish production. Inland fish production has accelerated from 5.14 per cent to 6.34 per cent from 1980-81 to 1989-90 to 1990-91 to 1998-99, while marine fish production decelerated from 3.73 to 2.50 percent during the same period. Culture fisheries comprising freshwater and brakishwater fish culture derived the growth in inland sector. The share of culture fisheries in inland sector increased tremendously from 43.33 percent in 1984-85 to 71.72 percent in 1989-90 and then to 84.07 per cent in 1994-95. At global level, fish production has increased from 86.53 million tones in 1985 to 112.91 million tones in 1995. Though marine fish production contributes more than 80 per cent of total fish production, but its share in total fish production has decreased during this period. On the other hand, share of inland sector has increased faster and increased from 12.40 per cent in 1985 to 18.61 percent in 1995. With increasing production levels and areas devoted to aquaculture and intensification of culture practices, environmental issues have come into focus.

Importance of fish germplasm conservation for the environment

The examination of fisheries conservation management indicates the advantages and limitations of existing measures to protect and enhance threatened freshwater fishes. Despite the global recognition to conserve fish biodiversity being only a recent phenomenon, good legal and institutional frameworks exist in many countries (Crivelli, 2002; Kirchhofer, 2002; Skelton, 2002). However, many of the institutions lack the resources to implement sound conservation plans (Impson, *et al.*, 2002) and representation of freshwater fishes is generally weak. Similarly, the resources to enforce legislation are lacking, thus freshwater fishes are not always afforded the protection they are designated under law. The main problems with existing measures are the lack of baseline information on which to manage the fishes, lack of public awareness and weak integration of conservation in water resource management planning. Although lack of baseline data is an issue, the urgency for direct management intervention is so great for many species that decisions should be based on the best available science and existing experience to support management options. Where information on which to make a decision remains inadequate the precautionary approach (FAO, 1996 and 1997) must be adopted. This is particularly important where development schemes are likely to impact on fish communities about which little is known. However, a networking mechanism for reporting the successes and failures in conservation management needs to be developed so that lessons can be transferred to all. One of the factors common to successful conservation projects appears to be involvement of people, as the general public act as excellent ambassadors to promote fish issues (Cambray and Pister, 2002). The biggest problem identified, however, is that the general population have poor awareness of the problems facing freshwater fishes, thus greater opportunity should be made of their willingness to support conservation campaigns by promoting education and extension programmes (Cambray and Pister, 2002). Increasing pressures on aquatic resources dictate that fish conservation can no longer be treated in isolation and an integrated approach to aquatic resource management is required (Cowx, 1998). Demands for sustainability grew out of the Rio Conference and have placed emphasis on the need to not only manage exploited resources but also promote biodiversity.

Fish Germplasm

Fisheries resources of India are highly diversified in type and nature. Nearly 22,000 species of finfishes are known worldwide (Warren and Burr, 1994). About 10 per cent (2,200 species) of the global fish diversity is found in Indian waters. Out of these, about 400 species are commercially important which include cultured, cultivable and wild species, however, only few are being presently utilized in aquaculture. These fishes are have acquired a wide varieties of forms and habitats which have been reflected in their adaptations to survive in markedly varying biotypes, ranging from cold torrential mountain streams to the dark byssal depth of the seas. Indian marine ecosystem is occupying 1440 (65.45 per cent) finfish species, Brakishwater 143 (6.50 per cent), Warmwater of plains 544 (24.73 per cent) and Coldwater 73 (3.32 per cent) (Anon, 1992-93; Das, 1994; Das and Pandey, 1995). Only a few species have been worked out in detail with regard to their biological attributes. Among different species available in inland water resources, only few have been domesticated under culture practices. Any analysis of the state of fisheries and their resources needs to be undertaken in its broader aquatic context. In that respect, most aquatic environments indicate a lack of stewardship, illustrated by growing degradation, loss of habitat, lack of coherence in aquatic science policy, inadequate management-oriented research, poor or inexistent long-term monitoring, lack of strategic, integrated planning of conflicting uses, etc.

Modern fisheries management, as practiced since the early 1940s, is strongly based on the ecosystem theory but focuses primarily on fishing activity and target fish resources. In inland waters, affected earlier and more strongly than marine waters by environmental problems, it developed as an extension of wildlife management (Laikre,1999) and involves a substantial amount of direct intervention on the habitat, species composition, etc. In marine ecosystems, however, because the possibility of direct intervention on the ecosystem is limited, management strategies concentrated on controlling human intervention (fishing) while observing proxies for the state of an otherwise opaque ecosystem and fugitive resources. It is defined as "the integrated process of information gathering, analysis, planning, decision making, allocation of resources and formulation and enforcement of fishery regulations by which the fisheries management authority controls the present and future behaviors of the interested parties in the fishery, in order to ensure the continued productivity of the living resources" (FAO, 1995b). It aims at optimizing the use of fishery resources as a source of human livelihood, food and recreation, dynamically regulating fishing activity, meeting resource-related objectives or constraints, mainly indirectly.

A thorough knowledge of fishery resources, their availability and distribution in a particular water body is essential for proper exploitation. Considerable work has been done on the availability and distribution of reservoir fishes in general (Reid, 1943; Parrish and Blaxter,1963; Mecombie, 1977).

Factors Affecting Fish Diversity

Habitat Shrinkage

Fish genetic resources, especially in the developing countries are under harsh environment attributable to the multipronged degradation, which may be due to over exploitation or water pollution or habitat shrinkage. As a result some of the prime fish germplasm are either depleting or gradually declining day by day. Factors responsible for habitat shrinkage are industrialization, urbanization and over harnessing of aquatic recourses for commercial purposes. It includes impoundment of lotic eco-system, destruction of luxuriant flora in the catchments area, reclamation of wetland for agriculture and allied activities, layout of roads along the river courses, natural calamities like prolonged drought can destruct the aquatic diversity to a great extent.

Habitat Modification

The developmental activities, whether it is concerned with water resources or agriculture or industry or flood control or navigation improvement, all conflict considerably with fisheries. While man-made lakes, diking of sea or river banks bring abrupt change in the ecology; large scale earth moving, clearance of forest disrupt partly or wholly the entire ecosystem. All these activities alter the flow pattern in the riverine system thus reducing the volume of water, flushing rate and increase the silt load. These changes affect the natural fish recruitment due to loss in breeding and feeding grounds resulting in catches decline and population shift.

Fish habitats are destroyed as a consequence of many factors. Headwater regions of streams have been altered by deforestation and watershed erosion, and siltation has destroyed the breeding habitats of many species that require clear oxygen-rich waters. Agricultural run-off, pesticides, fertilizers, sewage, and chemical pollutants add additional stresses to remnant fish populations. Impoundments for water retention and electrical generation create barriers to the natural dispersal pathways of migratory fishes, and eliminate opportunities for gene flow among populations of primary freshwater species. Canalization and diversion of streams have eliminated riparian zones and destroyed aquatic ecosystems that maintain water quality, nutrient recycling, and contribute to the nurture of fish populations. Introduced exotic species provide the *coup de grace* for many native fishes. Construction of dams and canals provides artificial latchstring and riverine habitats that often are stocked with non-native game fish and commercially important species. Introduced carp and tilapia flourish in impoundments and fishponds and compete with native fish for food and nesting sites. When these exotic species escape, they often reproduce in surrounding streams and quickly replace native fishes. Similarly, *Gambusia* has been introduced widely for mosquito control and has devastated natural populations of small fishes throughout the world (Meffe, 1985). Introduced species pose an additional threat if they transmit exotic diseases to native fishes (Leberg and Vrijenhoek, 1994). The potential impact of exotic helminthes, bacteria, and viruses should be assessed before government agencies become involved in the wholesale transport and release of exotic fishes.

Practices attributed to commercial fishing and hatcheries have altered native fish populations. Gillnet fisheries in Lake Victoria have contributed to the decline of endemic tilapias and greatly changed the composition of native fish communities (Fryer, 1972). In addition to the broad scale introduction of tilapia and carp throughout the world, sport and commercial fisheries based on translocated native species have contributed to the genetic adulteration of fish stocks. Genes from released sheepshead minnows *Cyprinodon variegates* Lace´pe'de 1803 used as bait for sport fishing have introgressed into native pupfish *C. pecosensis* Echelle and Echelle, 1978 populations (Echelle and Conner, 1989). Genes from hatchery-reared rainbow trout *Oncorhynchus mykiss* (Walbaum, 1792) have introgressed into many cutthroat trout *O. clarki* (Richardson, 1836) populations, and rainbow trout have completely replaced cutthroats in many streams on western slopes of the Rocky Mountains in the U.S.A. (Allendorf and Leary, 1988). Native Apache *O. apache* (Miller, 1972) and Gila trout *O. gilae* (Miller, 1950) have also been replaced by hatchery reared rainbow trout, or altered by introgressive hybridization (Loudenslager *et al.*, 1986). Competitive replacement and genetic swamping of native species is a serious problem contributing to homogenization of the world's freshwater fish fauna. Although released species may serve the nutritional and entertainment needs of many people, they alter local fish communities and may destabilize aquatic ecosystems.

Riverine Regulation in a Tropical Stream System

The structure and persistence of native biotic communities with riverine ecosystems is having a strongly influence of both spatial and temporal variation in environmental conditions (Poff and Ward, 1989; Stanford *et al.*, 1996). Spatially complex riverine environments present diverse habitats along longitudinal, lateral, and vertical dimensions (Ward, 1989; Stanford and Ward, 1992), offering the possibility for spatial segregation of species and guilds, size classes, and life stages (Schlosser, 1991; Stanford *et al.*, 1996; Poff *et al.*, 1997). Temporal variation in stream flow, water temperature, dissolved oxygen concentration, transport of sediment and organic matter, and other environmental conditions continually modify the suitability of particular aquatic habitats, imposing an 'environmental regime' on those habitats. Environmental regimes influence the composition and structure of aquatic communities in three important ways:

1. by shaping environmental conditions and their variation within particular habitats;
2. by shaping the distribution and evolution of the mosaic of habitats; and
3). by influencing the movements of organisms between habitats.

The stream flow regime is a driving force in river ecosystems (Stanford *et al.*, 1996; Poff *et al.*, 1997). Stream flow controls key habitat parameters such as flow depth, velocity, and habitat volume. The often strong connections between stream flow, floodplain inundation, alluvial ground water movement, and water table fluctuation mediate the exchange of organisms, particulate matter, energy, and dissolved substances along the four dimensions of river systems: upstream–downstream, channel–hyporheic (ground water), channel–floodplain, and the temporal dimension (Amoros and Roux, 1988; Ward, 1989; Stanford and Ward, 1992; Ward and Stanford, 1995a). Flow is often tightly coupled with other environmental conditions as well, such as temperature and oxygen, channel morphology and substrate particle sizes (Sparks, 1992, 1995; Allan, 1995; Ward and Stanford, 1995b; Stanford *et al.*, 1996; Poff *et al.*, 1997; Richter *et al.*, 1997). Alteration of natural stream flow regimes modifies the distribution and availability of riverine habitat conditions, with adverse consequences for native biota (Poff *et al.*, 1997).

Constructions of Dams

The first man-made lake is believed to have come into existence about 6000 years ago (Fernando, 1980). Since then many reservoirs have been constructed in different parts of the world. These water bodies were built, for power generation, irrigation, flood control, supply of drinking water etc.

According to UNESCO report (Anon, 1978), the total water spread of reservoirs of the world is 6,00,000 sq. km. of which the contribution of India is 20,000 Sq.Km. according to survey of IIMA (Anon, 1983) high dam reservoirs are expected to grow to 1.5 to 2.3 times by 2000A.D. and 2025 A.D. respectively from the 1974, level of 15 million hectare (Natarajan, 1985).

Unregulated rivers provide highly permeable (Wiens *et al.*, 1985; Wiens 1992) corridors for aquatic fauna, but high (tall) dams can reduce or eliminate this permeability. In a system dominated by animals requiring both fresh and salt water for life cycle completion, such impermeability can lead to major shifts in assemblage structure. Comparatively little is known about the life history and migratory patterns of tropical stream fauna (Clay 1995), and barriers posed by dams in the tropics are likely to differ from effects of dams on the well-known salmonid migrations in temperate regions (*e.g.*, Stuart 1962; Mills 1989; Linløkken 1993). Extrapolation to other systems is not necessarily warranted (McDowall 1993). Like McDowall's (1993) New Zealand fishes, many shrimps and fishes of Caribbean streams attempt to overcome obstacles by climbing (via walking legs, pelvic-fin suction discs, or wriggling) rather than jumping (Covich and McDowell 1996). Dams differ in many ways; in turn, the effects of these structures vary as well (Petts 1984; Clay 1995). For instance, Ibrahim (1962) and Lee and Fielder (1979) found that some shrimps (in India and Australia, respectively) could climb structures but only in conjunction with flowing water (Figure 15.1).

The stream macrofauna of the Caribbean is dominated by an assemblage of decapods crustaceans (*e.g.*, Abele and Blum 1977; Hunte 1978; Crowl and Covich 1990; Pringle *et al.*, 1993; Covich and McDowell 1996); fishes are present as well (*e.g.*, Erdman 1972; Penczak and Lasso 1991; Covich and McDowell 1996) but generally less abundant. Salt water may be essential for life cycle completion for many of these fauna. Although the eel *Anguilla bengalensis bengalensis* is catadromous and the freshwater crab *Epilobocera sinuatifrons* has direct development (Erdman 1972), there is evidence that at least some of these species may be washed to the sea as larvae or eggs and return to the streams as juveniles ("amphidromy" McDowall 1992, 1993). Support for this assumption has been provided by observations

Figure 15.1: High Dam Never Allow the Exchange of Population Between Pre-dam and Post-dam Regions (Nathsagar Dam (Godavari river), Arangabad, Maharashtra)

of larvae of the fish *Sicydium plumieri* (a goby) being carried down streams to the ocean (Bell and Brown 1995) and upstream migrations of shrimps and fishes (Erdman 1972, 1986; Felgenhauer and Abele 1983; Covich and McDowell, 1996) as well as shrimp rearing experiments (*e.g.*, Choudhury 1971; Hunte 1977; 1979; 1980). It is unclear, however, if this suspected amphidromy is obligate or uniform across all species. For example, Soltero, (1991) successfully raised shrimp in freshwater, and some related species in other systems are known to complete their life cycles in freshwater (*e.g.*, Hughes *et al.*, 1995). McKaye *et al.* (1979) and Erdman *et al.* (1984) reports successful reproduction in freshwater for some fishes (Eleotridae or "sleepers"), and there are reports of native shrimps and fishes in Puerto Rican reservoirs (*I. Corujo*). If access to salt water is not requisite for native fauna, then adult populations and shrimp larvae should be found in all stream categories, including reaches above long-established dams without spillway discharge. Alternatively, if access to the ocean is necessary for development, we should not find these fauna above dams lacking discharge. In addition, if shrimp larvae do not have a requisite marine phase, then we should find advanced larval stages in the streams, but if such a phase is necessary, we should find only first stage larvae. Complete blockage of faunal migration would clearly eliminate stream corridor permeability for the dominant macrobiota of these tropical streams. It can hypothesized that dams *with* spillway discharge might only reduce, rather than eliminate, such permeability. Finding native fauna present, but with reduced abundances and/or species richness, above such dams would be consistent with this hypothesis. If native taxa are absent above dams without spillway discharge, the relative importance of biotic resistance and intact disturbance regimes in limiting invasion can be compared. There would be native fauna and disturbance suppression below-dam and no native fauna and an intact disturbance regime above-dam. In such a scenario, presence of many exotic species and individuals below-dam but few above-dam would support the importance of intact disturbance regimes relative to biotic resistance in this system, whereas the opposite pattern would be consistent with greater importance of biotic resistance than intact disturbance regimes (Table 15.1).

Table 15.1: Fish Species Affected by Reservoirs in Various River System

Species	Resource
Mahaseers, Snow trout, Labeo dero, freshwater Eels and prawns,	Himalayan river systems
Labeo dyochelius	All major river systems
Mahaseers, L. calbasu, Cirrhian cirrhosa, Anguilla bengalensis Bengalensis, T. Khudree, Macrobrachium rosenbergii	Godavari river system
Puntius dobsoni, P. dubis, P. Carnaticus, Cirrhina cirrhosa, L. kontius	Cauvery river system
Puntius sarana, Tort or, T. mahanadicus, T. mosal, L. fimbriatus, L. calbasu, Rhinomugil corsula	Mahanadi river system
P. Kolus, P. dubius, P. Sarana, L. Fimbratus, L. Calbasu, L. pangusia, T. Khudree	Krishna river system
Mahaseers, Eels, Osteobrama belangeri	North-east rivers

Fish Barriers

Fish barriers are either natural or human-created obstacles that impede the passage of fish. Barriers include culverts, dams, waterfalls, logjams, and beaver ponds. This has obvious repercussions for anadromous, diadromous and amphidromous fishes. Under unobstructed streams all types of fish could move through the headwater streams to the mouth and beyond the rivers (Figure 15.2).

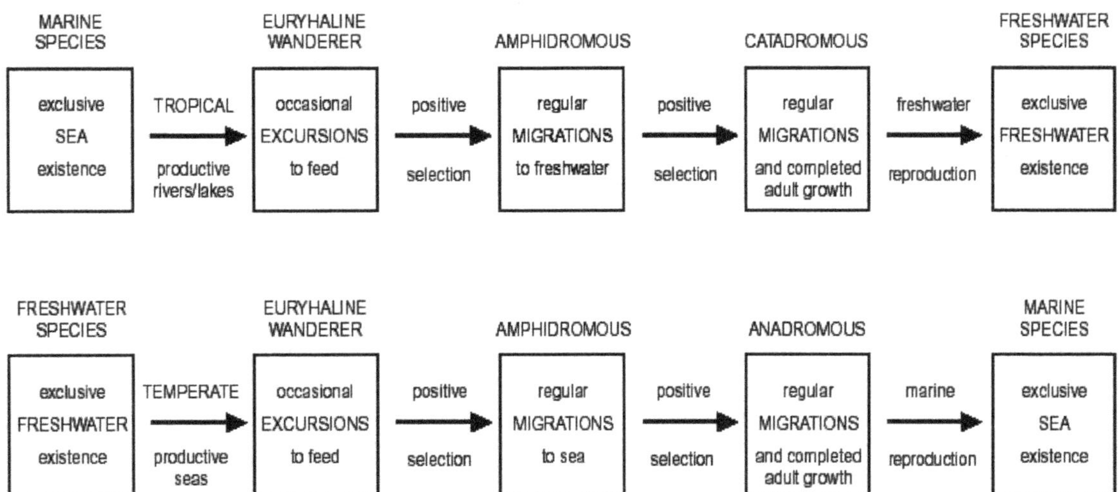

Figure 15.2: Patterns of Evolution of the Different Forms of Diadromy as Hypothesized by Gross (1987) (Redrawn from Gross, 1987)

Barriers can impact all aquatic species. Changes in habitat, population, or water quality conditions create pressures to relocate for more favorable conditions. Therefore, barriers are significant for all aquatic organisms including both anadromous and resident fish species. Major impacts of fish barriers on resident, non-migratory populations include:

1. Juvenile and resident adult fish must be able to move upstream and downstream to adjust to changing habitat conditions (*i.e.*, temperature fluctuations, high or low flows, competition for available food and cover);
2. Resident fish need continuity of stream networks to prevent population fragmentation, which decreases gene flow and genetic integrity;
3. Catastrophic events can displace entire resident fish populations. Barriers can prevent the recolonization of these habitats.

In U.S. culverts on state highways and national highways that act as fish barriers, their locations are reported on an ODFW database. These culverts were studied for the watershed and are all classified as low or medium priority. To survey a stream for salmon spawning, ODFW personnel walk the streams looking for evidence of fish presence and make notes on the condition of the habitat. They count live fish, spawned-out fish (mortalities), and "redds," or the gravel mounds created by fish at spawning sites. Numerous studies, including ones conducted in 1996 by the National Research Council, conclude that migration barriers have substantially impacted fish populations. The extent to which culverts impede or block fish migration appears to be substantial. During fish surveys conducted in coastal basins during 1995, 96 per cent of the barriers identified were culverts associated with road crossings (Figure 15.3).

Fish Passage Criteria

These culverts reported in the database are found on fish-bearing streams and were evaluated against established passage criteria for juvenile and adult salmonids. Parameters measured or estimated and recorded include:

Culvert diameter (inches) and length (feet); Culvert slope (percent); Generally, non-embedded metal and concrete culverts are considered impassable if the slope exceeds 0.5 to 1.0 per cent. At slopes greater than this, water velocities within the culvert are likely to be excessive and hinder passage; Presence/absence of a pool; Pool depth, if present, (in inches); Distance (inches) of drop, if any, to the streambed or pool at outlet; Conditions at the culvert outlet are evaluated for drop (distance from culvert invert to stream below) and the presence or absence of a jump pool. If a pool is present, its depth is recorded. The general criteria for pool depth are 1.5- to 2.0 times the height of the jump (drop) into the culvert; pools shallower than this depth are considered inadequate. If the height of the jump (pool surface to water level in the culvert) into a culvert exceeds 12 inches during the period of migration, the culvert is judged inadequate and included in the listing of culverts needing attention. If the jump is greater than 6 inches but less than 12, the culvert is judged to be a passage problem for juveniles only; Whether the culvert is embedded in the streambed and contained substrate; Whether water runs beneath the culvert at the upstream end of the culvert (a problem for downstream migration of juvenile fish in low water); Fish size (juvenile, adult or both) likely to be hindered (Figure 15.4). In India such type of studies were not undertaken, which can help in maintaining natural fish habitats.

Pollution of Fisheries Resources

Fish populations are vulnerable to pollution because the aquatic environment is the recipient of virtually every form of human waste (Moyle and Leidy, 1992). The IUCN (International Union for Conservation of Nature) Red List (1996) records 734 fish species as threatened and 92 species as extinct worldwide. Water pollution is one of several contributors to such declines in fish populations (Clark, 1992; Moyle and Leidy, 1992; Lawton and May, 1995; Maitland, 1995). Pollution can be defined as "the presence in the environment, or the introduction into it, of products of human activity which have harmful or objectionable effects" (Oxford English Dictionary). Studies of fishes have figured prominently in pollution research, particularly in sublethal physiological effects (review: Kime, 1995). Many studies focus on reproduction because this is one of the most vulnerable periods in the life cycle of fishes (Gerking, 1980; Little *et al.*, 1985). Recently there has been development towards the use of behaviour in toxicological research (Little *et al.*, 1985; D¢ving, 1991; Smith and Logan, 1997). This represents a fusion of the fields of behavior, ecology, toxicology and conservation biology. Effects of various factors threatening fish germplasm based on various studies is enumerated in Figure 15.5.

Secondary Effects of Pollution

A voluminous literature is available on the secondary effects of pollutants on reproductive behaviour. For example, silting arises from high sediment loading, and lowered oxygen levels often result from bacterial degradation under high organic pollution (Clark, 1992; Mason, 1996). Although such studies were not considered by most of the workers because they did not use pollutants directly, their results could still be useful in water quality management and conservation. Silting was found to affect parental care in the fifteen spine stickleback (Spinachia spinachia, Gasterosteidae) (Potts *et al.*, 1988). During parental care, males use their fins to fan water over their egg mass. Silting increased the number of times males fanned their eggs. Nest inspection and nest pushing also became more frequent. Lowered oxygen has also been found to affect paternal care in the three spine stickleback (Reebs *et al.*, 1984), with an increase in the length of egg-fanning bouts. When oxygen levels fall below 2.8 mg/l, males spent more time swimming outside the nest but still fanned at intervals. In the guppy, lowered oxygen levels increased the frequency of breathing at the surface by males at a cost of decreased courtship (Kramer and Mehegan, 1981). To date, such studies have not examined consequences for

Figure 15.3: Effective Fish Pass of Rocky Ramp Type Helps in Fish Movements Across the Stream

Figure 15.4: Weir of Hydropower Station Showing Fish Pass (Circle)

reproductive success. There is now a growing interest in understanding sublethal effects of pollutants on aquatic organisms (Alabaster and Lloyd, 1982; Muller and Lloyd, 1994). To evaluate the usefulness of developing this field further, it is helpful to integrate general theory from behavioral ecology to clarify the biological links between behavioral responses of individuals and population changes. Studies reviewed by Jackie *et al.* (1997) are concerned with reproductive behavior of fishes and their exposure to sublethal doses of the contaminant. Life history theory, supported by a large body of empirical research on a variety of taxa, suggests that commitment of resources to a given reproductive

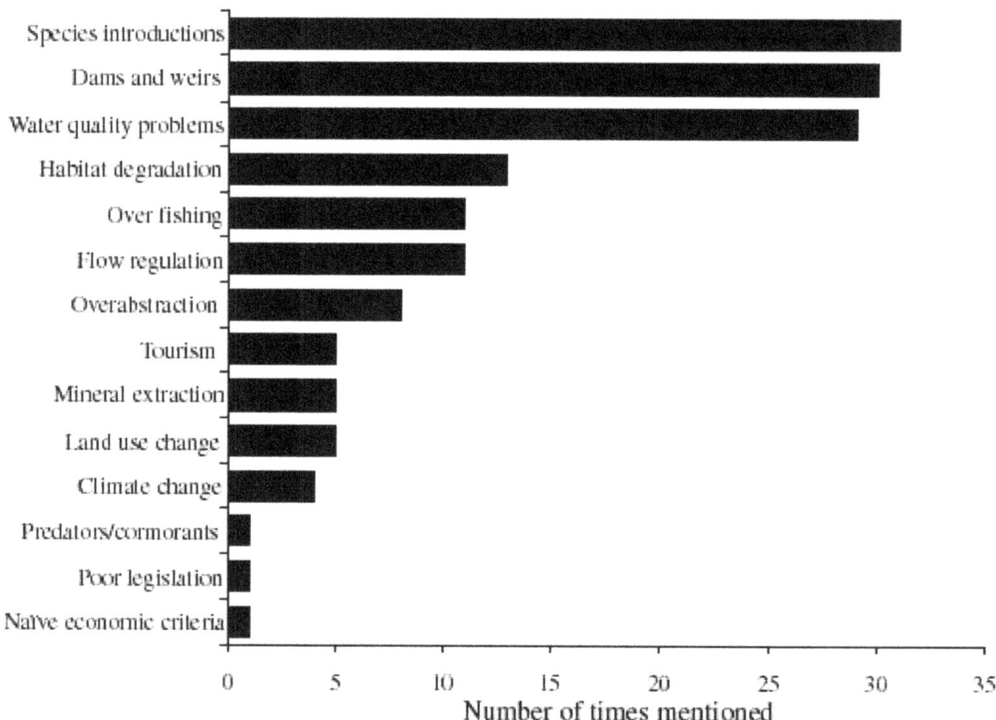

Figure 15.2: Principal Threats to Freshwater Fishes Based on Assessment of the Literature (from Cowx, 2002a)

bout should have costs for survival, growth, or reproduction in subsequent bouts (Williams, 1966; Roff, 1992; Stearns, 1992; Sargent and Gross, 1993). If this were not true, animals could make huge investments in reproduction forever (Partridge and Harvey, 1985). An example from fishes concerns male bluegill sunfish, which have been induced to ventilate their eggs by fanning water across them more frequently when given larger broods experimentally. Increased fanning reduces the time available to males for courting new females, and it reduces energy reserves through the loss of non-polar lipids (Coleman and Fischer, 1991). This may impair survival and future reproduction. Such reproductive trade-offs may be most evident under conditions of stress (Roff, 1992). For example, in the sand goby (*Pomatoschistus minutus,* Gobiidae), females that initially produced large clutches before receiving a low food ration produced smaller clutches in the next bout (Kvarnemo, 1997). This trade-off did not occur with high food. Thus, environmental stressors such as pollutants could have delayed effects on reproduction, which would be overlooked if only a single reproductive cycle were studied.

Links between individual behaviour and population biology have been much sought-after by ecologists (Sibly and Smith, 1985; Sutherland, 1996). In theory, any external force that alters behaviour of individuals from the optimum under natural and sexual selection could lead to reduced population sizes. But this effect may be far from straightforward, depending on the nature of density dependence in the population. For example, Lorenz and Taylor (1992b) showed that under conditions of low pH, convict cichlids were less able to rear their young in the presence of nonspecific. But if population densities of these fish were restricted by food or predation at a different stage of life, reduced survival of young being guarded by parents may have no impact on population size. The difficulty of showing relationships between reproduction and population size is well known to fisheries biologists who rarely find clear patterns between stock size and recruitment (Hilborn and Walters, 1992; King, 1995). Thus, if one is concerned with conservation, a major challenge is to demonstrate clear effects of sublethal doses of pollutants on population size.

Genetic Changes

Environmental impacts that change the reproductive behavior of individuals (*e.g.* choice of mate or oviposition site) could alter the genetic composition of populations. An example from a natural environmental parameter concerns the effects of light levels on behaviour of Trinidadian guppies (Reynolds, 1993). Under low light levels, large-bodied males have higher mating success than smaller ones, because of female choice. This pattern is reversed under higher light levels, when males may be hampered by greater risk of predation. Because male body size has a genetic basis, such changes in mating behaviour in response to an environmental change may have genetic consequences for the population. Genetic changes such as these have yet to be shown in the context of pollution, but given the numerous demonstrations of heritability of traits under natural selection (Falconer, 1989) and sexual selection (Pomiankowski and M¢ller, 1995; Alatalo *et al.,* 1997), researchers would be well advised to bear these possibilities in mind.

Long-term selection may lead to the evolution of resistance. This depends on the intensity of selection, additive genetic variance in resistance and the population size. Benton *et al.* (1994) provided an example in a comparison of populations of freshwater snails (*Helisoma trivolvis*, Planorbidae) and mosquito fish from a relatively clean site and a site contaminated with a variety of heavy metals and other elements. They found that in the populations exposed to the pollutants, both species may have evolved increased tolerance. In mosquito fish, this tolerant genotype seems to be linked to small body size. Particularly relevant here is the possibility that altered mate choice, courtship or parental care could be adaptive, leading to a buffering against environmental change. This could occur either through short-term, facultative changes in behaviour or through genetic responses to selection. This is a risky proposition, because changes in behaviour could be non-adaptive, and genetic adaptations may not evolve quickly enough. Furthermore, genetic bottlenecks and inbreeding are risks when the effective population size is reduced owing to fewer individuals reproducing (Meffe, 1986).

Over-fishing and pollution, in addition to their direct effects on natural fisheries, may also influence natural fish population genetically. Over-fishing drastically reduces population size and since the larger individuals are selectively removed, is equivalent to selection of smaller sized fish. Adaptation of natural fish populations to their environment is reduced by rapid environmental changes resulting from pollution and infestation. Inbreeding, negative selection and lack of adaptation are considered as genetic causes for decline off natural fisheries and lack of recovery. Over-exploitation of the fish resources coupled with habitat destruction, results in the shrinkage of fish population. Due to these factors, a number of fishes are declining rapidly in some conventional fishing grounds and some have even become endangered too (Williams *et al.,* 1989; Minckley and Deacon, 1991; Dehadrai, 1994; Warren and Burr, 1994; Das and Pandey, 1995). Since genetic variations are the raw material in species/population enabling them to adapt to the changes in their environment, any loss of genetic variation results in erosion of evolutionary flexibility. This leads to a poorer match of organisms to adopt to the environment increasing the probability to their extinction (Simpson, 1953; Franklin, 1980; Ralls and Ballau, 1983). The associated severe genetic problems in the small effective population take the form of genetic bottlenecks, genetic drift and accumulation of homozygosity (inbreeding depression) (Meffe, 1986; Das, 1989, 1992; Das *et al.*, 1990; Eknath and Doyle, 1990; Sinha, 1993).

Impact of Introduction of Exotics

Problems associated with exotic fishes are manifold and often exacerbated by river regulation (Courtenay and Stauffer 1990; Allan 1995; Courtenay and Moyle 1996; Vitousek *et al.*, 1996). Although much emphasis has been placed on biotic resistance to invasions (Elton 1958; Li and Moyle 1981; Case 1991; Vermeij 1991; Lodge 1993), Moyle and Light (1996 *a*, *b*) contend that biotic resistance to exotic fishes, though important, is often secondary to abiotic control. Regulated rivers and streams are viewed as particularly susceptible to invasion because of suppression of natural disturbance regimes (Walker 1985; Moyle and Light 1996 a,b; Vitousek *et al.*, 1996; and Meffe, 1986). In spite of an already rich and diverse fish genetic resource of India, more than 300 exotic species have been introduced into the country so far. While vast majority of them are ornamental fishes which remain, more or less confined to the aquaria and some others have been introduced in aquaculture and open water system with varying degree of success. Some of the important exotic species introduced are *Salmo gairdneri, Cyprinus carpio* var. *specularis, C. carpio.* var. *cummunis, Ctenopharyngodon idella, Hypophthalamichthys molitrix, Oreochromis mossambicus, Clarias gariepinus, Gambusia affinis, Labestes reticulates, Betta splendens, Xiphophorus helleri* and *Carassius auratus* (Das, 1989)

Introduction of exotics in many cases have been useful in aquaculture ponds to increase the fish production per unit area but their role in open water system and effect on indigenous germplasm has been more negative than positive. For example, the accidental introduction of silver carp in Govindsagar reservoir has no doubt increased the fish yield from the reservoir but has caused great damage by significantly reducing the stock of Catla and other indigenous species. Further, the damage to food chain at various trophic level due to this introduction has not been evaluated. Similar decline in native Schizothoracids is reported from Dal lake by introduction of common carp. There are many other examples of other river systems also.

The most destructive and voracious feeder catfish *Clarias gariepinus* commonly called as Sharpttoth catfish or Nile catfish was introduced in 1975 in Asia, which has made a significant impact on the Asian aquaculture scenario (Csavas, 1994). Asians do not appears to relish the flesh quality of *C.gariepinus* so much. However, its fast rate of growth and hardiness has made it popular amongst fish farmers. The problem of flesh quality and taste has been overcome by producing hybrids of female local clariids (primariy *C. macrocephalus* and *C. fiscus*) and male *C. gariepinus.* The hybrids so produced are reported to have improved meat quality, quicker growth and better resistance to diseases. The hybrids which are partially fertile, have virtually replaced the native cultured species almost everywhere in south-East Asia. While concerns are being shown about the negative impacts of introduction of the species or its hybrids on the indigenous fish fauna, the fact remains that their culture is spreading rapidly in Asia. *C. gariepinus* and its hybrids also started appearing in fish market in every cities of India. The fish perhaps gains access into the country through fish traders from Bangladesh. The day is not far when African catfish and its hybrids will establish their domain in India and in the South-East Asian countries (Thakur, 1998).

Over Exploitation

Over-exploitation of fishery resources due to its extraordinary economic value has been a causative factor exacerbating the vulnerability of the population in different ecosystems (Kirchhofer and Hefti, 1996). Tor spp., schizothorax spp. And Botia spp. In upland waters (Dhanze and Dhanze, 1994;

Nautiyal, 1994; Pathani, 1994), *Notopterus chitala, Ompak pabda, Pangasius pangasius, Eutropiichthys vacha, Semiplotus semiplotus, Glyphis gangeticus* etc. in warmwater (Pandey and Awasthi, 1994; Prasad, 1994; Copango, 1997) are declining at faster rate.

Wanton Destruction

Wanton killing by the use of dynamites, electric shocks and poisoning of brood fishes in spawning season and juveniles during post-monsoon periods have affected a number of food and game fishes of upland waters, especially in rivers and streams originating in Assam, Nepal, Bhutan, Garhwal, Kumaun and Himachal Pradesh (Joshi, 1994; Chauhan *et al.*, 1994; Nautiyal, 1994; Shreshtra, 1997). Mass killing of fishes of all the sizes during summer months in pools formed in river courses is an alarming situation, particularly in Indian rivers, which does not retain much water during summer now a day.

Habitat Selection (or Use) versus Preference versus Requirement

The concepts of habitat selection, preference, and requirement are sometimes confused in habitat studies, and information on habitat selection is frequently used to infer habitat requirement. Habitat selection (*i.e.*, differential occupancy) occurs when an organism avoids a particular habitat (negative selection) or uses a habitat in greater proportion than its availability in the environment (positive selection). Habitat selection can be demonstrated if fish occur at higher densities in particular habitats, or if fish occur at higher frequencies in particular microhabitats to relative frequency of that microhabitat in the environment; for instance, selection occurs when the frequency distribution of depths at which fish are observed differs from the distribution of depths available. Selective use of different hab- becomes the necessary challenge for fisheries scientists and managers. Solutions are usually limited to whole-scale ecosystem manipulations (*e.g.*, Hartman and Scrivener 1990; Carpenter *et al.*, 1995; Carpenter 1996) or to population models that link the dynamics and habitat requirements of individual life stages (*e.g.*, Minns *et al.*, 1996; Nickleson and Lawson, 1998).

References

Abele, L.G. and N. Blum, 1977. Panama. *Biotropica*, 9: 239–252.

Alabaster, J.S. and Lloyd, R., 1982. *Water Quality Criteria for Freshwater Fish*, 2nd edn. London: Butterworths.

Alatalo, R., Mappes, J. and Elgar, M.A., 1997. *Nature*, 385: 402–403.

Allan, J.D., 1995. *Stream Ecology: Structure and Function of Running Waters*. Chapman and Hall, London.

Allendorf, F.W. and Leary, R.F., 1988. *Conservation Biology*, 2: 170–184.

Anon, 1978. World Water balance and water resources of the earth Paris, UNESCO.

Anon, 1983. *Inland Fish Marketing in India: Overview Summary and Conclusions*. IIMA.

Anon, 1992–93. Annual report. National Bureau of fish genetic resources, Lucknow.

Bell, K.N.I. and J.A. Brown, 1995. *Marine Biology*, 121: 409–417.

Benton, M.J., Diamond, S.A. and Guttman, S.I., 1994. *Ecotoxicol. Env. Safety*, 29: 20–37.

Cambray, J.A. and Pister, E.P., 2002. The role of scientists in creating public awareness for the conservation of fish species: African and American case studies. In: *Conservation of Freshwater Fishes: Options for the Future*, (Eds.) M.J. Collares-Pereira, I.G. Cowx and Coelho M.M. Oxford: Fishing News Books, Blackwell Science, pp. 414–423.

Carpenter, S.R., 1996. *Ecology*, 77:677–680.

Carpenter, S.R., S.W. Chisholm, C.J. Krebs, D.W. Schindler and R.F. Wright, 1995. *Science*, 269: 324–327.

Chauhan, S.K., P. Singh, K.K. Pandey and M.S. Lal, 1994. In: *Threatened Fishes of India*, (Eds.) P.V.Dehadrai; P. Das and Verma, pp. 161–164. NATCON Pub. No. 4, Muzaffarnagar.

Choudhury, P.C., 1971. *Crustaceana*, 20:113–120.

Clark, R.B., 1992. *Marine Pollution*, 3rd edn. Oxford: Clarendon Press, 172 pp.

Clay, C.H., 1995. *Design of Fishways and Other Fish Facilities*. CRC Press, Boca Raton, Florida.

Coleman, R.M. and Fischer, R.U., 1991. *Ethology*, 87: 177–188.

Copango, L.J.V., 1997. *Environ. Biol. Fishes*, 49: 400.

Courtenay, W.R. Jr. and and Stauffer, J.R. Jr. (Eds), 1984. *Distribution, Biology and Management of Exotic Fishes*. Johns Hopkins University Press, Baltimore.

Courtenay W.R. Jr and Moyle, P.B., 1992. Crimes against biodiversity: The lasting legacy of fish introductions. In: *Biological Diversity in Aquatic Management*, (Eds.) J.E. Williams and R.J. Neves, pp. 365–372. Transactions 57th North American Wildlife and Natural Resources

Covich, A.P., and W.H. McDowell, 1996. The stream community. *The Food Web of a Tropical Rainforest*, (Eds.) D.P. Reagan and R.B. Waide. University of Chicago, Chicago, pp. 433–458

Cowx, I.G., 1998. Aquatic resource management planning for resolution of fisheries management issues. In: *Recreational Fisheries: Social, Economic and Management Aspects*, (Eds.) P. Hickley and H. Tompkins. Fishing News Books, Blackwell Science, Oxford, pp. 97–105.

Cowx, I.G., 2002a. Analysis of threats to freshwater fish conservation: past and present challenges. In: *Conservation of Freshwater Fishes: Options for the Future*, (Eds.) M.J. Collares-Pereira, I.G. Cowx and Coelho, M.M. Fishing News Books, Blackwell Science, Oxford, pp. 201–220.

Crivelli, A.J., 2002. The role of protected areas in freshwater fish conservation. In: *Conservation of Freshwater Fishes: Options for the Future*, (Eds.) M.J. Collares-Pereira, I.G. Cowx and Coelho, M.M. Fishing News Books, Blackwell Science, Oxford, pp. 373–388.

Crowl, T.A. and A.P. Covich, 1990. *Science*, 247: 949–951.

Csavas, Imre, 1994. Status and perspectives of culturing catfish in east and south-east Asia. Review paper presented at the International workshop on the "*Biological Base for Aquacultre of Siluriformes*", held at Montpellier, France on 23–27 May 1994.

D¢ving, K.B., 1991. Assessment of animal behaviour as a method to indicate environmental toxicity. *Comp. Biochem. Physiol.* 100C, 247–252. Behaviour. Oxford: Blackwell Scientific Publications. 620 pp.

Das, P., 1994. Strategies for conserving threatened fishes. In: *Threatened Fishes of India*, (Eds.) P.V. Dehadrai, P.Das and S.R. Verma. NATCON Pub.No. 4. Muzaffarnagar, pp. 307–310.

Das, P. and A.K. Pandey, 1995. Fish biodiversity of India: Present status and conservation measures. In: *Environmental Toxicology*, (Eds.) B.K.Dwivedi and G. Pandey. Bioved Research Society, Allahabad, pp.145–166.

Das, P., 1989. Exotic fish germplasm resources in India and their conservation.In: *Exotic Aquatic Species in India*, (Ed.) M.Mohan Joseph. Asian Fisheries Society: Indian Branch. Special Publication No.1. College of Fisheries, Manglore, pp.49–50.

Dehadrai, P.V., P.Das and S.R. Verma, 1994. *Threatened fishes of India*, 412 pp. NATCON Pub. No.4. Muzaffarnagar.

Dhanze, J.R. and R. Dhanze, 1994. An appraisal of depleting fish genetic resources of Himachal Pradesh. In: *Threatened fishes of India*, (Eds.) P.V. Dehadrai, P. Das and Verma . NATCON Pub. No. 4, Muzaffarnagar, pp. 197–204.

Echelle, A.A. and Conner, P.J., 1989. *Evolution*, 43: 717–727.

Eknath, A.E. and R.W. Doyle, 1990. *Aquaculture*, 85: 293–305.

Erdman, D.S., 1972. Inland game fishes of Puerto Rico. Dingell-Johnson final report F–1–20. Department of Agriculture, San Juan, Puerto Rico.

Erdman, D.S., I.N. Corujo-Flores, J. González-Azar, and W. Ortiz-Carasquillo, 1984. Los peces de agua dulce de Puerto Rico. Departmento de Recursos Naturales, San Juan.

Falconer, D.S., 1989. *Introduction to Quantitative Genetics*, 3rd edn. Harlow: Longman Scientific and Technical. 438 pp.

FAO, 1995b. *Report of the Expert Consultation on Guidelines for Responsible Fisheries Management*. FAO Fish. Rep. No. 519, 100 pp.

FAO, 1996. *Precautionary Approach to Capture Fisheries and Species Introductions*. FAO Technical Guidelines for Responsible Fisheries No. 2, Rome: FAO, 54 pp.

FAO, 1997. *Inland Fisheries*. Rome: FAO Technical Guidelines for Responsible Fisheries No. 6, Rome: FAO, 36 pp.

Fernando, C.H., 1980. Fishery potential of Man-made lake in South Asia and some strategies for its optimization. In: *BIOTROP Anniversary Publication BOGOR, BIOTROP*, pp. 25–28.

Franklin, I.R., 1980. Evolutionary changes in small population. In: *Conservation Biology: An Evolutionary-Ecological Perspective*, (Eds.) M.E.Soule and B.A. Wilcoz. California University Press, California, pp. 135–149.

Fryer, G., 1972. *Biological Conservation*, 4: 256–262.

Gerking, S.D., 1980. Fish reproduction and stress. In: *Environmental Physiology of Fishes*, (Ed.) Ali, M.A. Plenum Press, New York, pp. 569–587.

Gross, M.R., 1987. The evolution of diadromy in fishes. *Am. Fish. Soc. Symp.*, 1: 14–25.

Hartman, G. F. and J.C. Scrivener, 1990. *Canadian Bulletin of Fisheries and Aquatic Sciences*, pp. 223.

Hilborn, R. and Walters, C.J., 1992. *Quantitative Stock Assessment: Choice, Dynamics and Uncertainty*. Chapman and Hall, London.

Hughes, J.M., S.E. Bunn, D.M. Kingston, and D.A. Hurwood, 1995. *Australia. Journal of the North American Benthological Society*, 14: 158–173.

Hunte, W., 1977. *Aquaculture*, 11:373–378.

Hunte, W., 1978. *Zoological Journal of the Linnaean Society, Jamaica*, 64:135–150.

Hunte, W., 1979. *Crustaceana Suppl.*, 5:153–166.

Hunte, W., 1980. *Caribbean Journal of Science*, 16:57–61.

Ibrahim, K.H., 1962. *Indian Journal of Fisheries*, 9: 433–467.

Impson, N.D., Bills, I.R. and Cambray, J.A., 2002. A conservation plan for the unique and highly threatened freshwater fishes of the Cape Floral Kingdom. In: *Conservation of Freshwater Fishes: Options for the Future*, (Eds.) M.J. Collares-Pereira, I.G. Cowx and Coelho, M.M. Fishing News Books, Blackwell Science, Oxford, pp. 432–440.

IUCN. 1996. *1996 IUCN Red List of Threatened Animals*. Gland, Switzerland: IUCN, 368pp.

Jhingran, V.G., 1991. *Fish and fisheries of India*. Hindustan Pub.Co., New Delhi, 727p.

Joshi, P.C., 1994. Status of fish conservation in river Ramganga. In: *Threatened fishes of India*, (Eds.) P.V.Dehadrai; P. Das and Verma. NATCON Pub. No. 4, Muzaffarnagar, pp. 349–353.

Kime, D.E., 1995. *Rev. Fish Biol. Fisheries*, 5: 52–96.

King, M., 1995. *Fisheries Biology, Assessment and Management*. Blackwell Science Ltd., Oxford, 341 pp.

Kirchoffer, A. and D. Hefti, 1996. *Conservation of Endangered Freshwater Fish in Europe*. Birkhauser-Verlag, Basael, Switzerland.

Kirchhofer, A., 2002. The role of legislation, institutions and policy making in fish conservation in Switzerland: past, present and future challenges. In: *Conservation of Freshwater Fishes: Options for the Future*, (Eds.) M.J. Collares-Pereira, I.G. Cowx and Coelho, M.M. Fishing News Books, Blackwell Science, Oxford, pp. 389–401

Kramer, D.L. and Mehegan, J.P., 1981. *Env. Biol. Fishes* 6, 299–313.

Kvarnemo, C., 1997. *Behav. Ecol.*

Laikre, L. 1999. Conservation genetic management of brown trout (*Salmo trutta*) in Europe Report by the Concerted action on identification, management and exploitation of genetic resources in the brown trout (*Salmo trutta*) ("TROUTCONCERT"; EU FAIR CT97–3882) (ISBN 87–987732–0–8 1999) 91 pp.

Lawton, J.H. and May, R.M., 1995. *Extinction Rates*. Oxford University Press, Oxford, 233 pp.

Leberg, P. and Vrijenhoek, R.C., 1994. *Conservation Biology*, 8: 419–424.

Lee, C.L., and D.R. Fielder, 1979. *Crustaceana*, 37: 219–222.

Linløkken, A., 1993. *Regulated Rivers*, 8: 145–153.

Little, E.E., Flerov, B.A. and Ruzhinskaya, N.N., 1985. Behavioral approaches in aquatic toxicity investigations: a review. In *Toxic Substances in the Aquatic Environment: An International Aspect*, (Eds.) Mehrle, P.M. Jr, Gray, R.H. and Kendall, R.L. Bethesda, MD: American Fisheries Society, pp. 72–98.

Lorenz, J.J. and Taylor, D.H., 1992b. *Copeia*, pp. 832–839.

Loudenslager, E.J., Rinne, J.N., Gall, G.A.E. and David, R.E., 1986. *Southwestern Naturalist*, 31: 221–234.

Maitland, P.S., 1995. *Biol. Conserv.*, 72: 259–270.

Mason, C.F., 1996. *Longman*, 356 pp.

McDowall, R.M., 1992. *Copeia*, 1: 248–251.

McDowall, R.M., 1993. *New Zealand Journal of Marine and Freshwater Research*, 27: 453–462.

McKaye, K.R., D.J. Weiland, T. and T.M. Lim, 1979. *Copeia*, pp. 542–544.

Mecombie, A.M., 1977. *J. of Fish. Res. Board of Canada*, 20(3): 735.

Meffe, G. K., 1985. *Southwestern Naturalist,* 30: 173–187.

Meffe, G.K., 1986. *Fishes,* 11: 14–23.

Miller, P.S. and Hedrick, P.W. 1992. *Conservation Biology,* 5: 556–558.

Mills, D., 1989. *Ecology and Management of Atlantic Salmon*. Chapman and Hall, London.

Minckley, W.L. and J.E. Deacon, 1991. *Battle Against Extinction: Native Fish Management in the America*. Western University of Arizona Press, Tucscon and London.

Minns, C.K., R.G. Randall, J.E. Moore and V.W. Cairns, 1996. *Canadian Journal of Fisheries and Aquatic Sciences,* 53(Supplement 1): 20–34.

Moyle, P.B. and Leidy, R.A., 1992. Loss of biodiversity in aquatic ecosystems: Evidence from fish faunas. In: *Conservation Biology,* (Eds.) Fiedler, P.L. and S.K. Jain. Chapman and Hall, New York, pp. 129–169.

Muller, R. and Lloyd, R., 1994. *Sublethal and Chronic Effects of Pollution on Freshwater Fish.* Oxford: Blackwell Science Ltd. 371 pp.

Natrajan, A.V., 1985. *Soc. of Fish. Tech. of India,* pp.14.

Nautiyal, P., 1994. *Mahseer the Game Fish: Natural History, Status and Conservation Practices in India and Nepal.* Rachna publication, Srinagar, Garhwal.

Nickleson, T.E. and P.W. Lawson, 1998. *Canadian Journal of Fisheries and Aquatic Sciences,* 55: 2383–2392.

Pandey, K.D. and S.K. Awasthi, 1994. Endangered, threatened and rare fishes of U.P. In: *Threatened Fishes of India,* (Eds.) P.V.Dehadrai; P. Das and Verma. NATCON Pub. No. 4, Muzaffarnagar, pp. 13–15.

Parrish, B.B. and Blaxter, H.S., 1963. No.87 FAO. IInd World fishing gear Congress. *J. Fish. Res. Board of Canada,* 20(3): 735.

Partridge, L. and Harvey, P.H., 1985. *Nature,* 316: 20.

Coleman, R.M., Gross, M.R. and Sargent, R.C., 1985. *Behav. Ecol. Sociobiol.,* 18: 59–66.

Pathani, S.S., 1994. Trends of Mahaseer, snow trout and chital abundance in Kumaun Himalayas. In: *Threatened Fishes of India,* (Eds.) P.V.Dehadrai; P. Das and Verma. NATCON Pub. No. 4, Muzaffarnagar, pp. 155–160.

Penczak, T., and C. Lasso, 1991. *Hydrobiologia,* 215: 121–133.

Petts, G.E., 1984. *Impounded rivers.* John Wiley, Chichester, United Kingdom.

Poff, N.L., J.D. Allan, M.B. Bain, J.R. Karr, K.L. Prestegaard, B.D. Richter, R.E. Sparks and J.C. Stromberg, 1997. *Bio-Science,* 47: 769–784.

Pomiankowski, A. and M¢ller, A.P., 1995. A resolution of the lek paradox. *Proc. R. Soc. Lond.* 260B: 21–29.

Potts, G.W., Keenleyside, M.H.A. and Edwards, J.M., 1988. The effect of silt on the parental.

Prasad, P.S., 1994. Status paper on endangered, vulnerable and rare fish species of Bihar. In: *Threatened Fishes of India,* (Eds.) P.V.Dehadrai; P. Das and Verma. NATCON Pub. No. 4, Muzaffarnagar, pp. 25–29.

Pringle, C.M., G.A. Blake, A.P. Covich, K.M. Buzby and A. Finley, 1993. *Oceologia*, 93:1–11.

Ralls, K. and Ballau. 1983. Extinction : lesson from zoo. In: *Genetics and Conservation: A Reference for Managing Wild Animals and Plant Populations*, (Eds.) C.M. Schonewald-Cox, S.M.Chambers and L.Thomas. California University Press, Californias, pp. 164–184.

Reebs, S.G., Whoriskey, F.G. and FitzGerald, G.J., 1984. Diel patterns of fanning activity, egg respiration, and the nocturnal behavior of male three-spined sicklebacks, *Gasterosteus aculeatus* L. (*F. trachurus*). *Can. J. Zool.*, 62: 329–334.

Reid, G.K., 1943. *Ecology of Inland Waters and Esturies*, pp. 332.

Reynolds, J.D., 1993. *Amer. Nat.*, 141: 914–927.

Reynolds, J.D. and Coate, I.M., 1995. *Behav. Ecol.*, 6: 175–181.

Roff, D.A. 1992.*The Evolution of Life Histories*. Chapman and Hall, New York, 535 pp.

Sargent, R.C. and Gross, M.R., 1993. Williams' principle: an explanation of parental care in teleost fishes. In *Behaviour of Teleost Fishes*, 2nd edn, (Ed.) Pitcher, T.J. Chapman and Hall, London, pp. 333–361.

Shrestha, T.K., 1997. *The Mahseer in the Rivers of Nepal Disrupted by Dams and Ranching Strategies*. R.K. Printers, Kathmandu, Nepal.

Skelton, P.H. 2002. An overview of the challenges of conserving freshwater fishes in South Africa. In: M*Conservation of Freshwater Fishes: Options for the Future*, (Eds.) M.J. Collares-Pereira, I.G. Cowx and Coelho, M.M. Fishing News Books, Blackwell Science, Oxford, pp. 221–236.

Sibly, R.M. and Smith, R.H., 1985. *Behavioural Ecology: Ecological Consequences of Adaptive Behaviour*. Blackwell Scientific Publications, Oxford, 620 pp.

Simpson, G.G., 1953. *The Major Features of Evolution*. Columbia University Press, New York.

Sinha, N.K., 1993. *Cheetal*, 32(3): 10–14.

Smith, E.H. and Logan, D.T., 1997. Linking environmental toxicology, ethology, and conservation. In *Behavioral Approaches to Conservation in the Wild*, (Eds.) Clemmons, J.R. and Buchholz, R. Cambridge University Press, Cambridge, pp. 277–302.

Soltero, S. C., 1991. Hatching and larval rearing of *Atya scabra* (Leach) and *A. lanipes* (Holthuis) in the laboratory, including effects of different salinities, temperatures, and diets. *Ph.D. Thesis*, University of Puerto Rico, Mayagüez.

Stearns, S.C., 1992. *The Evolution of Life Histories*. Oxford University Press, Oxford, 249 pp.

Stuart, T.A., 1962. *Freshwater and Salmon Fisheries Research*, 28: 1–46.

Sutherland, W.J., 1996. *From Individual Behaviour to Population Ecology*. Oxford University Press, 213 pp.

Thakur, N.K. 1998. A biological profile of African catfish, Clarias geriepinus and impact of its introduction in Asia. In. *Fish Genetics and Biodiversity Conservation*, (Eds.) Ponniah, A.G., P. Das and Verma, S.R. Nature Conservator, Muzaffarnagar, pp. 315–322.

Warren, M.L. and Burr, 1994. *Fisheries* (Bethesda), 19(1): 6–18.

Wiens, J.A., 1992. Ecological folws across landscape boundaries : a conceptual overview. In: *Landscape Boundries: Consequences for Biotic Diversity and Ecological Flows*, (Eds. A.J.Hansen and F.di Castro). Springer-Verlag, New York, pp. 217–235.

Wiens, J.A., C.A.Crawford and J.R.Gosz, 1985. *Oikos*, 45: 421–427.

Williams, G.C., 1966. *Am. Nat.*, 100: 687–690.

Williams, J.D. and M.L. Warren, K.S. Cummings, J.L. Harris and R.J.Neves, 1993. *Fisheries (Bethesda)*, 18(9): 6–22.

Sustainable Environmental Management *Pages* **152–163**
Edited by: **Dr. L.V. Gangawane & Dr. V.C. Khilare**
Published by: **DAYA PUBLISHING HOUSE**

Chapter 16

Municipal Solid and Hazardous Waste Management: A Review

V.C. Khilare[1] and L.V. Gangawane[2]

[1]Department of Botany, Vasantrao Naik Mahavidyalya, Aurangabad – 431 003, MS, India
[2]Soil Microbiology and Pesticides Laboratory, Dr. Babasaheb Ambedkar Marathwada University, Aurangabad – 431 004, MS, India

ABSTRACT

Sources of hazardous waste include those from industrial processes, mining extraction and pesticides. The major hazardous waste-generating industries in India include petrochemicals, pharmaceuticals, pesticides, paint and dye, petroleum, fertilizers, asbestos, caustic soda, inorganic chemicals and general engineering industries. The main source of hazardous waste and cause of an adverse impact on the environment is the Indian chemical industry. Hazardous wastes from the industrial sectors contain heavy metals, cyanides, pesticides, complex aromatic compounds and other chemicals, which are toxic, flammable, reactive, and corrosive or have explosive properties. Municipal solid wastes in India increasing significantly over the last few decades due to rapid population growth in the country. The reasons for this trend could be our changing lifestyles, food habits and changes in the standard of living. The hazardous solid waste management emphasizes the need for scientific disposal of waste and policies to encourage waste minimization and adoption of cleaner technologies in this article.

Keywords: *Solid waste, Industrial waste, Policy.*

Introduction

Detection of traces of toxic chemicals in drinking water supplies, in polar ice caps, groundwater sources and in Minamata Bay (Japan) and Love Canal (USA) have focused the attention of the public worldwide on the risks posed by the inappropriate disposal of hazardous waste and accidental release of toxic chemicals into the environment. In India regulations to control and manage air and water related pollution were in place as early as 1974 and 1981 when the Water Act and Air Acts, were respectively, introduced in country. However, the concern and need to manage the hazardous waste generated in the country in a scientific manner was felt only in the mid-eighties after the occurrence of the famous Bhopal gas tragedy on 2nd and 3rd December 1984. The Government's attention was then drawn towards environmental damage and the casualties that hazardous chemical substances and toxic wastes can cause. In order to protect the environment from different pollutants the Environment (Protection) Act in 1986 was passed by parliament. Subsequent to this Act, in order to prevent indiscriminate disposal of hazardous waste, the MoEF (Ministry of Environment and Forests) promulgated the Hazardous Wastes (Management and Handling) Rules in 1989, and efforts to inventories hazardous waste generation were initiated. Though the hazardous waste rules were introduced in 1989, the response towards their implementation has remained very poor. Also, due to the liberalized policy the pace of industrialization has been accelerated, which has resulted in increasing amounts of hazardous wastes every year. This along with a growing amount of municipal solid waste due to rapid urbanization and hospital waste due inadequate policy and technological measures continues to remain a daunting issue of environmental concern to India.

Pressure of Industrial and Hazardous Waste

Sources of hazardous waste include those from industrial processes, mining extraction, tailings from pesticide based agricultural practices, etc. Industrial operations lead to considerable generation of hazardous waste and in a rapidly industrializing country such as India the contribution to hazardous waste from industries is largest. Hazardous waste generation from industries is also critical due to their large geographical spread in the country, leading to region wide impacts. The annual growth in hazardous waste generation can be directly linked to industrial growth in the country. States such as Gujarat, Maharashtra, Tamil Nadu, and Andhra Pradesh, which are relatively more industrialized, face problems of toxic and hazardous waste disposal far more acutely than less developed states. The major hazardous waste-generating industries in India include petrochemicals, pharmaceuticals, pesticides, paint and dye, petroleum, fertilizers, asbestos, caustic soda, inorganic chemicals and general engineering industries. During the last 30 years, the industrial sector in India has quadrupled in size. The main source of hazardous waste and cause of an adverse impact on the environment has been the Indian chemical industry. Hazardous wastes from the industrial sectors mentioned above contain heavy metals, cyanides, pesticides, complex aromatic compounds (such as PCBs), and other chemicals, which are toxic, flammable, reactive, and corrosive or have explosive properties.

Municipal Solid Wastes

There has been a significant increase in the generation of MSW (municipal solid wastes) in India over the last few decades. This is largely a result of rapid population growth in the country. The daily per capita generation of municipal solid waste in India ranges from about 100 g in small towns to 500 g in large towns. Although national level data do not exist for municipal solid waste generation, collection and disposal, for the lack of a nation wide inventory, the growth of solid waste generation over the years can be studied for a few selected urban centres. The population of Mumbai increased from around 8.2 millions in 1981, to 12.3 millions in 1991, a growth of around 49 per cent. The municipal waste generation however grew from 3,200 tonnes per day to 5,355 tonnes per day in the same period, a growth of around 67 per cent. The waste quantities are estimated to increase from 46 million tonnes in 2001 to 65 million tonnes in 2010 (Kumar and Gaikwad, 2004). This clearly indicates that the growth in municipal waste generation in our urban centres has outpaced the growth in population in recent years. The reasons for this trend could be our changing lifestyles, food habits and changes in the standard of living. MSW in cities is collected by the municipalities and transported to designated disposal sites normally a low-lying area on the outskirts of the city for disposal. The choice of a disposal site is more a matter of what is available than what is suitable.

Status of Industrial and Hazardous Waste

The first few attempts to quantify hazardous waste generation in the country remain limited to indirect estimations. For instance, using the correlation between economic activity and hazardous waste generation established by the Organization for Economic Cooperation and Development (OECD), the reported generation of hazardous waste was about 0.3 million tonnes per annum in 1984. World Bank estimates place this at approximately 4 million tonnes annually for the year 1995. These scattered inventories were not very useful in designing hazardous waste strategies for the country since hazardous waste generation is very dynamic owing to the intense growth in industrial activities taking place. In order to generate an updated inventory for hazardous waste in the country, an exercise in different states of India was initiated by the CPCB (Central Pollution Control Board) in the year 1993. The present information on total hazardous waste generated from industries and facilities available for its disposal in Indian states has been collected by the MoEF through the respective SPCBs (state pollution control boards). At present, around 7.2 million tonnes of hazardous waste is generated in the country of which 1.4 million tonnes is recyclable, 0.1 million tonnes is incinerable and 5.2 million tonnes is destined for disposal on land (MoEF 2000). As per the information provided by the MoEF, there are 323 hazardous waste recycling units in India, and of these 303 recycling units use indigenous raw material while 20 depend on imported recyclable wastes. The status of hazardous waste imported for recycling and recovery of mostly metallic constituents in country. The major types of hazardous waste imported by the country include battery scrap, lead and zinc dross, ash, skimmings and galvanized zinc.

The major generators of non-hazardous industrial solid wastes in India are thermal power stations producing coal ash, steel mills producing blast furnace slag and steel melting slag, non-ferrous industries such as aluminium, zinc and copper producing red mud and tailings, sugar industries generating press mud, pulp and paper industries producing lime sludge and fertiliser and allied industries producing gypsum. Since these wastes are generated in huge quantities in the country the recycle/reuse potential of these wastes should be explored, otherwise a huge land area would be required for disposal.

As stated earlier, the daily per capita generation of MSW in India ranges from about 100 g in small towns to 500 g in large towns. The recyclable content of waste ranges from 13 per cent to 20 per cent (CPCB 1994/95). A primary survey in 1971 estimated that the urban population generated 374 g/capita/day of solid waste (Bhide and Sundersan, 1983). In another survey conducted by NEERI the quantity of waste produced has been found to vary from 200 to 600 g/capita/day. A survey in 1995 at 456 g/capita/day (EPTRI 1995). A survey conducted by ORG in 1989 places total MSW generation for 33 Indian cities at 14,934 tonnes a day. The EPTRI estimates of the survey in 1995 for 23 Indian cities places it around 11 million tonnes a year. The survey conducted by CPCB puts total municipal waste generation from class I and II cities to around 18 million tonnes in 1997 (CPCB 2000a). The present annual solid waste generated in Indian cities has increased from 6 million tonnes in 1947 to 48 million tonnes in 1997 and is expected to increase to 300 million tonnes per annum by 2047 (CPCB 2000a). The characteristics of MSW collected from any area depends on a number of factors such as food habits, cultural traditions of inhabitants, lifestyles, climate, etc. The changes in the characteristics and the relative share of different constituents of waste in the past several decades. The percentage of recyclable waste is increasing in the municipal waste streams. This can be largely attributed to changing lifestyles and increasing consumerism. The disposal of plastic bags along with other types of waste streams is a major problem in the strategy leading with municipal solid waste management in the country, should therefore target maximizing recycling/reuse efforts so that dependence on landfills for final waste disposal can be minimized.

Only few cities follow such good practice of waste disposal as tipping of waste using mechanized equipment for leveling and compacting and placing a daily cover of soil on top of it before compacting it further. Some municipalities also practice composting the organic fraction of the waste. Vermi-composting being practiced at one of the dumpsites. However, overall, the average waste collection efficiency of the total generation in Indian cities is around 72 per cent (NIUA 1989) and 70 per cent of Indian cities do not have adequate waste transportation facilities. Lots of littering usually takes place while waste is stored in collections centres and also during its transport.

In addition, till date, biomedical waste generated from clinics, hospitals, nursing homes, pathological laboratories, blood banks and veterinary centres, in absence of any legislation till very recently, and a lack of awareness of impacts due to its indiscriminate disposal, was also being disposed along with municipal waste in dumpsites. Assuming a waste generation factor of 250 g/bed/day for infectious biomedical waste, the Directorate General of Health Services has estimated the total infectious biomedical waste generated from different states in India at 54 404 tonnes per annum as on 1 January 1993 (CPCB 2000b). A WHO study on health care waste has estimated that of the total waste generated in health care facilities, about 85 per cent of the waste is non-infectious, 10 per cent infectious but non-hazardous and 5 per cent hazardous (CPCB 2000b). Based on these estimates, the total health care waste generated as per the 1993 data in the country can be taken as 544 040 tonnes per annum and hazardous waste generation from health care facilities can be taken as 27,202 tonnes per annum. A proper waste segregation scheme for separating hospital waste into infectious and non-infectious categories is therefore desired. This should be coupled with separate and dedicated treatment facilities for infectious waste categories so that co-disposal of infectious waste with municipal waste can be avoided.

Impact of Industrial and Hazardous Waste

Improper storage, handling, transportation, treatment and disposal of hazardous waste results in adverse impact on ecosystems including the human environment. When discharged on land, heavy metals and certain organic compounds are phytotoxic and at relatively low levels can adversely affect soil productivity for extended period of times. For example, uncontrolled release of chromium contaminated wastewater and sludge resulted in contamination of aquifers in the North Arcot area of Tamilnadu. These aquifers can no longer be used as sources of freshwater. Discharge of acidic and alkaline waste affects the natural buffering capacity of surface waters and soils and may result in the reduction of a number of species. This has adverse impacts on not only the ecosystem but also on the human environment. Unscientific disposal practices leave waste unattended at the disposal sites, which attract birds, rodents, fleas, etc. to the waste and creates unhygienic conditions (odour, release of airborne pathogens, etc.). The plastic content of the municipal waste is picked up by rag pickers for recycling either at primary collection centres or at dumpsites. Plastic are recycled mostly in factories, which do not have adequate technologies to process them in a safe manner. This exposes the workers to toxic fumes and unhygienic conditions. Moreover, since the rag picking sector is not formalized, not all the recyclables, particularly plastic bags, get picked up and are found littered everywhere, reaching the drains and water bodies ultimately and choking them. Policy intervention to strengthen administrative structures can help in mitigating the adverse impacts of the waste on public health. The efforts of the Surat Municipal Corporation after the plague epidemic in 1994 have resulted in a complete metamorphosis of the city. This successful example has streamlined the management of solid waste and has helped in creating an atmosphere where the urban local bodies and citizens can discuss the gravity of the problem and share responsibilities with a more positive attitude. Most biomedical waste generated from health care facilities are at present, collected without segregation into infectious and non- infectious categories and are disposed in municipal bins located either inside or outside the facility premises. Sanitary workers pick this waste from here along with other MSW and transport and dispose it at municipal dumpsites. Since the infectious waste gets mixed with municipal solid waste, it has potential to make the whole lot infectious in adverse environmental conditions. Moreover, biomedical waste also contains sharp objects (scalpels, needles, broken glasses/ampoules, etc.) the disposal of which poses a risk of injury and exposure to infection to sanitary workers and rag pickers working at these dumpsites. Since most of these dumpsites are unscientifically managed, the chances of pathogens contained in infectious waste becoming airborne and getting released to nearby water bodies or affecting the local resident population cannot be ruled out.

Projections

As stated earlier, the present hazardous waste generation in the country is around 7.2 million tonnes out of which 1.4 million metric tones is recyclable, 0.1 million tonnes is incinerable and 5.2 million tonnes are destined for disposal on land. This indicates that discounting the recyclable fraction of hazardous waste, total of around 5.3 million tonnes of hazardous waste requires some treatment and disposal. Taking the unit average cost of treatment and disposal of hazardous waste at Rs 3,000 per tonne of the waste, this requires an investment of around Rs 15,900 million every year for treatment and disposal of the hazardous waste in a scientific way. The land required to dispose this waste in an engineered landfill, assuming the average density of waste to be around 1.2 tonnes/m3 and the depth of the landfill 4 m, would be around 1.08 Km^2 every year. This data can be applied to future waste projections to arrive at future land requirements for the disposal of hazardous waste. In addition to hazardous waste, industries also generate around 147 million tonnes of no-hazardous wastes every year at present which is mostly disposed on open, low lying land. A study conducted by the CPCB on management of municipal solid waste in the country estimates that waste generation from the present 48 million tonnes is expected to increase to 300 million tonnes per year by the year 2047 (490 g per capita to 945 g per capita). The estimated requirement of land for disposal would be 169.6 sq km in 2047 as against 20.2 sq km in 1997 (CPCB 2000a).

Responsibility of Existing Policy

The MoEF, Government of India is the nodal agency at the central level for planning, promoting and co-coordinating environmental programmes, apart from policy formulation. The executive responsibilities for industrial pollution prevention and control are primarily executed by the CPCB at the central level, which is a statutory authority, attached to the MoEF. The CPCB was constituted in September 1974, for implementing provisions of the Water (Prevention and Control of) Pollution Act, 1974. The State Departments of Environment and SPCBs and Pollution Control Committees (PCCs) are the agencies designated to perform these functions at the state and union territory level.

Policies for Hazardous Waste Management

The Hazardous Wastes (Management and Handling) Rules, 1989 was introduced under Sections 6, 8, and 25 of the Environment (Protection) Act of 1986 (referred to as HWM Rules 1989). The HWM Rules, 1989 provide for control of generation, collection, treatment, transport, import, storage and disposal of wastes listed in the schedule annexed to these rules. Implementation of these rules is done through the SPCBs and pollution control committees in respective states and union territories. Besides these rules, in 1991, the MoEF issued Guidelines for Management and Handling of Hazardous Wastes for (a) generators, (b) transport of hazardous waste, and (c) owners/operators of hazardous waste storage, treatment and disposal facility. These guidelines also established the mechanisms for the development of a reporting system for the movement of hazardous waste (the manifest system) and for the first time established procedures for closure and post-closure requirements for landfills. In 1995, these were followed by publication of Guidelines for Safe Road Transport of Hazardous Chemicals that established basic rules for Hazardous Goods Transport and provided for the establishment of a Transport Emergency Plan and for provisions on Identification and Assessment of Hazards. In addition to these direct rules dealing with issues of hazardous waste management, the Government has moved to enact into legislation, additional incentives for industries to comply with environmental provisions and bring market forces out into the business of environment. In this vein, the Public Liability Act 1991 was adopted to require industries dealing with hazards to ensure against accidents or damages caused by release of pollutants. The National Environmental Tribunal Act 1995 provides provisions for expeditious remedies to parties injured by environmental crimes. Legislation on a Community Right to Know 1996 has been adopted to provide more access to information regarding potential hazards from industrial operations. India is also a signatory to the Basel Convention, 1989 on control of transboundary movement of hazardous wastes and their disposal. There were few inherent limitations observed in implementation of HWM Rules, 1989. To remove these limitations, the MoEF notified Hazardous Wastes (Management and Handling)

Initiatives Taken for Hazardous Waste Management

Emerging policy directions in the field of hazardous waste management emphasize the need for scientific disposal of waste and policies to encourage waste minimization and adoption of cleaner technologies. Various activities initiated by the Government of India to meet these objectives are listed and discussed below:

1. MoEF has initiated task of hazardous waste inventory in various states to gather updated information. State governments are in the process of identifying hazardous waste disposal sites based on EIA of the potential sites.
2. The CPCB has prepared a ready reckoner in 1998 providing technical information on sources of hazardous wastes, their characteristics, and the methods for recycling and Disposal.
3. Training programmes have been organized for concerned personnel in ports and customs and in pollution control boards so as to familiarise them with precautionary measures and testing methodologies for hazardous waste constituents.
4. It has been decided to impose a ban on import of hazardous wastes containing beryllium, selenium, chromium (hexavalent), thallium, pesticides, herbicides and their intermediates/residues based on recommendations by an Expert Committee constituted at the national level for advising in matters related to hazardous Wastes.
5. In order to control movement of Basel Wastes, cyanide wastes and mercury- and arsenic-bearing wastes have been prohibited for export and import from December 1996. 6. Import of

waste oil and metal bearing wastes such as zinc ash, skimmings, brass dross and lead acid batteries for processing to recover resources would be regulated by MoEF and allowed only by environmentally acceptable technologies

Policies for Municipal Solid Waste Management

The MoEF, Government of India has now issued the Municipal Solid Wastes (Management and Handling) Rules in the year 2000. These rules identify the CPCB as the agency that will monitor the implementation of these rules and municipalities will be required to submit annual reports regarding municipal waste management in their areas to the CPCB. For management of biomedical waste, the MoEF has notified Bio-Medical Waste (Management and Handling) Rules in 1998 under sections 6, 8 and 25 of Environment (Protection) Act of 1986.

Initiatives Taken for Municipal Solid Waste Management

Apart from notification of rules for management of municipal solid wastes in 2000 by the MoEF, several attempts are underway to improve the management of municipal solid waste. Some of the initiatives taken at the national level and efforts made by various ministries at the central level are as follows:

NWMC (National Waste Management Council).

The NWMC was constituted in 1990 and one of its objectives was municipal solid waste management. The council is at present engaged in a survey of 22 municipalities to estimate the quantity of recyclable waste and its fate during waste collection, transportation, and disposal. NWMC in 1993 constituted a national plastic waste management task force to suggest measures to minimize the adverse environmental and health impacts arising out of plastic recycling. Based on the recommendations of this task force, the MoEF in 1998, came out with draft Recycled Plastic Usage Rules, 1998 which bans storing, carrying and packing of food items in recycled plastic bags. It also specifies the quality standards for manufacturing recycled plastic bags.

Strategy Paper

The MoUAE engaged NEERI (National Environmental Engineering Research Institute) for formulating a strategy paper on municipal waste management and also for preparing a manual on solid waste management. These documents highlight various critical issues relating to the management of solid wastes and have offered a number of suggestions for improving management practices.

Policy Paper

The CPHEEO (Central Public Health Environmental Engineering Organization) of MoUAE has prepared a policy paper on promoting the integrated provisions of water, sanitation, solid waste management and drainage utilities in India.

Master Plan for MSW

The MoEF and CPCB organized an interaction meet on March 1995 with municipal authorities and other concerned ministers to evolve a strategy for the management of municipal solid wastes. CPCB also formulated guidelines for safe disposal of hospital wastes.

Realizing the potential and the need for proper treatment of wastes and resultant recovery of energy, the MNES, in June 1995, launched a National Programme on Energy Recovery from urban–municipal and industrial wastes, with a view to promoting the adoption of appropriate technologies.

Various fiscal and financial incentives are offered by the MNES under this programme for energy recovery from wastes.

High Powered Committees

A high powered committee on urban waste was constituted by the Government of India during 1975. The committee, in its report made 76 recommendations, covering eight important areas of waste management. Another high-powered committee was constituted in 1995. The committee has given number of recommendations covering issues like segregation, door-to-door collection, proper handling and transportation, waste composting and treatment and use of appropriate technologies for waste treatment and disposal.

Judicial Interventions

Failure in implementation of existing legislation to check the environmental damage caused by non-conforming industrial units has resulted in issue of directions in the year 1996 from Supreme Court (SC) of India ordering closure/shifting of industrial units using hazardous processes and hazardous chemicals from Delhi region to regions identified by government in the National Capital Region. In addition, SC has ordered closure of 200 tanneries in Tamilnadu, and 35 foundries in Bengal.

Policy Gaps in Hazardous Waste Management

1. The rules promulgated by the MoEF in the year 2000 dealing with hazardous waste management fail to provide any incentive for waste reduction/minimization efforts.

   Industries are therefore reluctant to adopt such measures.
2. In absence of standards for clean up of contaminated sites and limits for disposal of waste on land, those industries which are causing contamination of land and water bodies through inappropriate waste disposal are not legally bound to clean the site unless ordered by judicial interventions.
3. Though draft rules for the management of municipal waste were notified as early as 1998, the final rules could be notified only in the year 2000. These rules along with rules for biomedical waste management do not clearly identify the role and responsibilities to be undertaken by the CPCB and SPCBs.

Knowledge/Information Data Gaps in Hazardous Waste Management

1. The hazardous waste inventory carried out by different states has been a one-time exercise. But since the growth of the industrial sector is dynamic in the country, there is a need to constantly update this waste inventory so that appropriate waste management strategies can be incorporated in waste management plans.
2. In absence of a reliable waste inventory, there is very little practice at present for using tools such as EIA for hazardous waste problems. This has led to very little research on exploring the risks and health impacts of hazardous waste disposal on surrounding ecosystem and communities.
3. Apart from some dedicated facilities at large chemical industries, India lacks the sort of infrastructure that is required for proper treatment and disposal of hazardous waste largely due to the inability of regulatory authorities to achieve strict enforcement of rules. This is

also partly due to inadequate infrastructure including staff in different SPCBs assigned for hazardous waste management in the state.

4. Although attempts have been made at the city level in some selected pockets of the country to identify and quantify municipal waste and biomedical waste, there are no state/nation-wide waste inventories available in both the cases. It becomes very difficult in the absence of such an inventory to prepare waste management plans.
5. Most of the waste whether municipal or biomedical, is at present dumped in open low lying areas with no provisions for liners, leachate collection and treatment system or gas collection system.
6. In absence of segregation of waste at source, waste treatment alternatives such as recycling, waste-to-energy projects and or composting become uneconomical to operate.
7. Most infectious biomedical waste segregated at the source of generation gets disposed at municipal waste dumpsites in absence of dedicated waste disposal facilities for biomedical waste generators.

Policy recommendations for industrial and hazardous waste management

1. The strategy required to ensure scientific management of hazardous waste, which is expected to increase over the years due to our liberalized economic policies and related growth in industry should encompass all the aspects of waste management cycles starting from generation of waste to its handling, segregation, transportation, treatment, and disposal.
2. In addition, the strategy should also target waste minimization/reduction as its primary focus. This becomes particularly important in view of stricter environmental standards being enforced on industries. This results in increased cost of treatment and disposal to meet the stricter standards. Any waste minimization/reduction effort would thus result in less waste generation and lesser waste to be managed thus reducing the cost of waste management. In addition, any recycle/reuse effort may in fact earn net revenue on the waste generation.
3. Although the Government of India recognizes the localized nature of hazardous waste generators and while significant progress has been made in identifying large concentrations of hazardous waste, further efforts are required to quantify and characterize the volume of waste residues generated by industries. As discussed above, there is need to constantly upgrade this waste inventory so that appropriate waste management strategies can be incorporated in waste management plans.
4. Although substantial progress has been made in imparting training and capacity building to SPCB officials, additional capacity at SPCB is needed to deal with analytical and monitoring requirements regarding tracking of hazardous waste movement and management. In addition training is also required for critical industrial sectors generating hazardous waste to address their responsibility in handling, storage, transportation, treatment and disposal of hazardous waste. This becomes particularly important in light of new amended hazardous waste rules introduced in country in January 2000. The amended hazardous waste rules expand the definition of hazardous waste from previous one incorporating the hazardous waste streams identified in Basel Convention.
5. It is suggested to incorporate comprehensive approaches such as EIA to carry out environmental and social assessments of hazardous waste management operations. This

will help us assessing the risks and health impacts of inappropriate disposal of hazardous waste on surrounding ecosystem and communities.

6. Environmental emergencies and accidental spillage or indiscriminate disposal of chemicals or waste on land causes contamination of soil and groundwater. Use of any treatment or cleanup option requires cleaning of soil and groundwater to some acceptable level of contaminants. Most of the time, in dealing with contaminated soil or groundwater, it is neither economically nor technologically feasible to achieve the zero level of cleanup. It is, therefore, necessary for the Government to set standards not only for disposal of waste on land but also for clean up of contaminated soils and groundwater.
7. Apart from some dedicated facilities at large chemical industries, India lacks the sort of infrastructure that is required for proper treatment and disposal of hazardous waste. Opportunity of setting such facility at the state level, addressing the willingness-to-pay issue by participating industries, type of ownership, financial mechanisms to finance such ventures and extent of private sector participation need to be addressed/explored to ensure that such facilities come into existence.

Municipal Solid Waste Management

In order to have a satisfactory, efficient, and a sustainable system of solid waste management, proper planning, implementation, and management systems must be incorporated in framing the national policy for solid waste management for the country. Present and future ways to manage solid waste stream need consideration of the following aspects (Singhal and Pandey, 2001).

Setting Targets for Waste Reduction

Reduction at source can be accomplished in three ways (1) fees and tax incentives to promote market mechanisms to effect source reduction, (2) mandatory standards and regulation, and (3) education and voluntary compliance with policies by business and consumers, (Marcin, Durbak, and Ince 1994). However, these strategies need to be sensitive to the concerns of possible loss of business and jobs in affected industries. Reduction in the quantity of municipal solid waste could affect employment, taxes/revenues, and economic activity in unpredictable ways (Marcin, Durbak, and Ince 1994).

Technological Interventions

India has lagged behind in adopting technologies for solid waste management. In particular, three technical components, collection, transportation, and treatment and disposal of waste need urgent attention.

Collection of Waste

One immediate measure to revamp the existing collection service structure is to provide community waste bins conveniently placed for the people to deposit domestic waste. As a first step, this will ensure that people do not throw their garbage on the roads and hence do not create open dumpsites. The second measure should entail separation of waste at source into biodegradable and non-biodegradable components.

Transportation of Waste

Waste should be carried in covered vehicles. For the narrow lanes in the congested Old City where a dumper placer cannot move and where the waste has to be carried longer than 1 km to the nearest municipal bin, small, covered vehicles built over a three-wheeler scooter, preferably with a tipping arrangement, may be used. Infectious and hazardous waste from health care facilities should be carried strictly in separate covered vehicles. Hospital waste of some categories, *e.g.* biomedical waste consisting of human body parts, body fluids, etc., has to be incinerated but for other categories of waste, methods like microwaving and autoclaving are possible.

Treatment and Disposal

Proper segregation would lead to better options and opportunities for scientific disposal of waste. Recyclables could be straightaway transported to recycling units, which, in turn, would pay the corporations for it, thereby adding to their income. The organic matter could be disposed of either by aerobic composting, anaerobic digestion or sanitary land filling. Depending upon land availability and financial resources, either of these methods could be adopted. However, it appears that land filling would continue to be the most widely adopted practice in India in the coming few years, in which case certain improvements will have to be done to ensure that it is sanitary land filling and not merely dumping of waste.

Institutional and Regulatory Reforms

The municipalities are the primary institutions responsible for solid waste management in India, but most of the urban local bodies, barring a few progressive ones, are unable to provide the desirable level of conservancy services. The 12th Schedule in 74th Amendment Act 1992, (Entry 6 in Schedule 12 (Article 243-W) empowers the local bodies by giving them independence, authority, and power to impose taxes, duties, tolls, and fees for services including public health, sanitation, conservancy, and solid waste management.

NEERI's Contribution Towards Municipal Solid Waste Management in India

Since last thirty years the Solid Waste Management Division of NEERI has been developed a extension programme to improve the status of waste management in the country. The prominent best practices evolved during the last five years, are described below.

1. Preparation of strategy paper on SWM in India.
2. Long term planning of Solid Waste Management.
3. Biomethanation of vegetable market wastes.
4. Greenhouse gas inventory estimation for waste sector, its uncertainty analysis and formulated measures to mitigate the same.
5. Utilization of landfill site for construction of Rail Car Depot.
6. Site selection criteria for sanitary landfills.
7. Utilization of residue from destruction of soiled currency notes.

References

Agarwal, R., 1998. *Basel Action News*, 1(1).

Anjello, R. and Ranawana, A., 1996. 'Death in Slow Motion- India has become the dumping groun for the west's toxic waste'. *Asiaweek.*

Bhattacharya, A. and Shrivastava, R., 1994. 'The night air turned poisonous'. *Down to Earth*, December.

Bhide, A.D. and Sundaresan, B.B., 1983. *Solid Waste Management in Developing Countries.* New Delhi, Indian National Scientific Documentation Centre, 222 pp.

CPCB, 2000a. *Management of Municipal Solid Waste.* Central Pollution Control Board, Delhi.

CPCB, 2000b. *Manual on Hospital Waste Management.* Central Pollution Control Board, Delhi.

EPTRI, 1995. *Status of Solid Waste Disposal in Metropolis Hyderabad.* Environmental Protection Training and Research Institute, Hyderabad, 46 pp.

Greenpeace, 1997. *The Waste Invasion.* http:/www.greenpeace.org/~comms/no.nukes/p970131.html.

Greenpeace, 1998. *Dutch PVC Waste Still Exported to Asia: call for an end to delayed dumping.*

Kumar, S. and Gaikwad, S.A., 2004. Municipal solid waste management in Indian urban centres: An approach for betterment. In: *Urban Development Debates in the New Millennium,* (Ed) K.R. Gupta. Atlantic Publishers and Distributors, New Delhi, pp. 100–111.

Marcin, T.C., Durbak, I.A. and Ince, P.J., 1994. *Source Reduction Strategy and Technological Change Affecting Demand for Pulp and Paper in North America.* Centre for International Trade in Foreign Products (Proceedings), pp.146–164. Seattle: 3rd International Symp., Sept. 13-14, 1994.

MoEF, 2000. *Draft on Status of Implementation of the Hazardous Waste Rules, 1989.* Ministry of Environment and Forests, New Delhi.

NIUA, 1989. *Upgrading Municipal Services: Norms and Financial Implications.* National, New Delhi.

Shaleen Singhal and Sunil Pandey, (2001). TERI Information Monitor on Environmental Science, 6 (1): 1–4.

www.greenpeace.org/pressrelease/toxics/1988feb4.html.

Sustainable Environmental Management *Pages* ***164–170***
Edited by: **Dr. L.V. Gangawane & Dr. V.C. Khilare**
Published by: **DAYA PUBLISHING HOUSE**

Chapter 17

Pesticide Threat and Biotechnological Remedies

G.K. Kulkarni

Department of Zoology, Dr. Babasaheb Ambedkar Marathwada University, Aurangabad – 431 004, India

ABSTRACT

Pesticides are indispensable in modern agriculture. Their indiscriminate and injudicious use may prove inimical to interests of man. Some of the pesticides quite persistence and contaminate agricultural and animal commodities. This article reviews pesticides in crop commodities, animal products and their effects on human health. Management of pesticide residue through bioremediation is also discussed with the help of microbes like *Pseudomonas, Achromobacter, Flavobacterium, Nocardia, Arthobacter and other forms like Vibrio, Bacillus, Micrococcus, Acinetobacter, arbuscular mycorrhizal, Candida, Rhodotorula, Aurobasidium* and *Sporobolomyces.*

Keywords: *Pesticide residue, Bioremediation, Microbes.*

Introduction

The pesticides are undoubtedly the largest group of potentially toxic chemicals that are introduced purposefully into the environment. They are defined as any substance or mixture of substances used for preventing, destroying, repelling or mitigating any pest. Chlorinated hydrocarbons, organophosphates, carbamates, chlorophenoxy acids and pyrethroids are the main categories of pesticides. 80 per cent of all pesticides are used on cotton, rice, sugarcane, vegetables, and fruits, and these pesticides remain in soils with low organic matter and microbial biodiversity (Bindra, 1971). It is widely accepted that extensive and non-judicious use of such harmful chemicals has led to contamination of soil, crops, resulting food products and drinks. Water used for agriculture gets contaminated with pesticides. This leads to contamination of groundwater. Surface run-off from contaminated agricultural land pollutes rivers and other water bodies that they culminate in.

About 220 kinds of registered pesticides are used worldwide of which 90 are manufactured in India. As a consequence of Green Revolution the pesticide use in India has risen from 2000 tons/year in the 1950s to over 3,00,000 tons in 2002-03. India accounts for nearly 1/3rd of the pesticide poisoning cases in the third world. Where such enormous amounts of pesticide were used? Is every thing been effectively used to control pests? No doubt that about 50 to 60 per cent of it is finally running off the land by various methods and contaminating the aquatic ecosystem.

Millions of humans are either killed or disabled annually from pesticide borne diseases. To control pests in our hot and humid climate, frequent and indiscriminate use of various pesticides are common, and thus environment is loaded with heavy amounts of pesticides. The use of pesticides at one end helping in the increase in agricultural production and at the other end the indiscriminate use of these chemicals is toxic to human beings and domestic animals and their products (Bindra and Karla, 1971).

Table 17.1: Pesticide Residues in Market Samples of Crop Commodities

Commodities	*Sampling Sites*	*No. of Samples*	*Pesticides Detected (%)*	*Incidence Range (ppm)*	*Residue*
Cereals and Cereal Products	A.P., Delhi Haryana,	1670	DDT	53	Tr.–12.0
	Maharashtra, Pubjab, U.P.	BHC	31		Tr.–20.0
Pulses	A.P., Karnataka,	DDT	25		Tr.– 10.0
	Punjab, U.P.	289	BHC	20	Tr.–10.0
Vegetables and Fruits	A.P., Delhi, Karnataka,	2244	DDT	27	Tr.–35.0
	Maharashtra	BHC	28		Tr.–60.0
Vegetable Oils	A.P., Delhi	83	DDT	95	Tr.–25.7
	Punjab, U.P.	BHC	95		Tr.–06.4

DDT: MPL–42 ng/L; Allowed Daily Intake–0.005 mg/kg; BHC: MPL–53 ng/L; Allowed Daily Intake–0.007 mg/kg.

MPL: Maximum Permissible Limit.

Table 17.2: Pesticide Residues in Animal Products in India

Commodities	*Sampling Sites*	*No. of Samples*	*Pesticides Detected (%)*	*Incidence Range (ppm)*	*Residue*
Meat	A.P., Delhi	157	DDT	96	Tr.–20.8
	Punjab, U.P		BHC	90	Tr.–00.20
Fishes	A.P., Punjab		DDT	92	Tr.–02.2
	West-Bengal	90	BHC	85	0.02–1.1
Poultry and Eggs	A.P., Delhi, Gujarat, Maharashtra	110	DDT	60	Tr.–8.0
			BHC	20	Tr.–2.5
Milk	A.P., Delhi	1010	DDT	95	0.19–216.0
	Punjab, Gujarat Maharashtra		BHC	90	0.12–40.0
Butter	A.P., Delhi,	360	DDT	100	1.72-9.81
	Gujarat, Maharashtra, Karnataka, T.N., U.P.		BHC	100	0.58-7.63
Desi Ghee	A.P., U.P.,	140	DDT	100	2.1-13.0
	Rajasthan, Punjab.	BHC	100	2.0-09.8	

DDT: MPL–42 ng/L; Allowed Daily Intake–0.005 mg/kg; BHC: MPL–53 ng/L; Allowed Daily Intake–0.007 mg/kg.

Table 17.3: Class of Pesticides by their Target Organisms

Class	*Pesticides*	*Target Organisms*
Acaricide and Miticides	Sevin, Malathion	Ticks and Mites
Algaecides	Copper oxychloride	Algae
Fumigant	Methyl bromide	Soil pest and weeds
Fungicides	Quinomethoate	Fungi
Insecticides	Sevin, Malathion, BHC, DDT	Insects
Molluscicides	Mercuric chloride, Copper sulphate	Predatory snails, slugs, oyster
Nematicides/Helminthicides	Heterophos, Dazomet	Nematodes, tiny round worms, etc.
Ovicides	Heptachlor, Deltametrion	Eggs of insects, mites, nematodes etc
Rodenticides	Zinc phosphate, Sodium fluoroacetate	Rats, Mice and other rodents.
Herbicides	Dinoseb, Sinbar, Nitrofen	Herbs, weeds, grass etc.

Table 17.4: Pathological Effects of Heavy Metal Pesticides

Metal Salt	*Effects on Human*
Mercuric chloride etc.	Abdominal pain, headache, diarrhea, brain damage
Lead nitrate etc.	Anemia, nausea, vomiting, damage to brain, kidney, liver, convulsions, loss of appetite etc.
Arsenic chloride etc.	Mental disturbance, liver cirrhosis, lung cancer, ulcers in GI tract, kidney damage etc.
Cadmium chloride etc.	Diarrhea, bone deformation, stunted growth, kidney damage, testicular atrophy, anemia etc.
Copper sulphate etc	Hypertension, uremia, coma, sporadic fever etc.
Barium chloride etc.	Vomiting, excessive salivation, diarrhea, colic pain.
Manganese chloride etc.	Chronic exposure leads to neurological disorders
Zinc phosphate etc.	Renal damage, cramps, vomiting etc.
Selenium salts	Liver, kidney and spleen damage, low BP, vomiting, carcinogenic and blindness effects etc.
Chromium (hexavalent salts)	Nephritis, GI tract ulcers, CNS impairment, Cancerous.
Cobalt chloride etc	Diarrhea, low BP, lung irritation, bone deformities etc

Effects of Some Organic Pesticides on Human

1. DDT causes liver cirrhosis, endocrine and reproductive organ damage.
2. Endosulfan suppresses the immune system, affects the kidneys, causes fetal abnormalities and increases the chance of cancer.
3. Lindane causes chronic liver disease.
4. Dibromochloropropane (DBCP), an agriculture fumigant, is a known spermotoxin inducing infertility in men.
5. Chropyrene induces abortions in pregnant women.

All of them cause learning disabilities and diminished mental capacity in children. Increased incidences of abnormal development of testes and reproductive incapabilities have aroused a great concern. The pesticides have adverse effects on spermatogenesis and cause testicular atrophy and fetotoxicity.

Biotechnological Remedies

Insects cause millions of dollars damage to cotton, grains, fruit and vegetable crops every year. They also have a significant impact on meat and poultry production. By using the genetic approach, researchers are minimising the use of traditional pesticides.

Pesticide Residue

Pesticides can still be found on some unprocessed fruit and vegetables on the supermarket shelf. While residues can be useful in extending the period a chemical is exposed to insects, longer term, residues can enter the broader environment with harmful effects. Biotechnology is helping reduce the level of pesticide residues (Lichtenstein and Sculz, 1965).

Pesticide Resistance

Repeated use of pesticides on plants and animals increases the risk of insect pests developing resistance to these chemicals, thereby reducing their effectiveness. This can create a situation where it becomes harder to manage insects using chemicals that are currently available. This can lead to an increase in the frequency of chemical application. Biotechnology is helping reduce the level of pesticide resistance in insects.

Effect on Non-target Insects

When used on cotton, grains, fruit and vegetable crops, broad spectrum insecticides also kill non-target insects, including those that normally hunt problem insects. The incidental death of these 'beneficial' insects adds to the efforts required to keep insect pest numbers within a level that does not cause economic damage to crops. Biotechnology is helping maintain the levels of non-target (beneficial)insects while effectively managing insect pest numbers. It includes a variety of compatible technologies, which includes methods for preventing pollution and effective recycling of wastes, breaking down of recalcitrant compounds etc. For example, the two common methods for cleaning a polluted site (via bioremediation) include Biostimulation and Bioaugmentation. The general approaches to bioremediation are to enhance natural biodegradation by Native Organisms (intrinsic bioremediation), to carry out environmental modification by applying nutrients, aeration and water to accelerate degradation by organisms occurring naturally at the site (Biostimulation) or through addition of microorganisms (Bioaugmentation).

Bioremediation

It employs microorganisms and/or other living organisms to destroy pollutants or to facilitate their removal from the environment. This is a natural process whereby bacteria or other microorganisms alter and break down organic molecules into other substances, eventually producing fatty acids and carbon dioxide. Of the 132-bioremediation activities with clearly identified target compounds reported to the USEPA, 75 dealt with petroleum and related material.

The microorganisms transform contaminants to less harmful compounds through aerobic and anaerobic respiration, fermentation, etc. Diverse micro- organisms including bacteria and fungi have evolved the metabolic capacity to degrade the environmental pollution. They demobilize contaminants in different ways:

1. Microbial biomass can absorb hydrophilic organic molecules. Biomass can stop or slow the movement of the contaminant. This concept is sometimes called Biocurtain.
2. Microorganisms can produce reduced or oxidized species that cause metals to precipitate.
3. Microorganisms can biodegrade organic compounds that bind with metals and keep the metals in solutions.
4. The most prevalent bacterial degraders belong to the decreasing order, to the genera *Pseudomonas, Achromobacter, Flavobacterium, Nocardia, Arthobacter and other forms like Vibrio, Bacillus, Micrococcus* and *Acinetobacte*r.
5. The fungi that can be used in bioremediation are the members of Zygomycetes *e.g.* the mucaraceous fungi and the *arbuscular mycorrhizal* fungi.
6. Among the yeast-like forms of fungi, the most often reported genera are *Candida, Rhodotorula, Aurobasidium* and *Sporobolomyces.*
7. The filamentous fungi include *Penicillium, Aspergillus* and *Cladosporodium* are frequently used as degraders.

Microbial Biodiversity

Microbial diversity can be regarded as one of the largely unexplored natural resources contributing substantially for biotechnological applications. It can be considered on one hand as a problem due to the large variety of infectious microorganisms causing many diseases and on the other hand as a solution due to the rich biotechnological potential including disease control.

As they are also the movers and shakers of global cycles of elements, microorganisms contribute significantly to the stability and functioning of ecosystems, which can be threatened by man made disturbances including such originating from biotechnological applications. Thus the sustenance of the environment requires a thoughtful management of microorganisms.

Microorganisms are somewhat vaguely defined by their size and not by their position in the tree of life. In marine and freshwater, they are viruses, prokaryotes (bacteria), protests (mainly flagellates) and algae, whereas in soil they are mainly viruses, prokaryotes, protests (mainly amoebae) and fungi. One specific characteristics of them is of their short generation time.

Functions of Microbes

Ecosystem functions provided by microbes are

1. The transformation of inorganic carbon into biomass by primary producers, nutrient generation and cycling.
2. Conversion of organic matter including humus that would otherwise be lost from the food web into living biomass.
3. Regulation of biogeochemical cycles and consequently climate. Degradation of organic matter (including oil or pesticides) requires complex metabolic pathways.
4. The succession of microbes such as viruses, bacteria, fungi, microflagellates and protozoans in various communities and ecosystems are of great importance.

Pesticide Degradation

Pesticide degradation, or the breakdown of pesticides, is usually beneficial. The reactions that destroy pesticides change most pesticide residues in the environment to inactive, less toxic, and harmless compounds (Mookherjee *et al* 1965). However, degradation is detrimental when a pesticide is destroyed before the target pest has been controlled. Three types of pesticide degradation are microbial, chemical, and photodegradation.

1. Microbial degradation is the breakdown of pesticides by fungi, bacteria, and other microorganisms that use pesticides as a food source. Most microbial degradation of pesticides occurs in the soil. Soil conditions such as moisture, temperature, aeration, pH, and the amount of organic matter affect the rate of microbial degradation because of their direct influence on microbial growth and activity. The frequency of pesticide applications can also influence microbial degradation. Rapid microbial degradation is more likely when the same pesticide is used repeatedly in a field. Repeated applications can actually stimulate the buildup of organisms effective in degrading the chemical. As the population of these organisms increases, degradation accelerates and the amount of pesticide available to control the pest is reduced.
2. Chemical degradation is the breakdown of pesticides by processes that do not involve living organisms. Temperature, moisture, pH, and adsorption, in addition to the chemical and physical properties of the pesticide, determine which chemical reactions take place and how quickly they occur. Because of lack of light, heat, and oxygen in the water-saturated layers of the soil profile below the surface, chemical breakdown is generally much slower than at the surface. In northern states, the season influences groundwater temperatures from 5 to 10 feet below the ground surface, varying from 39° F to 41° F during the coldest part of the winter to 59° F to 61° F during the hottest part of the summer. Groundwater below 10 to 15 feet maintains a constant temperature of 50° F to 53° F. These low temperatures greatly reduce tile rate of chemical breakdown. One of the most common pesticide degradation reactions is hydrolysis a breakdown process where the pesticide reacts with water. Depending on the pesticide, this may occur in both acid and alkaline conditions. Many organophosphate and carbamate insecticides are particularly susceptible to hydrolysis under alkaline conditions. Some are actually broken clown within a matter of hours when mixed with alkaline water.
3. Photodegradation is the breakdown of pesticides by light, particularly sunlight. It can destroy pesticides on foliage, on the soil surface, and even in the air. Factors that influence pesticide photodegradation include the intensity of the sunlight, properties of the application site, the application method, and the properties of the pesticide.

Conclusion

According to an eminent environmentalist Prof. William Longowood, while man is poisoning his pest enemies, he is also poisoning himself by being forced to eat traces of these substances that remain as residue in his food. Therefore, indiscriminate use of pesticides is stopped to conserve the precious non-target organisms. Instead eco-friendly alternative biocides be selected as a part of integrated pest management (IPM) programme, which will help to save the biodiversity of non-target economically beneficial organisms. For this purpose the use of novel biotechnological tools available at hand must be employed.

References

Bindra, O.P., 1971. Proc. 1st All-India Symp. '*Progress and Problems in Pesticide-Residue Analysis*', 18-19 November, PAU, Ludhiana, India, Pp. 41-50.

Bindra, O.P. and Karla, R.L., 1971. Proc. 1st All-India Symp. '*Progress and Problems in Pesticide-Residue Analysis*', 18-19 November, PAU, Ludhiana, India, Pp. 1-336.

Lichtenstein, E.P. and Sculz, K.R., 1965. *J. Agric. Food Chem.*, 13: 57.

Mookherjee, P.B., Beri, Y.P., Sharma, G.C. and Dewan, R.S., 1965. *Ind. J. Ent.*, 27: 476–480.

Sustainable Environmental Management *Pages* **171–179**
Edited by: **Dr. L.V. Gangawane & Dr. V.C. Khilare**
Published by: **DAYA PUBLISHING HOUSE**

Chapter 18

Importance of VAM Fungal Diversity on Forest Tree Species in a Deciduous Forest

H.C. Lakshman

Post Graduate Dept. of Botany and Microbiology Laboratory, Karnataka University, Dharwad – 580 003, Karnataka, India

ABSTRACT

Vesicular-arbuscular mycorrhizal (VAM) fungi are known to be significantly beneficial to the nutrition of host plants and are ubiquitous in distribution and physiologically unspecailised. Now-a-days serious degradation in the quality of forest and significant reduction in the area of the forest cover. A survey was carried in forty-five tree species growing in Dharwad moist deciduous forest to study the effect of VAM fungi. Emphasis has been laid on nutrient uptake, host growth response and host-pathogen interactions. This paper leads that there is a great diversity among VAM fungi. The significant increase of biomass production was observed in three plants by the inoculations of *Glomus fasciculatum*.

***Keywords**: Deciduous forest, VAM fungi, Glomus fasciculatum.*

Introduction

Forest sustains vast populations and is also store house of mineral resources and genetic Diversity. Survey carried by UNEP estimated that at least 225 hectare of tropical forest area will be degraded by the end of this century. There is growing concern in Dharwad district in Karnataka about serious degradation in the quality of forest and significant reduction in the area of forest cover. Vesicular-arbuscular mycorrhizal (VAM) fungi, mutually beneficial symbiotic soil fungi, is ubiquitous in distribution and physiologically unspecailised. It is most widespread in plant kingdom that one can list the families in which it does not occur. Often, the agronomists, agriculturists, plant physiologist, foresters and plant pathologists are unaware of the fact that the plant species they study are normally VA-Mycorrihzal.

VA mycorrizae are known to be significantly beneficial to the nutrition of host plants, many workers in different disciplines of the basic and applied sciences are doing a wide variety of researches in mycorrhizae. Studies are wide ranging and the literature and several effects of VAM fungi on host are now ample and diverse. Emphasis has been laid on nutrient uptake, host growth response and host-pathogen interactions.

Since three decades VAM (Vesicular arbuscular mycorrhizal) fungi has gained prominence in view of there role in improving nutrient uptake, percent survival, growth performance and biomass yield of plant species (Thaper and Khan, 1973; Janos, 1980; Borges and Willium, 1988; Reena and Bhagyaraj,1990; Rachel *et al*, 1993; Durga and Gupta,1995).Very little work has been carried out on the mycorrhizal association of forest trees (Rahangdale and Gupta,1999) an indigenous and introduced flora of humid tropical forest in particular. The present investigation on important tree species from moist deciduous forest of Dharwad district for exploiting them in energy/biomass plantation programme.

Materials and Methods

A survey was carried in forty-five tree species growing in Dharwad moist deciduous forest. The geologically it is located in between (14° 15¹ and 15° 50¹ north longitude and 74° 48¹ and 76° 20¹ East latitude). Rhizosperic soil and root samples from different tree species were collected from depth of 0 to 25 cm. The roots were freed from the adhering soil, gently washed and then cut in to 1cm segments. The degree of colonization was calculated by using the slide method (Giovannetti and Mossae,1980).The colonization of roots were determined after clearing roots with 10 per cent KOH and staning with 0.05 per centtryphan blue in lactophenol (Phillips and Haymann,1970). For assessment of occurrence of spores in the rhizosphere, soil was subjected to the wet-sieving and decanting technique following (Gedermann and Nicolson, 1963).The VAM spores were identified using the mannual of (Schenck and Perez,1991).Three plants *i.e., Pterocarpus marsupiun, Lagerstroeium lanceolatus* and *santalum album* were grown for 120 days in green house conditions at 30 + 5°c. Each treatment was replicated three times. Growth parameters such as plant height, dry weight, percent VAM colonization, spore number, macro and micro elements was determined according to (Jackson,1973) at the time of harvest. This possibility of inoculating green house level with selected strains of VA mycorrhiza more effective than those of (non-inoculated) control plants.

Result and Discussion

Result of experiments given in the Tables 18.1–18.3 revealed that biodiversity of tree species are correlated with diversified VAM spore population. Altogether forty-five tree species have been screened. In general most of the trees showed varied percentage of colonization spore number. Least percent root colonization from 14-17 per cent was observed in *Grewelia robusta* and *Santalum album*. However, forty trees harbored with 44-86 per cent of root colonization. The highest percentage of root colonization was noted down in member of Rutaceae and Combretaceae.Similarly least number of spores 98-99/50g soil was recovered in the members of Fabaceae.Tweny five trees had spore number from 101 to 135/50g soil, Seventeen trees had spore number from 104 to 198/50g soil. Highest spore number of spores 203/50g soil was recorded in Rutaceae (Table 18.1). Altogether thirty-four different

VAM spores were isolated from fifteen places where experimental trees are growing (Table 18.2). These spores are arranged in the gradation in ascending order. Among the recovered spores *Glomus* species were most predominated over *Acaulospora, Gigaspora, Sclerocystis* and *Scutellospora* species shown in (Table 18.3). Growth response of three tree species *Pterocarpus marsupiun, Lagerstroeium lanceolatus* and *santalum album* revealed that the plant height were 5 to 6 fold increased over the non-mycorrhizal plants. Plant biomass, percent of root colonization, spore number and uptake of macro and micronutrients was significantly higher over the non-inoculated plants shown in (Tables 18.4–18.6).

Table 18.1: Per cent Root Colonization Spore Number of AM Fungi in Forty-five Tree Species of Dry Deciduous Forest of Dharwad District

Plant Species	*Family*	*% of VAM Colonization*	*Spore Number/ 50g of Soil*
Aegle maronelos Corr.Ex.Roxb.	Rutaceae	92.5	106
Acacia Arabica (Lam.)Willd.	Mimosae	53.2	101
Acacia melonoxylon Willd.	Mimosae	50.2	162
Acacia niolotica Lam.	Mimosae	57.3	133
Artocarpus heterophylls Gaertn.	Moraceae	72.8	124
Azadiracta indica Juss.	Meliaceae	81.4	175
Barringtonia acutangula	Lecythidaceae	61.3	102
Bauhinia purpurea L.	Fabaceae	52.2	172
Bauhinia varigata L.	Fabaceae	54.1	99
Bombax ceiba L.	Bombacaceae	67.2	104
Butea manosperma Lam.	Fabaceae	51.7	109
Cassia fistula Lam.	Fabaceae	52.5	107
Cassia siamea Lam.	Fabaceae	45.4	192
Casurina equisetifolia Forst	Casurinaceae	44.3	103
Citrus aurantifolia L.	Rutaceae	95.1	209
Dalbergia sissoo Roxb.	Fabaceae	54.2	114
Dendrocalamus strictus Nees.	Bombacaceae	69.6	111
Erythrina indica Lam.	Fabaceae	55.5	152
Eucalyptus tereticornis Wt.	Myrtaceae	75.7	143
Ficus religiosa L.	Moraceae	58.5	135
Garcinia indica Choiss.	Guttiferae	66.4	147
Gravellia robusta A.cunn.	Proteaceae	14.3	184
Gmelia arborea Roxb.	Rubiaceae	71.2	115
Hardwickia binata Roxb.	Fabaceae	51.5	104
Lagerstroemia lanceolat Wall.	Lythraceae	—	—
Leucaena leucocephala Link.	Fabaceae	52.6	172
Madhuca longifolia Gmel.	Sapotaceae	69.3	151
Mangifera indica L.	Anacardiaceae	77.2	132

Contd...

Table 18.1–Contd...

Plant Species	Family	% of VAM Colonization	Spore Number/ 50g of Soil
Mesue ferrea L.		61.3	105
Michelia champaca L.	Magnolaceae	62.2	128
Myristica malabarica	Moringaceae	60.4	141
Pongamia pinnata Vent.	Fabaceae	47.3	106
Pterocarpus marsupium Lam.	Fabaceae	46.6	101
Pithocelobium dulce (Roxb).Benth.	Fabaceae	48.1	102
Santalum indica L.	Santalaceae	17.8	132
Syzygium cumni L.	Myrtaceae	59.7	140
Syzygium jambolana DC.	Myrtaceae	63.2	123
Tamarindus indica L.	Fabaceae	48.4	98
Tectona grandis L.	Verbenaceae	57.3	106
Terminalia bellarica Roxb.	Combretaceae	81.7	104
Terminalia chebula Retz.	Combretaceae	86.6	122
Terminalia tementosa W.andA.	Combretaceae	92.5	199
Thespesia macrophylla BL.	Malvaceae	78.1	176
Zizyphus glabrata Wt.	Rhamnaceae	67.2	153
Zizyphus zujuba Lam.	Rhamnaceae	69.3	144

Isolation of VAM spores from respective rhizosphere soil of the screened species indicated that altogether five genera were *Acaulospora, Glomus, Gigaspora, Sclerocystis* and *Scutellospora*. The *Sclerocystis* were represented by 2-3 species. *Glomus* was more predominated.

The higher level of root colonization in 43 tree species indicated its deep susceptibility towards VAM fungi as well as infectiveness of VAM fungi. Variation in the percent of colonization might be regulated at the species level. Variation of spore population in rhizospheric soil showed that the multiplication of spores depends on species to species level (Rahangdale and Gupta, 1999; Lakshman *etal.*, 2003).

Experiments on growth response with the most indigenous fungi in the rhizosphere of selected three tree species reveal that the fungi differ in there potential to colonize plant roots under similar condition (Tables 18.4 and 18.6).The root colonization was found to be very high with the inoculation of *Glomus fasciculatum* compared to the other species *Glomus mossae*. This showed there preference towards host species. A lot of variation in colonization was observed among AM fungi by (Reddy *etal.*, 1996).and this was attributed to the ability of AM fungi to colonize the host roots quickly and extensively. The high level of root colonization may be the prime determinant of the efficiency of symbiosis (Menge, 1983; Lakshman and Geeta Patil, 2004) and the fungus *Glomus fasciculatum* was found to be the efficient fungus for colonization than *Glomus mossae*. Difference in the response of AM fungi within host suggested that under certain condition selection should occur to favor certain host-fungus combinations.

Table 18.2: The Association of Arbuscular Mycorrhizal Fungi with Three Species of Dharwad District

Mycorrhizal Species	*Host Species*		
	P. masupium	*L. lanceolatus*	*S. album*
Acaulospora appendicula (Spain.)Siev and Sch.	–	+	–
Acaulospora delicata (Walker.)Pte.and Bloss	+	+	–
Acaulospora denticulata Sieverding and Toro.	–	–	+
Acaulospora elegans Trappe and Gerdermann.	–	+	+
Acaulospora lacunosa Morton.	–	–	+
Acaulospora rugusa Morton.	–	–	+
Acaulospora spinosa Walker and Trappe.	+	–	–
Acaulospora sporocarpia Berch.	+	+	–
Acaulospora tuberculata Janos and Trappe.	–	–	+
Entrophospora infrequens (Hall.)ames and Schneider.	+	+	–
Gigaspora caudida Bhattacharjee and Mukerji	–	–	–
Gigaspora ramisporaphora Spain.	–	+	–
Gigaspora rosea Nicolson andSchenck.	+	+	–
Glomus aggregatum Schenckand Smith.	–	–	+
Glomus albidum Walker and Rhodes.	+	+	+
Glomus boreale (Tnaxter) Trappe and Gerdermann.	+	+	–
Glomus citricolum Tang and zang.	–	–	+
Glomus clarum Nicolson and Schenck.	+	+	–
Glomus dimorphicum Boyetchko and Tiwari.	–	–	+
Glomus clunicatum Becker and Gerdermann.	+	+	–
Glomus fasciculatum (Tnaxter) Gerd. and Walker.	+	+	+
Glomus fragile Berk andBrome.	+	+	–
Glomus geosporum Nicol. and Gerd.(Walker.)	+	–	+
Glomus hoi Berch and Trappe.	–	–	+
Glomus macrocarpum Tul andTul.	+	–	+
Glomus microcarpum Tul andTul.	+	–	+
Glomus mosseae Gerd.and Trappe.	+	+	+
Glomus multicanle Gerd.and Trappe.	+	–	–
Glomus tenerum Tandy	–	–	+
Sclerocystis clavispora Trappe.	–	+	+
Sclerocystis indica (Pat.)Honn.	–	+	–
Sclerocystis rubiformis Gerd.and Trappe.	+	–	–
Scutellospora aurigloba (Hall.)Wal. and Sanders.	–	–	+
Scutellospora calospora Walker and Sanders.	+	+	–
Scutellospora gilmorea (Trappe. and Gerd.) Walker and Sanders.	–	+	–

Table 18.3: Distribution and VAM Spores Diversity in Fifteen Places of Dharwad District

Yerikoppa	G.ci,	G.eta,	A.tu,	A.lo,	S.gil.
Kiruwatti	G.ma,	G.te,	G.fa,	Sc.ru,	G.al
Mottagi	G.ag,	G.ta,	G.ma,	S.tri,	Gi.con.
Yeriguppe	G.mo,	A.spi,	G.mi,	G.ci,	Sl.ru.
Devanur	Sc.in,	G.ag,	G.ma,	A.tu,	A.spi.
Badrapur	S.cal,	G.mi,	G.fa,	Sc.ru,	Sl.ru
Aravatti	S.tri,	G.al,	G.ma,	A.tu,	G.al.
Navalgund	G.ag,	Gi.con,	G.ci,	G.ag,	S.tri.
Kudrikeri	A.lo,	G.te,	G.mi,	Sc.in,	G.ag.
Hebsur	G.ma,	A.spi,	G.ta,	G.te,	A.de.
Nigadi	G.mo,	G.ci,	A.tu,	A.lo,	G.mo.
Amblikoppa	A.de,	Gi.can,	G.ag,	S.tri,	A.tu.
Narendra	G.fa,	G.mi,	Sc.in,	G.ag,	G.te.
Mugad	A.lo,	G.ci,	S.cal,	G.fa,	G.m.
Alnavar	G.fa,	Sc.in,	S.tri,	G.ag,	G.ma.

G.ci: *Glomus citricola*; G.etu: *Glomus etunicatum*; A. tu: *Acaulospora tuberculata*; A.de: *Acaulospora denticulate*; G.ag: *Glomus aggregatum;* A.lo: *Acaulospora longula*; G.fa: *Glomus fasciculatum*; G.ma: Glomus *macrocarpum*; G.mo: *Glomus mossae*; Sc.in: *Sclerocystis indica*; Sc.ru: *Sclerocystis rubiformis*; S.cal: *Scuttelospora calospora;* S.tri= *Scuttelospora tricalypta*; G.te: *Glomus tenubricum*;G.al: *Glomus albida;* A.spi: *Acaulospora spinosa*; Gi.can: *Gigaspora candida*; G.mi: *Glomus microcarpum.*

Table 18.4: Effect of Arbuscular Mycorrhizal Fungi on the Growth Biomass Production, Per cent AM Colonization, Spore Number, Macro and Micronutrient Status in *D. marsupium* for Sixty Days

Treatment	*Plant Height (cm)*	*Total Plant Dry Weight (g)*	*% AMF Coloni-zation*	*Spore Number/ 50 g Soil*	*Nutrient Status*					
					Macro %			*Micro %*		
					N	*P*	*K*	*Zn*	*Cu*	*Mg*
Control	19.71	2.92	–	–	1.51	0.11	1.07	1.8	2.2	2.3
G.Fasciculatum	46.53	8.42	53.2	114	3.73	0.27	2.12	3.3	4.1	2.2
C.D.@%	7.15	3.45	11.3	10.2	0.08	0.005	0.05	0.002	0.01	0.001

Table 18.5: Effect of Arbuscular Mycorrhizal Fungi on the Growth Biomass Production, Per cent AM Colonization, Spore Number, Macro and Micronutrient Status in *L. lanceolatus* for Sixty Days

Treatment	*Plant Height (cm)*	*Total Plant Dry Weight (g)*	*% AMF Coloni-zation*	*Spore Number/ 50 g Soil*	*Nutrient Status*					
					Macro %			*Micro %*		
					N	*P*	*K*	*Zn*	*Cu*	*Mg*
Control	27.64	3.87	–	–	1.73	0.12	1.12	2.1	1.7	2.1
G.Fasciculatum	83.41	14.55	86.62	263	3.82	0.32	2.62	3.1	3.8	1.6
C.D.@ 5%	9.16	4.41	15.30	13.2	0.17	0.005	0.05	0.001	0.05	0.001

Table 18.6: Effect of Arbuscular Mycorrhizal Fungi on the Growth Biomass Production, Per cent AM Colonization, Spore Number, Macro and Micronutrient Status in *Santalum album* for Sixty Days

Treatment	Plant Height (cm)	Total Plant Dry Weight (g)	% AMF Coloni-zation	Spore Number/ 50 g Soil	Nutrient Status					
					Macro %			Micro %		
					N	P	K	Zn	Cu	Mg
Control	17.42	2.61	–	–	1.55	0.05	1.07	1.2	1.9	2.2
G.Fasciculatum	39.53	7.47	48.30	103	2.92	0.19	2.16	2.6	3.1	1.8
C.D.@ 5%	6.33	3.51	13.10	08.0	0.06	0.005	0.05	0.002	0.003	0.002

AM fungal inoculation has increased the growth characteristics of the selected three trees. There was increased dry weight, macro and micro nutrient in *Pterocarpus marsupial, Lagerstroeium lanceolatus* and *santalum album*. Similar results were observed by (Sitaramaiah and Khanna, 1997).Correlation between total biomass production and percent root colonization was recorded in the present study. This type of relation was reported by (Bhat *et al.*, 1993). However, a difference in effectiveness was observed among AM fungi on inoculations of host plants. Although AM fungal species are not considered to have any speciality towards different taxa of hosts under favorable conditions, there are many differences in there effectiveness in a particular host (Kotuari and Singh, 1996).The enhanced photosynthetic rate associated with increased N.P.K.and Cu, Zn uptake may be attributed directly to the growth improvement of the plants (Schwob *et al*, 1998). Different fungi stimulate plant growth to different extents and therefore tree productivity can be increased by inoculation with selected mycorrhizal fungi. Increases in nutrient content of the plant may be due to increased absorption of the limiting element in that soil or to increased absorption of several other elements. Although much of the research emphasis of mycorrhizas has been P, they also increase uptake to many other nutrients like K, S, Zn, Sr-90 (a Ca-tracer) Cu and NH_4 –N (Bowen, 1980,1984,TimmerandLeydon, 1980). The increased uptake of poorly mobile ions by mycorrhizas is by the growth of hyphae into soil, which absorbs nutrients and then transfers back to the plant. The increased inflow of phosphates in mycorrhizal plants in relation to mycelial growth has been demonstrated by inflow analysis for both VAM and ectomycorrhiza (Sanders *et al*, 1977, Sander and Tinker, 1973).

Mycorrhizal development depends on the status of nutrients in soil. Plants in rich fertile soils do not usually need mycorrhizal association for adequate nutrition and growth. It is now well established that increased phosphorus in the substrate results in decreased root exuviate formation and decreased VAM formation (Graham *et al.*, 1981; Soil fertility does not always decrease ectomycorrhizae formation.

Microbes solubilize non-available nutrients for plants and thus help reclamation of disturbed sites (Menge, 1983). Mycorrhizal plants, which can also fix nitrogen, are efficient in reclamation purposes (Osonubi *et al*, 1991). The nutrient status of the NM plants was clearly inferior to that of mycorrhizal plants. Non-mycorrhizal plants under low nutrient condition may develop strategies to acquire more nutrients from the soil by the production of cluster roots. There for, in the present investigation, mycorrhizal dependency was found higher in forty three examined plants out of forty five trees. This may be due to its dense but more branched root system with short root hairs.

Conclusion

VA mycorrhizal dependence among the tree species was highly variable. Part of their dependency differences was due to the variation in the root structure and the ability of roots to absorb. It has been proposed that the length of the root hair is indicative of degree of mycorrhizal dependency (Baylis, 1970). Short root hair indicates a relatively high degree of mycorrhizal dependency, while long hairs indicate a low degree of mycorrhizal dependency. VA mycorrhizal inoculum can be dispersed naturally or with the assistance of nursery managers. Wind, rain and animals naturally disperse propagules. Historically mycorrhizal inoculum was introduced to tree seedling nurseries as forest soil or litter. Therefore nursery cultural practices should be designed to foster a healthy rhizosphere environment for tree seedlings. Results indicated there is a great diversity among VAM fungi. Growth response has brought significant increase of biomass production in three plants by the inoculations of *Glomus fasciculatum* preference mycosymbiot.

Acknowledgments

Author is indebted to U.G.C. New Delhi, for sanctioning Major Research Project and financial assistance.

References

Bhat, M.N., Jeyarajan, R. and Ramraj, B., 1993. *Indian J. For.*, 16: 309–312.

Bowen, G.D., 1984. Tree roots and the use of soil nutrients. In: *Nutrion of Plantation Forests*, (Eds.) Y. Waisel, A. Esnel and U. Kafkafi. Marced and Dekker Inc. , New York, pp. 309–330.

Bowen, G.D., 1980 Mycorrhizal roles in tropical plants and ecosystems. In: *Tropical Mycorrizal Research*, (Ed.) P. Mikola. Oxford University Press, Oxford, pp. 165–190.

Durga, V.V.K. and Gupta, S., 1995. *The Ind. Forest*, 121: 518–527.

Gerdermann, J.W. and Nicolson, T.H., 1963. *Trans. Brit. Mycol. Soc.*, 46: 235–244.

Giovanneti, M. and Mossae, B., 1980. *New Phytol.*, 84: 489–500.

Graham, J.H. Leonard, R.T. and Menge, J.A., 1981. *Plant Physiology*, 68: 548–552.

Jackson, M.L., 1973. *Soil Chemical Analysis.* Printice Hall Inc., New Delhi, pp. 199.

Kothri, S.K. and Singh, U.B., 1996. *Plant Soil*, 178: 231–237.

Lakshman, H.C., Inchal, R.F. and Mulla, F.I., 2003. *Ind. J. Eco. Plan.*, 179: 161–169.

Lakshman, H.C. and Geeta B. Patil, 2004. *Envi. Ecol. Conser.*, 19: 21–29.

Menge, J.A., 1983. *Can J. Bot.*, 61: 1015–1024.

Osonubi, O., Mulongoy, K., Awokoye, O.O., Atayese, M.O. and Okali, D.U.U., 1991. *Plant and Soil*, 136: 131–216.

Phillips, J.M. and Hayman, D.S., 1970. *Trans. Brit. Micol. Soc.*, 55: 158–161.

Rahagdale, R. and Nibha gupta, 1999. *Ind. J. Forestry*, 22: 62–65.

Reddy, B., Bhagyaraj, D.J. and Mailesha, B.C., 1996. *Indian J. Microbiol.*, 36: 13–16.

Reeva, J. and Bhagyaraj, D.J., 1990. *J. Microb. Biotech.*, 6: 59–63.

Sanders, F.E. and Tinker, P.B., 1973. *Pesticide Science*, 4: 383–395

Sanders, F.E., Tinker, P.B., Black, R.L.B. and Palmerley, S.M., 1977. *New Phytol.*, 78: 257–268.

Schenck, N.C. and Perez, Y., 1990. *Manual for Identification of VA Mycorrhizal Fungi*. Gainsville. USA; Synergistic Publications, 286 pp.

Sitaramaiah, K. and Khanna, R., 1997. *J. Mycol. Plant Pathology*, 27: 21–24.

Thaper, H.S. and Khan, S.N., 1973. In: *Proc. Nat. Sci. Acad.*, 39: 687–694.

Timmer, L.W. and Leydon, R.F., 1980. *New Phytol.*, 85: 15–24.

Sustainable Environmental Management *Pages* ***180–188***
Edited by: **Dr. L.V. Gangawane & Dr. V.C. Khilare**
Published by: **DAYA PUBLISHING HOUSE**

Chapter 19

Distribution, Density and Diversity of Primary Producers in Thane Creek Waters

V.W. Lande and J.P. Kotangale

National Environmental Engineering Research Institute (NEERI),
Nehru Marg, Nagpur – 440 020, MS, India

ABSTRACT

Discharges of wastewater and its serious consequences in Thane creek waters affecting physico-chemical parameters and phytoplankton community studies were carried out during summer-2001. Daily discharge of waste water drastically changes the water quality in upper narrow zone of the creek. Physicochemical parameters *viz.* water temperature 29.5-32.5°C, pH 7.0-8.0, dissolved oxygen (DO) 0.8mg/l to 4.4mg/l and biochemical oxygen demand (BOD) 2.8mg/l to 24mg/l, ammonical nitrogen (NH3-N) 0.38mg/l to 2.26mg/l and high concentration of orthophosphate (PO4-P) 0.10mg/l to 0.67mg/l corresponding to phytoplankton numbers 12-119/ml were observed in upper narrow zone of creek (transect 1-4). Middle and outer zones of creek showed zone of recovery of water due to influx of sea water during high tide and fairly good water quality was observed in creek. Total 76 phytoplankton genera were recorded in samples from transects 1-10. Tolerant forms of ulotricales, euglenales, pinnales, zygnematales were observed in creek waters. Significant correlation coefficient between BOD and PO4-P 0.555165 in upper zone (transect 1-4), 0.709512 in middle zone (transect 5-7) and between NH3-N and PO4-P 0.685464 in outer zone (Transect 8-10) had indicated pollution load in creek waters.

Keywords: *Primary producers, Pollution, Physico-chemical parameters, Correlation coefficient.*

Introduction

Thane creek waters receive fresh water from Ulhas river in upper narrow zone. Serious consequences of chemicals, high biochemical oxygen demand, phenolic compound, heavy metals discharges in effluents of various industries located at Thane–Belapur road further deteriorated water quality in creek during summer season. Several studies on primary producer and their potentials have been carried out in estuarine water. Algae are known as indicators of pollution as they respond to both toxicants and growth stimulants in aquatic bodies. Paerl *et al.* (2003) reviewed distinguishing and integrating impacts of natural and human stressors which are essential for understanding temporal changes in microbial diversity and function. Laws *et al.* (1999) studied point and nonpoint sources of freshwater inputs in the estuaries which function as buffer zones by trapping source of the sediment and nutrients that would otherwise enter the coastal ocean during the period of 1 year in Mamala bay, Hawaiian islands. Wang *et al* (1999) showed phosphorus enrichment and nitrogen limited algal growth when the internal nutrient cycling and transport exceeded external loadings in Tampa bay, Florida. Generally phytoplankton dominated by diatoms in spring and autumn, chlorophytes in summer and irregular occurrence of cyanophytes which characterized a eutrophic river environment (Noppe *et al.*, 1999). Kannan and Vasantha (1992) observed high species richness of phytoplankton in the Pitchavaram mangals. Ketchum (1967) indicated characteristic features of the estuaries (i) when it receives large amount of organic matter/plant nutrients leached from soil and (ii) pollutants from domestic sources enhances the water organic contents and increase phytoplankton population, (iii) followed by enrichment of nutrient concentration in subsurface layer of estuaries mixing to sea water was higher than the surface waters. Sivani and Rao (2001) studied the abundance of primary producer resulting from enrichment of nutrients due to organic nature of waste from the nearby Jute mill in Gosthani estuary near Visakhapatnam. Mukhopadhayay *et al.* (1998) showed that physicochemical characteristics such as salinity, hardness, BOD and phytoplankton, zooplankton populations significantly varied in relation to industrial waste discharged in to the Hooghly estuary. Prasad (2003) pointed out that trace metals act as limiting factors for growth of plankton population and productivity in mangrove waterways of Coringa, east coast of India. Ram *et al.* (2003) reported highly toxic Hg levels in suspended particulate matter of waters and sediment of the Ulhas Estuary. Pollutants discharges in terms reduce pH, dissolved oxygen, nitrite, nitrate, primary productivity and increase ammonia, phosphate, silicate, heavy metals, biochemical oxygen demand, chlorophyll pigment, particulate organic carbon (POC), suspended solids and bacteria may not be overlooked in heavily polluted Adyer river at the mixing site during low tides (Subramanian *et al.*, 2003). Patrick (1967) studied pollution effects on the distribution of algae biomass, which comprised of few species, changing its normal flora and fauna while curtailed its usefulness as food and the biomass under certain circumstances may be as high as in a naturally eutrophic estuary. Phytoplankton photosynthesis is apparently phosphorus limited in the freshwater portion, nitrogen and/or silica limited toward sea reaching portion and further exogenous influx of major plant nutrients by means of trace elements, *e.g.* iron and copper, skewed the primary production of the creek, (Zhang, 2000). Studies on surface water quality in relation to pollution by Paradeep phosphate ltd. (PPL) in river Atkerabakni which ultimately joins the Mahanadi river near the confluence point of the bay of Bengal were carried out (Patnaik *et al.*, 2002). Organic pollutants showed impact on quality of diatoms of Gopalpur coast in Orissa, (Mallick and Padhi, 1999). Cloern (1999) studied that there is large difference among coastal ecosystems in the rates and patterns of nutrient assimilation and cycling. Considering the earlier studies, present observation of physicochemical parameters and phytoplankton from Thane-road bridge on upstream to Trombay region upto the location of 19.02° N outer region of creek was studied during summer-2001 (Figure 19.1).

Materials and Methods

Water samples were collected from thirty sampling locations of ten transects. Four transects are located in upper narrow zone covering the area Thane Kalwa road bridge to Airoli bridge, approximate distance 4.5 km. Next to this zone middle zone of creek, the stretch of 9 km comprised of three transect and outer zone of the distance of 5 km, another three transect. From each transect of creek, three sampling locations (East, Center and West) samples for hydrobiological studies were collected. Physicochemical parameters like water temperature, pH, dissolved oxygen (DO), biochemical oxygen demand (BOD), ammonical nitrogen (NH3-N), orthophosphate (PO4-P) and phytoplankton distribution were analyzed following methods given in Standard methods, (APHA, 1999), A manual for sea water analysis (Strickland and Parsons, 1968). Identification and enumeration of phytoplankton were carried out in the laboratory (Prescott,1978).

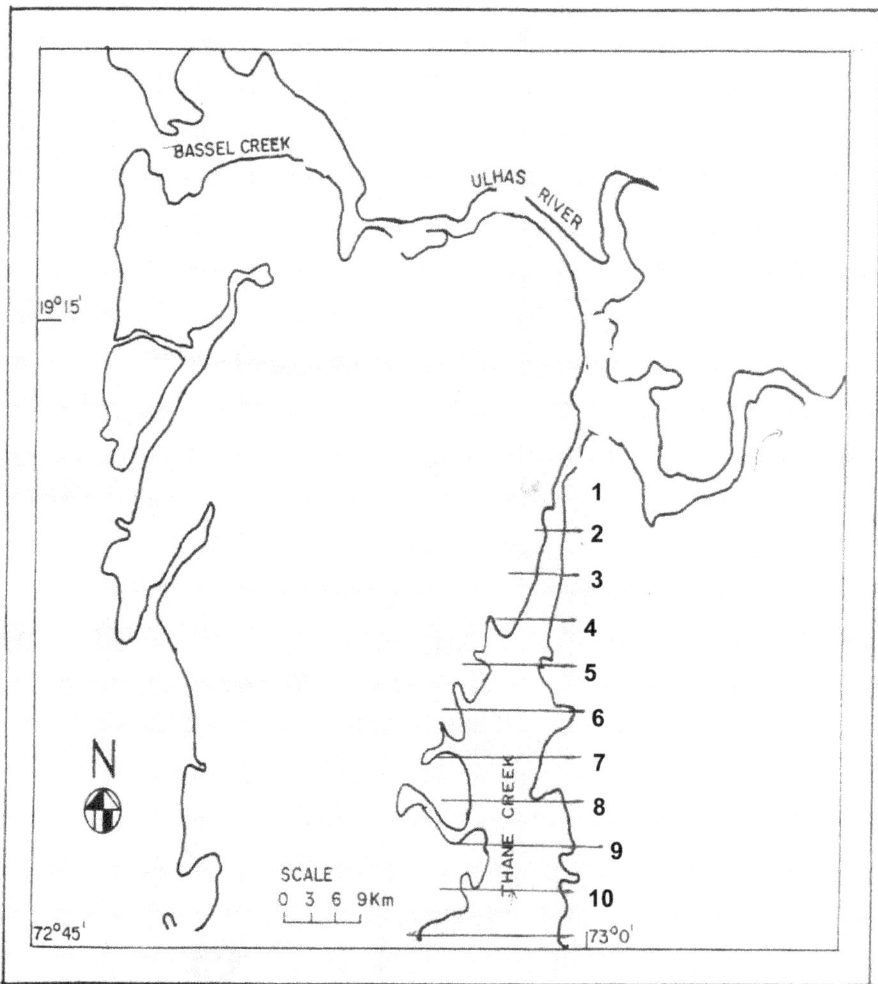

Figure 19.1: Sampling Locations in Thane Creek (Transects 1 to 10)

Results

Thane creek's upper narrow region receives four discharges of waste water, one from treatment facility of Thane and three nallas carrying large quantity of waste discharges. This zone is heavily stressed due to waste discharges. The upper region of zone also has fresh water source, which directly influence physicochemical and biological parameter of creek waters during monsoon season. Surface water temperature ranged from 29.5 to 32.5 °C in day time. pH was ranged from 7.0 to 8.0 neutral to alkaline range, did not show much variation. DO directly affected due to degradable organic waste, ranging from 0.8mg/l to 4.4mg/l corresponding to higher BOD 2.4mg/l to 24mg/l was recorded in the region. Ammoniacal nitrogen 0.38 mg/l to 2.26 mg/l was found which incidentally may be toxic to plankton (Table 19.1). Several times higher phosphate content 0.10mg/l to 0.67 mg/l indicated preference of single species dominance in creek waters during observation period. The region has rocky bottom at transect 1 and 2 during low tide. Mangrove mud flats are also seen in this region. Forty six genera of phytoplankton ranging 12-119/ml numbers were recorded in upper region of creek. Dominant genus *Ulothrix* sp. and subdominant genera *Eremosphaeria* sp., *Chlorella* sp., *Phacus* sp., several diatoms genera and *Cryptomonas* sp., *Mallomonas* sp., *Prorocentrum* sp., *Nautococcus* sp. and *Ceratium* sp. were recorded in transect 1 to 4. Many species of desmids like *Desmidium* sp., *Netrium* sp., *Gonatozygon* sp., *Euastrum* sp. as the indicators of chemical pollution (Sulfites, sulfides of iron) were also seen in nanoplankton of this zone, (Table 19.2). In middle region transect 5 to7, this stretch receives mainly effluent from Bhandup on western bank and domestic waste water discharge from Airoli, Koparkhairane, Vashi and industrial waste water through Common Effluent Treatment Plant (CETP) disposal point at eastern bank. The water is deeper at west end whereas considerable area at eastern bank gets dried up during low tide. Water quality observed to be substantially better than upper zone due to increase of width and volume of water in creek. Water temperature 1°C higher than the upper zone and pH ranged from 7.0 to 8.2 were observed. Clear water indicates reduction of suspended organic matter by settling and diurnal tidal variation. DO was ranged from 1.7 to 5.5 mg/l, low BOD ranged from 1.0-13.0 mg/l was observed in this zone. Phosphate and ammoniacal nitrogen values were 0.00 to 0.56 mg/l and 0.20 to 3.12 mg/l respectively; they support the good growth of phytoplankton. Transect 5-7 comprised of phytoplankton numbering 4-627/ml with dominant genera *Ulothrix* sp., *Chlorella* sp., *Chaetoceras* sp. whereas uncommon forms *Treubaria* sp., *Chrysidiastrum* sp. and several diatom genera including *Biddulphia* sp. *Skeletonema* sp., *Nitschia* sp., *Cocconeis* sp., *Thalossiosira* sp., *Gyrosigma* sp., *Navicula* sp., *Gamphonema* sp., *Cyclotella* sp., *Cymatopleura* sp., *Synedra* sp., *Achenanthes* sp., *Amphipleura* sp., *Caloneis* sp, *Scolipleura* sp. *Neidium* sp., *Amphora* sp., *Pinnularia* sp., *Chlorobotrys* sp., *Monomastix* sp. were recorded. Fifty one species of phytoplankton have identified in middle zone of the creek. The outer region of sampling locations 8 to 10 receives discharges from Ghatkoper at western bank and Nerul STP on eastern bank. Shallower area on eastern bank and water depth more than 8 m on western bank was observed. Temperature ranged from 30.5°C to 33°C, pH had shown increased values of 7.2 to 8.3 mg/l in outer creek waters. Sea water enters in creek and increases the pH values more than those observed in upper and middle zone. DO was recorded 3.0 to 5.6 mg/l in creek waters. Nutrient levels phosphate (0.05 to 0.23 mg/l), ammoniacal nitrogen (0.20 to 2.40 mg/l) and phytoplankton counts 20-403/ml indicated water suitability for development of not only flora but fauna also in outer zone of creek. Forty three phytoplankton genera comprised of pinnales forms *Licmophora* sp.,*Gomphonema* sp., *Synedra* sp., *Nitzschia* sp., *Cocconeis* sp., *Skeletonema* sp., *Surirella* sp., *Asterionella* sp., *Gyrosigma* sp. and *Coscinodiscus* sp., *Melosira* sp., *Triceratium* sp. were identified.

Correlation coefficient between various physico-chemical parameters and phytoplankton species were studied. Significant correlation coefficient between BOD_5 and PO_5-P (0.555165) and water temperature with phytoplankton counts (0.516531) were observed in upper zone, (transect 1 to 4).

BOD_5 and PO_4-P (0.709512), PO_4-P and total count of phytoplankton (0.514147) were noticed in middle zone (transect 5 to 7). Water temperature and DO (0.523144), water temperature and ammoniacal nitrogen (0.760071), pH and BOD (0.694658), PO_4-P.

Table 19.1: Variation of Physico-chemical Parameters and Phytoplankton in Thane Creek Waters

Physico-chemical Parameters and Phytoplankton	*Transects 1-4 (Minimum–Maximum)*	*Transects 5-7 (Minimum–Maximum)*	*Transects 8-10 (Minimum–Maximum)*
Water temperature °C	29.5–32.5	30-33.5	30.5–33.0
pH	7.0-8.0	7.0-8.2	7.2-8.3
Dissolved oxygen mg/L	0.8-4.4	1.7-5.5	3.0-5.6
Biochemical oxygen demand mg/L	2.8-24.0	1.0-13.0	1.0-8.0
Ammonical nitrogen mg/L	0.38-2.26	0.20-3.12	0.20-2.40
Orthophosphate mg/L	0.10-0.67	0-0.56	0.05-0.23
Phytoplankton counts/ml	12–119	4-627	20-403

Table 19.2: Occurrence of Phytoplankton Genera in Thane Creek Waters

Identified Phytoplankton	*Transect 1-4*	*Transect 5-7*	*Transect 8-10*
Cyanophyceae			
Nostocales			
Anabaena sp.	+	–	–
Nodularia sp.	+	+	–
Oscillatorales			
Oscillatoria sp.	+	+	+
Phormidium sp.	+	–	–
Lyngbya sp.	–	–	+
Chrococcales			
Merismopedia sp.	–	+	–
Gloethece sp.	–	–	+
Euglenales			
Phacus sp.	+	+	+
Lepocinclis sp.	–	–	+
Chlorophyceae			
Tetrasporales			
Nautococcus sp.	–	–	+
Volvacales			
Chloromonas sp.	–	–	–
Chlamydomonas sp	+	+	+
Ulotrichales			
Ulothrix sp.	+	+	+
Stichococcus sp.	–	+	+
Rhaphidonema sp.	–	–	+

Contd...

Table 19.2–Contd...

Identified Phytoplankton	Transect 1-4	Transect 5-7	Transect 8-10
Chlorococcales			
Chodatella sp.	–	–	+
Chlorella sp.	+	+	+
Closteriopsis sp.	+	–	–
Treubaria sp.	–	+	–
Eremosphaeria sp.	+	–	+
Coelastrum sp.	+	+	–
Ankistrodesmus sp.	+	+	–
Zygnematales			
Spirotaenia sp.	+	+	+
Desmidium sp.	+	+	+
Netrium sp.	+	+	+
Gonatozygon sp.	+	–	+
Cosmarium sp.	+	+	–
Euastrum sp.	+	–	–
Hyalotheca sp.	+	–	–
Bacillariophyceae			
Pinnales			
Diatoma sp.	+	+	+
Synedra sp.	+	+	+
Cymatopleura sp.	+	+	–
Navicula sp.	+	+	+
Amphora sp.	+	+	+
Asterionella sp.	–	+	+
Cocconeis sp.	+	+	+
Bacillaria sp.	–	+	–
Caloneis sp.	+	+	–
Neidium sp.	–	+	–
Licmophora sp.	–	–	+
Gyrosigma sp.	+	+	+
Mastogloia sp.	+	+	–
Gomphonema sp.	+	+	+
Eunotia sp.	+	+	–
Nitzschia sp.	+	+	+
Cocconeis sp.	–	+	+
Skeletonema sp.	+	+	+
Amphiprora sp.	–	+	+

Contd...

Table 19.2–Contd...

Identified Phytoplankton	*Transect 1-4*	*Transect 5-7*	*Transect 8-10*
Surirella sp.	–	–	+
Thalassiosira sp.	+	+	+
Taballaria sp.	+	–	–
Achnanthes sp.	–	+	–
Scoliopleura sp.	–	+	+
Pinnularia sp.	–	+	+
Rhoicosphenia sp.	+	–	–
Amphipleura sp.	+	+	–
Centrales			
Melosira sp.	+	+	+
Staphanodiscus sp.	–	+	+
Biddulphia sp.	–	+	–
Cyclotella sp.	–	+	–
Thalassiothrix sp.	–	–	–
Chaetoceras sp.	+	+	+
Coscinodiscus sp.	+	+	+
Rhizosolania sp.	–	+	+
Triceratium sp.	+	+	+
Desmokontae			
Desmomonadales			
Prorocentrum sp.	+	+	+
Ochromomadales			
Mallomonas sp.	+	–	–
Dinophyceae			
Dinokontae			
Ceratium sp.	+	–	–
Cryptophyceae			
Cryptomonadacae			
Monomastix sp.	–	+	–
Cryptomonas sp.	+	+	–
Rhizochrysidales			
Chrysidiastrum sp.	–	+	–
Xanthophyceae			
Heterokontae			
Gloeachloris sp.	+	–	–
Chlorallantus sp.	–	–	+
Chlorobotrys sp.	+	+	–

+: Present; –: Absent.

Discussion

Ecomarphic pollution impact is likely to occur in creek upper zone. Even natural or anthropogenic eutrophication is the enrichment of water body in terms of nutrients utilizable by algae (Voltera and Conti, 2000). Algae is known to conjugate broad range of electrophilic compounds which are used in agriculture and industries. *Gloeotrichia echinulata* conjugate and metabolizes pentachloronitrobenzene by specific glutathione reductase in lake water (Field and Thurman, 1996). Mixotrophic species may dominate over strict autotrophs *e.g.* in the waters of poor in inorganic nutrients or under low light (Graneli *et al.*, 1999).

Earlier a case study of acidification of lake water showed a shift of pH from 6.5 to 5.0-5.5 and indicated that under pollution stress ecosystem processes in lake water appear quite stable while the functional redundancy develop in communities *e.g.* appearance of resistant species and succession in species composition allowing functions to continue as normal (Pratt and Cairns Jr., 1996). Ecological processes energetics get altered under stress as the community respiration increases and P/R (photosynthesis/respiration) become unbalanced, P/B and R/B (maintenance/biomass) ratio increase, importance of auxiliary energy source and stress influences consumers and primary production increase, nutrients cycling, community structure alters significantly. It is expected that resting spores of algae insure integrity of community as diversity of algae seems not to be affected while the numbers of phytoplankton affected very much in creek waters. In general, algal species, which reproduced faster, are more susceptible to external energy influx. The consequences resulted in increasing population size of one species among many species due to large amount of energy flowing into that portion of the ecosystem from internal and external sources and the consumers increase in number while decreasing the number of the species they consume. Fortunately, it appears that pollutants detected in sediments and carnivores well before any effects on individual communities or ecosystems are discernible (Chesser and Sugg,1996). Similar trends had observed in Thane creek during summer. Physicochemical parameters and phytoplankton correlation clusters indicated the following features (*i*) the correlation coefficient of physicochemical parameters and phytoplankton, negative values were observed under pollution impact (*ii*) Certain physicochemical parameters had shown negative correlation *e.g.* DO to ammoniacal nitrogen, pH to BOD and PO_4–P to NH_3–N. Cyanophycean organisms and diatoms are highly adaptive and known to colonize water polluted with heavy metal. Water quality severely affected secondary trophic level and benthos in the sediments of Thane Creek.

Acknowledgement

Authors are thankful to Director, for kind permission to publish this paper and Dr. Mrs. A. Chandorkar, for necessary assistance during field trips and samples collection and analysis.

References

APHA, 1999. *Standard Methods for the Examination of Water and Wastewater*, 20th ed., APHA, AWWA, and WEF, p. 10–15, index 1–I, Washington D.C.

Chesser, R.K. and D.W. Sugg, 1996. Toxicants as selective agents in population and community dynamic. In: *Ecotoxicology*: A Hierarchical Treatment, (Eds.) Michael C. Newman and C.H. Jagae, CRC, Lewis Publishers, London, 10: 293–317.

Cloern, J.E., 1999. *Aquatic Ecology*, 33(1): 3–16.

Field, J.A. and E.M. Thurman, 1996. *Enviro. Sci. and Technol.*, 30(10): 1413–1498.

Graneli, E., Carisson, P. and C. Legrand, 1999. *Aquatic Ecology*, 33(1): 17–27.

Kannan, L. and K. Vasantha, 1992. *Hydrobiologia*, 247(1): 77–86.

Ketchum Bostwick H., 1967. Phytoplankton Nutrients in Estuaries. In: *Estuaries*, (Ed.) George H. Lauff. Pub. No. 83, AAAS,Washington D.C.

Laws, E.L., Ziemann, D. and D. Schulman, 1999. *Marine Envir. Res.*, 48(1): 1–21.

Mallick, A.K. and S.B. Padhi, 1999. *Ind. J. Envi. and Ecoplanning*, 2(3): 289–293.

Mukhopadhyay, M.K., Vass K.K., Krishna Mitra and M.M. Bagchi, 1998. *Enviro. Ecol.*, 16(4): 790–797.

Noppe, K., Prygiel, J., Coste, M. and Lepretre, A., 1999. *J. Freshwater Ecol.*, 4(2): 167–178.

Paerl, H.W., Dyble, J., Pinckney, J.L., Steppe, T.F., Twomey, L., and L.M. Valdes, 2003. *Aquatic Microbial Ecol.*, 46: 23–246.

Patnaik, K.N., Satyanarayana, S.V. and Swoyam P. Rout, 2002. *Ind. Jr. Environ. Hlth.*, 44(3): 203–211.

Patrick, R., 1967. Diatom communities in Estuaries In: *Estuaries*, (Ed.) George H. Lauff. Pub. No. 83, AAAS, Washington, D.C.

Prasad, N.V., 2003. *Indian J. Environ and Ecoplan.*, 7(3): 465–469.

Pratt, J.R. and J. Cairns (Jr.)., 1996. Ecotoxicology and the redundancy problem: Understanding effects on community structure and function. In: *Ecotoxicology: A Hierarchical Treatment*, (Ed.) Michael C. Newman and C.H.Jagae. CRC, Lewis Publishers, London, 17: 347–370.

Prescott, G.W., 1978. *How to Know the Freshwater Algae*, 3rd Edition. The pictured key nature Series, USA, p. 293.

Ram, A., Rokade, M.A., Borole, D.V. and Zingde, M.D.2003. *Marine Pollution Bulletin*, 46(7): 846–857.

Sivani, G. and L.M. Rao, 2001. Influence of Jute mill effluents on the physico-chemical characteristics of Gosthani Eatuary Visakhapatnam. In: *Aquatic Pollution and Toxicology*, (Ed.) R.K. Trivedy. ABD Publishers, Jaipur, India, p. 139–144.

Strickland, J.D.H. and T.R. Parsons, 1968. *A Manual for Sea Water Analysis*. Fisheries Research Board of Canada, Ottawa, Canda, p. 398.

Subramanian, B., Mohan, N., Gandhiappan, J. and A. Mahadevan, 2003. *Indian J. Environ. Hlth.*, 45(4): 275–280.

Voltera, L. and M.E. Conti, 2000. *International Journal of Env. and Pollution*, 13(1–6): 92–125.

Wang, P.F., Martin J. and Morrison, G., 1999. *Estuarine, Coastal and Shelf Science*, 491: 1–20

Zhang, Jing, 2000. *Limnology and Oceanography*, 45(8): 1871–1878.

Sustainable Environmental Management *Pages 189–197*
Edited by: **Dr. L.V. Gangawane & Dr. V.C. Khilare**
Published by: **DAYA PUBLISHING HOUSE**

Chapter 20

Assessment of Biomass as Energy Resource

S. Mande[1] and S.B. Wagh[2]

[1]*Tata Energy Research Institute (TERI), Habitat Place, Lodi Road, New Delhi – 110 003*
[2]*Department of Environmental Science, Dr. Babasaheb Ambedkar Marathwada University, Aurangabad – 431 001, MS, India*

ABSTRACT

Energy use facilitates all human endeavor as well as social and economic progress. Energy is used for heating and cooling, illumination, health, food, education, industrial production and transportation. 'Biomass' is an important resource for energy production. The article reviews the types of biomass and the application in production of energy.

Keywords: *Biomass, Energy resources, Agro residues.*

Introduction

Worldwide photosynthetic activity is estimated to store 17th times as much energy as is annually consumed by all the nations in the world. Taking into account, the energy requirements of collection, processing and conversion to convert forms of that, biomass still assures a bright future from energy point of view. It holds the promise to supply complete energy needs of the world if managed and used effectively and sustainably.

The term biomass generally refers to renewable organic matter generated by plants through photosynthesis in which the solar energy combines the carbon dioxide and moisture to form carbohydrates and oxygen. Materials having combustible organic matter is also referred to as biomass. Coal is the end product of a sequence of biological and geological processes undergone by biomass itself. Biomass contains C, H and O, *i.e.*, they are oxygenated hydrocarbons. Biomass generally has high moisture and volatile matter constituents and has low bulk density and calorific value.

Biomass Sources

Biomass can be classified as woody and non-woody biomass. Non-woody biomass comprises of agro-crop and agro-industrial processing residue. Municipal solid wastes, animal and poultry wastes are also referred as biomass as they are biodegradable materials. Thus the main biomass sources are

1. *Wood:* Sawdust
2. *Residues:* Rice Husk, Bagasse, Groundnut shells, Coffee husk, Straws, Coconut shells, Coconut husk, Arhar stalks, Jute sticks etc.
3. *Aquatic and Marine Biomass:* Algae, Water hyacinth, Aquatic weeds and plants, Sea grass beds, Kelp, Coral reep etc.
4. *Wastes:* Municipal solid waste, Municipal sewage sludge, Animal waste, Industrial waste etc.

Being so versatile and scattered nature of biomass, the sufficient database and documentation in terms of its availability and consumption/utilization pattern are not always available. Although, the biomass meets a major part of the total energy requirements, yet, it does not find appropriate place in the overall energy balances of India, probably due to lack of proper documentation and database.

Potential Availability of Agro-residues

Agricultural residues can be divided into two groups: crop residues, and agro-industrial residues. Crop residues are plant materials remaining in the farm after removal of the main crop produce. The remaining materials could be of different sizes, shapes, forms and densities *i.e.* straw, stalks, sticks, leaves, haulms, fibrous materials, roots, branches, twigs etc. The agro-industrial residues, are by-products of post harvest processes of crops *viz.* cleaning, threshing, delinting, sieving, crushing, etc. and could be in the form of husk, dust, straws etc. The major crop residues produced in India are straws of paddy, wheat, millet, sorghum, pulses, and oil-seed crops, maize stalks and maize cobs, cotton stalks, jute sticks, sugar cane trash, mustard stalks etc. The agro-industrial residues are groundnut shells, rice husk, bagasse, cotton waste, coconut shell and coir pith (Dhingra *et al.*, 1996).

The quantity of agricultural residues produced, differs from crop to crop and is also affected by seasons, soil type, irrigation conditions etc. Production of agricultural residues is directly related to the corresponding crop production and ratio between main crop produce and residues, which varies from crop to crop and at times with the variety of the seeds in one crop itself. Thus for known amounts of crop production, it may be possible to estimate the amounts of agricultural residues produced using the residue to crop ratio (Vimal and Tyagi, 1984). It may be noted that with the improved agricultural farming techniques, the production of crops has been increasing consistently in the past three decades. Correspondingly, the availability of agricultural residues has also been changing with time (NPC, 1987). Some present and projected future data for various agro-residues at national level is given in Table 20.1 (Iyer *et al.*, 2002).

Most of the agricultural residues are not available throughout the year and become available at the time of harvest. This makes collection easy, but creates storage problems if residues have to be saved for use during other months of the year especially due to its low bulk density. The amount available depends upon harvesting time, their storage-related characteristics, the storage facility etc.

Table 20.1: Agricultural Crop Residues Production in India

Crop Residues	*(Million tones)*	
	1994	*2010*
A. Field-based residues		
Rice straw	214.35	284.99
Wheat straw	103.48	159.98
Millet stalks	19.42	17.77
Maize stalks	18.98	29.07
Cassava stalks	0.36	0.40
Cotton stalks	19.39	30.79
Soybeans (straw + pods)	12.87	34.87
Jute stalks	4.58	1.21
Sugar cane tops	68.12	117.97
Cocoa pods	0.01	0.01
Groundnut straw	19.00	23.16
Sub total–A	**480.55**	**700.22**
B. Processing based residues		
Rice husk	32.57	43.31
Rice bran	10.13	13.46
Maize cob	2.59	3.97
Maize husks	1.90	2.91
Coconut shells	0.94	1.50
Coconut husks	3.27	5.22
Groundnut husks	3.94	4.80
Sugar cane bagasse	65.84	114.04
Coffee husk	0.36	0.28
Sub total–B	**121.53**	**189.48**
Total (A+B)	**602.08**	**889.70**

Source: Biomass: Thermo-chemical characterization, 3rd Edition, 2002

Analysis to Estimate Monetary Value of Biofuels

Biomass, both woody as well as agricultural residues, have acquired considerable importance as biofuels for variety of energy enduses such as domestic cooking, industrial process heating, electrical power generation etc. and are used directly as well as in briquetted form. In order to formulate and implement long-term strategies for efficient and economic utilization of biofuels as energy source for energy conversion and utilization, it is important to estimate their monetary value for the end user. Two approaches can be used for determining the monetary value of biofuels.

1. Supply side approach or production cost method in which the contribution of each step in biofuel production, harvesting, processing, collection, transportation etc. is taken into account,

2. Demand side approach or an opportunity cost estimation based on the amount and cost of fuel(s) that is likely to be substituted by biofuels.

Fuel Woodas Feedstock for Energy Production

India is the world's largest consumer of fuel wood and about 70 per cent wood produced from forest is used as fuel wood. A large proportion of rural domestic energy demand is supplied by biomass fuels, which includes fuelwood, branches, twigs, and leaves (collectively termed as firewood). It is mostly collected as zero private cost from all types of areas, such as forests and other public lands, *panchayat* lands, roadsides, along railway lines and canals, private farmlands, and home gardens both for personal consumption and for sale. Firewood is also obtained from operations like fallings, cleanings, thinning, pruning, and pollarding from forests, horticultural areas, and farmlands, and illicit removals from forest areas and other public lands. In the scarcity hit pockets local weeds and shrubs are also collected as firewood. In India, the total round-wood production was around 188 Mt, out of which 172.5 Mt was used as fuel wood. Contribution of various sources of wood fuels in India is given in Table 20.2 (Majumdar, 1998).

Table 20.2: Land Availability Estimates for Biomass Production from Different Sources

Author/sources	*Category and land available (Mha)*	*Total (Mha)*
Planning Commission (1992); Degraded land quoted in Eighth Five Year Plan	Degraded forest-36 Degraded non-forest-94	130
Chambers *et al.* (1989); Land available for tree planting	Cultivated land-13 Strips and boundaries-2 Uncultivated, degraded land-33 Degraded forest land-36	84
Kapoor (1992); Land available for tree planting	Agricultural land-45 Forest land-28 Pasture land-7 Fallow land (long and short)–25 Urban land-1	106
Ministry of Agriculture (1992)	Forest land with <10% tree crown cover-11 Grazing land-12 Tree groves-3 Cultivable land-15 Old fallow-11 Current fallow-14	66
Sudha (1996)	Cultivable land under AEZ-26.1 Land not suitable for cultivation-13.6 Pasture land-2.9	42.6
NRSA (1995)	Forest degraded land-16.27 Wasteland-38.11 Other category-11.07	65.45

*Fuel wood*consumption grew at a rate of about 2.4 per cent per annum between 1980 and 1997 (MoEF, 1999) and is estimated at about 260 million cubic meters in 1997 as against a sustainable supply of only 52.6 million cubic meter (TERI, 1998). By 2010, the demand for fuel wood in India is projected to be 226 Mt and the total round-wood consumption is estimated to be around 283 Mt.

India's Forest Profile

According to FSI (Forest Survey of India) 2000, the recorded forest area is 76.52 Mha (million hectare), which accounts for 23.28 per cent of India's total geographical areas of 328.73 Mha. Though National Forest Policy 1952 and 1988 stipulates that, at least one-third of the land areas should be under forest or tree cover, the actual forest cover is far lower at 63.73 Mha (19.39 per cent) according to FSI, 2000. This consists of 37.74 Mha of dense forest (crown density > 40 per cent), 25.51 Mha of open forest (crown density 10-40 per cent) and 0.487 Mha of mangroves constituting 11.48 per cent, 7.76 per cent and 0.15 per cent of the geographic area respectively. In addition, there is another 5.19 Mha (1.58 per cent) of shrub cover. Map in Figure 20.1 shows the forest profile of India.

Forests meet nearly 40 per cent of the energy needs of the country of which more than 80 per cent is utilized in rural areas (MoEF, 1999). FSI (1998) estimates that the growing stock of the country is 4740.8 MCM (million cubic meters) with an annual increment of 87.62 MCM. Of this about 40 per cent (35 MCM) is estimated to be fuel wood and 60 per cent (52.62 MCM) timber. This represents an incremental annual growth of 1.36 m^3/ha/yr that compares poorly with global average annual growth of 2.1 m^3/ha/yr. Thus, it is clear that the forest resource of the country is not being managed to its full potential.

Land Availability and Fuel Wood Production Potential

The availability of land for biomass production, potential productivity and cost of production of biomass are critical issues in any assessment of biomass production for energy. These issues are particularly relevant for developing countries like India, which is experiencing high population growth rate and land shortages. The first step in analysing the bioenergy potential for electricity generation is to assess the current and future availability of land for biomass production.

Woody Biomass for Energy could be Obtained From

1. Clearing off or extraction from natural forests
2. Sustainable extraction from forests and from residues of the timber processing industry
3. From dedicated energy plantations.

Tropical forests are already under pressure, primarily to meet the land demand for food production and infrastructure. A reduction in forest cover in many tropical countries like India has led to soil erosion and loss of biodiversity, adversely affecting the ecological system. There is need to conserve primary forest for their biodiversity and their role as watershed and thus first option of clear felling of or extraction from natural forest can not be considered as a source of woody biomass for sustainable energy production. Given the fact, that the forests are already subjected to non-sustainable extraction, second option sustainable extraction from existing forest may not be feasible (Evans, 1993). Availability of timber processing waste is estimated to be low at 9.5 Mt in 1991 (Ravindranth and Hall, 1995). Thus, the only potential option for large-scale biomass supply for energy production is dedicated plantation forestry. The degraded forests, pastures and croplands available in most tropical countries like India could be used for biomass production for energy. Biomass from dedicated plantations, short rotation and intensive-culture plantations of trees is likely to be the major source of feedstock for energy (Woods and Hall, 1994).

It is estimated that there would not be any increase in land area for food production due to increased population. The food production in India has tripled between 1950 and 1990 and it is estimated that India will achieve a 40 per cent gain during next 40 years without increase in area

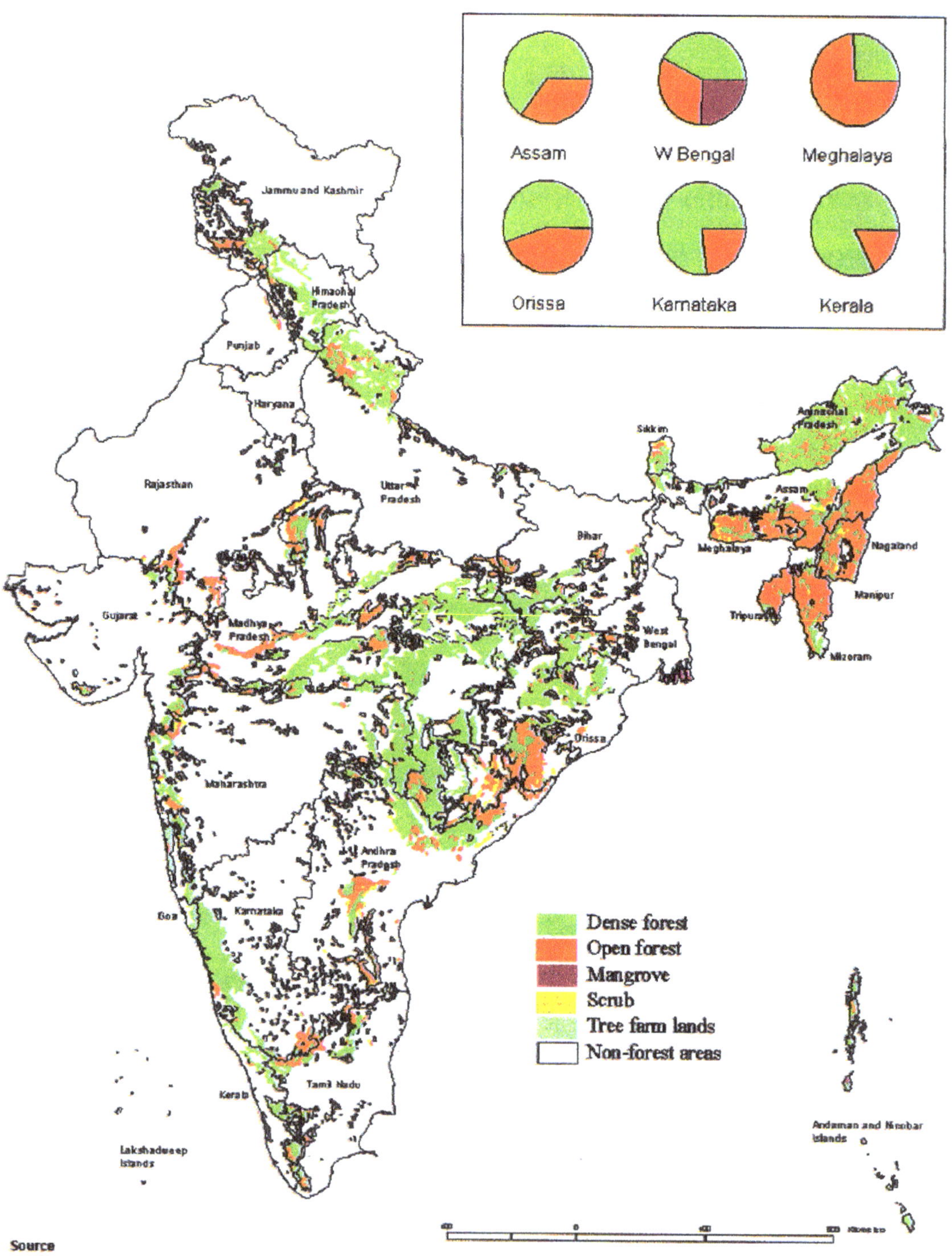

Source

Forest Survey of India, Ministry of Environment and Forests, Government of India

Figure 20.1: Natural Vegetation Map of India

under food crops (Brown and Kane, 1995). Many authors have estimated the land potentially available for biomass production in India, which is summarized in Table 20.3.

Table 20.3: Cost of Fuel Wood Production from Energy Plantation

Total Land (ha)	*Land Under Plantation (ha/yr)*	*Duration of Plantation (X years)*		
		10	*15*	*20*
100	20	820	753	731
200	40	720	678	664
300	60	687	653	642
400	80	670	641	631
500	100	660	633	624

Here, the total potential for biomass production is calculated using a conservative (lowest) estimate of 42.6 Mha of available land, according to the estimate made in AIT's report submitted to ARRPEEC-SIDA, which was done using the FAO/IIASA classification (Sudha, 1996). Here, an assessment of the area available for biomass production has been made under various agro-ecological regions (AEZ) of India (Fischer, 1994). The land under human settlements, infrastructure, protected areas, and forest areas were deducted to arrive at the actual cultivable area. Further, area estimated to be under agriculture in India by year 2010 was deducted to arrive at potentially available area for biomass production (Brown and Kane, 1995). Only 50 per cent and 25 per cent of NS (not suitable for cultivation) area and pastureland was considered for potential biomass production respectively (Ravindranath and Hall, 1995). Thus the total land available for biomass production is 42.6 Mha consisting of 26.1 Mha cultivable land under various agro-ecological regions, 13.6 Mha of NS area and 2.9 Mha of pasture land (only 50 per cent is considered for biomass production).

Following three different scenarios of plantation management have been considered for estimating the quantity of biomass that can be produced from potential available areas.

S-l: Low productivity scenario with no inputs and no improvement genetic stock of seeds, no fertilizer application and no extra water or irrigation

S-2: Moderate productivity scenario with genetic improvement of planting by way of seed and clonal selection but no supplement of fertilizer and irrigation

S-3: High productivity scenario where genetic stocks of improved plants are chosen and fertilizers used with no extra irrigation.

Utilizing corresponding productive levels in various agro-ecological zones, the range of potential biomass production under these three different input level scenarios is calculated. Under S-l (least input and at low productivity scenario) the total biomass production potential works out in the range of 214-395Mt. Under S-2, where there is an improvement of genetic stock but the fertilizer and irrigation is not used, the potential biomass production range is from 321 to 657 Mt. Under S-3, high productive scenario using genetic improved seeds and plants along with fertilizers with no irrigation; the biomass production in the range of 474-860 Mt can be achieved. Thus, depending on input levels annual biomass production from available 43 Mha areas can vary from 214 to 860 Mt.

Realizing the importance of producing biomass for fuel/energy, MNES launched a national level programme of energy plantation with substantial financial support to encourage the participation of farmers and private entrepreneurs. The MNES has also set-up Biomass Research Centers in different agro-climatic regions in India with the main goal of identifying productive tree species having short rotational cycle and to develop package of practices for their adoption in India. So far, about 70 species have been identified for the energy plantation purpose, and package of practices has been developed for 35 fast growing *fuel wood*tree species in marginal and degraded land (Singh, 1996). However, only few species such as *Eucalyptus, Acacia alurina* (Pahari kikar), *Acacia auriculiformis* (Bengali babul), *Acacia nilotica* (Babul), *Azadirachta indica* (Neem), *Butea monosperma* (Dhak), *Lucaena lucccephala* (Subabul), *Terminalia arjuna* (Arjun) etc. have been widely adopted (MNES, 2001). A financial analysis of fuel wood production has been carried out in this section to estimate the cost of fuel wood production using such plantations.

Inputs required for the energy plantation are seedlings, fertilizers, pesticides, irrigation, soil preparation, and human labour. The soil type, tree species, and agro-climatic conditions also affect the yield from plantation. In addition to this, some related factors, like tree spacing, planting time, season, rotational cycle, survival rate, lopping and harvesting etc. also play a significant role in the energy plantation. Although at present, the plantation programmes in India do not involve any significant level of soil preparation, fertilizer applications, irrigation, etc. however, these inputs may be required to improve the yield.

The cost of fuel wood production, obtained using Equation 5.8 for different plantation periods and plantation areas as per input parameters is calculated. Results obtained are summarized in Table 5.14. It may be seen that large area of plantation for longer duration is economical, as compared to plantation in smaller area for short duration. The cost of *fuel wood*production varies from Rs 624 to Rs 820 per tonne and is in close agreement with the present cost of *fuel wood*in India (Mande, 2002).

References

Brown, L.R. and Kane, H., 1995. *Reassessing the Earth's Population Carrying Capacity.* Earthscan Publishers Limited, London, 261 pp.

Chambers, R., Saxena, N.C. and Shaw, T., 1989. *To the Hands of the Poor-water and Tress,* Oxford and IBH Publication, New Delhi.

Dhingra, S., Mande, S., Kishore, V.V.N. and Joshi, V., 1996. Briquetting of Biomass: Status and Potential. In: *Proceedings of the International Workshop on Biomass Briquetting, New Delhi,* RWEDP (Regional Wood Energy Development Programme in Asia) Report No. 23, FAO (Food and Agriculture Organization) of United Nations, Bangkok, pp 24-30.

Evans, J., 1993. *Plantation Forestry in the Tropics,* Oxford Science Publications, Clarendon Press, Oxford.

FSI (Forest Survey of India), 1999. *State of Forests Report 1998,* Forest Survey of India, Dehradun, 72 pp.

FSI (Forest Survey of India), 2000. *State of Forests Report 1999,* Forest Survey of India, Dehradun, 113 pp.

Iyer, P.V.R., Rao, T.R. and Grover, P.D. (Eds), 2002. *Biomass: Thermo-chemical characterization, Revised Third Edition,* Published under MNES sponsored Gasifier Action Research Project (GARP), IIT Delhi.

Kapoor, R.P.A. and Aggarwal, 1992. *The Price of Forests,* (Eds.) Kapoor, R.P.A. and Aggarwal Center for Science and Environment, New Delhi, 173 pp.

Mande, S., 2002. *Ph.D. Thesis,* Dr. B.A. Marathwada University, Aurangabad, MS, India.

Majumdar, N.B., 1998. Integrating wood energy subjects into forest management curriculum. In: *Proceedings of National Experts Consultation on Wood Energy in Curricula of Forestry Training and Education in India,* Indira Gandhi National Forest Academy, Dehradun.

MNES (Ministry of Non-Conventional Energy Sources), 2001. *MNES Annual Report 2000–2001*, Govt. of India, New Delhi.

MoA (Ministry of Agriculture), 1992. *Land use statistics,* Ministry of Agriculture, Government of India, New Delhi

MoEF (Ministry of Environment and Forests), 1999. *National Forestry Action Programmes-India* (*Vol. 1*), Ministry of Environment and Forests, Government of India, New Delhi, 179 pp.

NPC (National Productivity Council), 1987. *Improvement on agricultural residues and ago by-products utilization,* A technical report, National Productivity Council, New Delhi.

NRSA (National Remote Sensing Agency), 1995. *Report on area statistics of land use/land cover generated using remote sensing techniques,* Department of Space, Hyderabad.

Planning Commission, 1992. *Planning Commission Report of the Eighth Five Year Plan,* Government of India, New Delhi.

Ravindranath, N.H. and Hall, D.O., 1995. *Biomass Energy and Environment: A Developing Country Perspective from India,* Oxford University Press, Oxford.

Sudha, P., 1996. *Plantation Forestry: Land Availability and Biomass Production Potential in Asia,* Report submitted to ARRPEEC (Asian Regional Research Programme in Energy Environment and Climate), SIDA (Swedish International Development Cooperation Agency) Energy Programme, AIT (Asian Institute of Technology), Bangkok, Thailand.

Vimal, O.P. and Tyagi, P.D., 1984. *Energy from Biomass.* Agricole Publishing Academy, New Delhi.

Sustainable Environmental Management *Pages* **198–202**
Edited by: **Dr. L.V. Gangawane & Dr. V.C. Khilare**
Published by: **DAYA PUBLISHING HOUSE**

Chapter 21
Mitosporic Fungi and their Importance

C. Manoharachary and B. Bhadraiah

Department of Botany, Osmania University, Hyderabad–500 007, A.P. India

ABSTRACT

Deuteromycetous fungi are occurred on diversified habitats and is a fascinating group.Their identification is possible through morpho-taxanomic criteria, conidial ontogeny and molecular techniques such as RFLP, RAPD and PCR usage has become an additional tool in fungal research. It is also important to mention that several members are of much significance in biotechnology.

Keywords: *Deuteromycetous fungi, Diversified habitats, Conidial ontogeny, Morpho-taxanomic criteria, Conidia/Spores, Molecular techniques and biotechnology.*

Introduction

Fungi-imperfecti is a fascinating group of fungi colonizing diversified habitats. They colnize, multiply and survive in/on water, litter, soil, excreta, seed, fruits and vegetables besides floating in the air and occurring as parasites on plants and as saprophytes on diversified habitats.

Mitosporic fungi are the asexual phases of Ascomycotina and Basidiomycotina. Deuteromycotina constitute an artificial group reproducing only through asexual spores or conidia produced by mitotic division. These fungi can live and multiply seemingly without a sexual phase. Technically these asexual or imperfect forms are known as anamorphs.

In fungi-imperfecti, the fungi are characterized in possessing asexual multiplication by mitosis. The multiplication occurs by mitotic spores or conidia. Conidia are formed on hyphae called conidiophore. Conidiophore is formed on a hypha. The conidiophore(s) are single or may be grouped into various structures (Conidiomata, Acervulus, Pycnidia, Synnema, Sporodochia). The development of conidium is called conidiogenesis and this helps in the segregation of genera. (Cole and Samson,

1979). The conidiogenesis proposed by Hughes (1953) enable us to classify and identify them. Mycelia sterilia or Agonomycetes do not produce conidia but multiply through hyphal fragments and sclerotia, the vegetative fruit bodies.

In a life cycle of a fungus, a phase of sexual reproduction (teleomorph) can co-exist with a phase of asexual multiplication (anamorph). In such circumstances the term holomorph is used to designate the fungus considered in all its phases. For example *Stemphylium botryosum* is the anamorph of *Pleospora herbarum,* an ascigenous teleomorph.

The sexual forms are massive while the asexual forms are small and even microscopic. As per Muller (1981), the asexual phases (anamorphs) of Asconmycetes care haploid and that of Basidiomycetes can be haploid or dikaryotic. Even if dolipore septum and clamp connections are observed and such fungi can also be grouped under Basidiomycotina.

Taxonomy and Classification

The taxonomy of Deuteromycotina is artificial. Several mycologists have proposed classifications. Saccardo's (1880, 1884) classification was based on grouping of conidial apparatus, the colour, form and septation of the conidia produced. Hughes (1953) classified conidial fungi into eight groups *viz.*, Blastospores, terminal spores, solitary chlamydospores and successive ones on annellophores, phialospores, meristem arthrospores, porospores, arthrospores and on basauxic conidiophores. Ellis (1971, 1976) classification broadly divides the conidia into thallic and blastic based on conidial ontogeny. They are further divided into holothallic, enterothallic, holoblastic and enteroblastic and these are further divided into subsections. Kiffer and Morelet (1999) classified the conidial fungi into thalloconidia which is further divided into chlamydospores, Arthroconidia, Meristemarthroconidia and Holothallic conidia. The second group includes Blastosporae which is further divided into Aeroblastospores, Sympodulospores, Botryblastospores, Porospores, Aleuriospores, Holoblastic, Annellospores, Phialospores and Anellospores. Mycelia sterilia or Agonomycetes are such group of fungi that do not produce conidia. Though several classifications were proposed, still lacunae and gaps exist in understanding these fungi and some difficulties do arise in classifying conidial fungi and other fungi-imperfecti. Hawksworth *et al.* (1995) have recognized Mastigomycotina, Zygomycotina, Ascomycotina and Basidiomycotina only. Deuteromycotina, which is an artificial single group has not been considered due to the fact that conidial fungi are nothing but the asexual stages of Ascomycotina or Basidioycotina.

Hughes (1953) proposed a coherent classification and divided Deuteromycotina into eight sections. Conidial ontogeny, relationship between the conidium and conidiogenous cells, evolution of the conidium and conidiophore and the mode of secession of spores form the base for modern classification. The light microscope is insufficient for the study of these phenomena. Electron Microscopic studies will further strengthen the morphotaxonomic criteria. Hennebert and Sutton (1994) proposed 20 parameters. Ellis (1971, 1976) has simplified the taxonomic concepts and proposed a classification based on conidial ontogeny. Subramanian (1962) has also proposed a classification of Hyphomycetes. Sutton (1980) has brought out a monograph on Coelomycetes and classified several fungi based on the morphotaxonomic criteria of asexual fruithbodies and conidial ontogeny. Tubaki (1963) has also made exhaustive taxonomic study of Hyphomycetes.

The fungi imperfecti may occur in water, wood, coprophiloous habitats, on insects, grains, seed, on lichens, nematodes, keratin substrates, in soil, as parasites on humans and animals besides being parasites on vascular plants.

Criteria

Morphotaxonomic criteria such as hyphal characters, septation, nuclear condition, aggregation of hyphae, coloration and branching help in the identification of fungi. Conidial development, septation, coloration, ornamentation, shape, size and other parameters including the type of asexual fructification or conidiomata aid in the identification of fungal genera and species.

Methodology

Sampling, slide preparation, mounting in appropriate medium, microscopic observation, biometry, microphotography, camera lucida drawing and observation of conidial ontogeny in culture on slide are the important steps to be followed. In complex and critical cases, it becomes essential to use a Scanning Electron Microscope (SEM) and Transmission Electron Microscopic (TEM) characters of the cell wall, septa, ornamentation and development of the conidium.

It is well known that fungal nuclei are small. In recent times the molecular techniques such as RFLP, RAPD and PCR usage has become an additional tool in fungal research.

These techniques will help in resolving taxonomic problems especially at the species and intra species level including strain differentiation. Ribosomal analysis (18s and 26s) can be used to ascertain the taxonomic position of fungi with uncertainity. Around 90 fungal genera of Deuteromycotina are worked out using molecular tools. Immunological techniques (ELISA), physiological needs, chemotaxonomy and other characters assist the mycologists studying taxa that are difficult to identify.

Anamorph and Teleomorphs

Anamorph indicates the asexual phase of fungus. For example *Aspergillus* (Conidial status) state is the anamorphic state of *Emericella* and *Eurotium*. Similarly *Penicillium* state is the asexual phase of *Talaromyces*. The teleomorph is the perfect state of a fungus belonging to Ascomycotina or Basidiomycotina. Holomorph represents complete fungus represented both by anamorphic state and teleomorph. Hughes (1979) has coined another term namely synanamorph, which means that the fungus can produce two different anamorphs. (*e.g.*, *Venturia inacqualis* teleomorph with two anamorphs– *Sclerophoma pythiophila* and *Hormonema dematioides*). Approximately 6500 species are reported in this group.

Biodiversity

The fungi include moulds and yeasts including edible mushrooms. It is estimated that around 15,000,00 fungal species exist in nature growing on diversified habitats. All over 76,000 fungal species are known in the literature which are reported around the world. Around 27,000 fungal species are recorded from India indicating that 1/3 global diversity of fungi exists in this country. 5-7 per cent fungi remain growing in culture. Around 50 per cent fungi remain as hidden wealth. Members of Ascomycotina and Basidiomycotina including their asexual phases (Deuteromycotina) form bulk of the fungal wealth followed by Zygomycotina and Mastigomycotina.

Biotechnological Importance

Fungi-imperfecti contribute to the biodegradation and recycling of organic matter being cellulolytic, lignolytic, pectiolytic etc., It is an important group in fungal succession that cause degradation. Members of Deuteromycotina play a role in bioremediation, waste treatment and pesticide degradation.

Fungi such as *Penicillium* are used in cheese production. *Botrytis cinera* which is a well known pathogen also helps in the production of sweet wines. A number species of *Aspergillus* and *Penicillium* produce organic acids, enzymes and pigments. Bisol, a biofertilizer is a product of *Penicillium chrysogenum*. Fungi like *Aspergillus niger* and *Penicillium* spp. are known for heavy metal tolerance and extraction of uranium and radium from atomic industrial waste water. Since the discovery of Pencillin from *Penicillium notatum* and *P. chrysogenum*, a revolution has taken place in biomedicine through the discovery of griseofulvin (*P. griseofulvum*) cephalosporin (*Cephalosporium*) and cyclosporines, the immunoregulators (*Tolypocladium*). Around 40 per cent of enzymes including the commercialized enzymes are mostly from the members of Deuteromycotina.

Fungi such as *Beauvaria bassiana* is a well known entomopathogen employed in the control of coffee berry borer in Karnataka. CASST, a product of *Alternaria cassiae* is well known weed killer. Nematode killing fungi which occur abundantly in nature are represented by *Arthrobotrys. Dactylaria,* and *Monacrosporium. Dactylaria Trichodema viride, T. harzianum, Verticillium biguttatum, Scytaladium* and others are the well known fungal antagonists and are used as biocontrol agents against many pathogenic fungi. *Colletotrichum gloeosporiodes* and *C.truncatum* are also used as mycoherbicides. Single cell protein namely quorn is secreted by *Fusarium graminearum*.

Besides the uses mentioned above some members of Deuteromycotina cause harmful effects such as *Aspergillosis* (*Aspergillus*), histoplasmosis (*Histoplasma*) *onchomycosis* (*Scytalidium, Graphium*), ring worm and allergies besides causing serious plant diseases including leaf spot of rice, the epidemic by *Helminthosporium oryzae*. Aflatoxins and some mycotoxins which are health hazardous and carcinogenic are also elaborated by *Aspergillus, Penicillium* and others.

Conclusions

Deuteromycetous fungi is a fascinating group of fungi forming an artificial group or dustbin group of fungi. Their identification is possible through morphotaxanomic criteria, conidial ontogeny and molecular techniques. Several complexities exist in the identification and classification of fungi imperfecti. It is also important to mention that several members are of much significance in biotechnology. These fungi occur on diversified habitats and form asexual phases of Ascomycotina and Basidiomycotina.

References

Cole, G.T and Samson, R.A., 1979. *Patterns of Development in Conidial Fungi*. Pitman, London, 190 p.

Ellis, M.B., 1976. *More Dematiaceous Hyphomycetes*. CMI, Kew, 608D.

Ellis, M.B., 1971. *Dematiaceous Hyphomycetes* CMI, Kew, 507 p.

Hawksworth, D.L., Kirk, P.M., Sutton, B.C. and Pegler, D.N., 1995. *Ainsworth and Bisby's Dictionary of the Fungi* 8th edition, CAB Int. Walling ford, pp. 616.

Hennebert, G.L. and Sutton, B.C., 1994. Unitary parameters in conidiogenesis. In: *Ascomycetic Systematics: Problems and Perspectives in the Ninties,* (Ed.) Hawksworth, Plenum Press, N.Y., pp. 65–76.

Hughes, S.J., 1953. *Can. J. Bot.*, 31: 577–659.

Kiffer, E. and Morelet, M., 1999. *The Deuteromycetes*. Sci. Publication, USA, 273p.

Muller, E., 1981. Relations between conidial anamorphs and their teleomorphs. In: *Biology of Conidial Fungi*, Vol. I. (Eds.) G.T. Cole and B.W. Kendrick. Academic Press, New York, p. 43–84.

Saccardo, P.A., 1880. *Mycologicum*, 2: 1–38.

Saccardo, P.A., 1884. Sylloge fungorum, Omnium hucusque cognitorum, Vol. 3, *Pavia*, 860 p.

Subramanian, C.V., 1962. *Curr. Sci.*, 31: 409–411.

Sutton, B.C., 1980. *The Coelomycetes*, CMI, Ken, 696 p.

Tubaki, K., 1963. *Ann. Rep. Inst. Ferm, Osalca,* 1: 25–54.

Sustainable Environmental Management *Pages* **203–207**
Edited by: **Dr. L.V. Gangawane & Dr. V.C. Khilare**
Published by: **DAYA PUBLISHING HOUSE**

Chapter 22

Environment and Its Global Issues

B.N. Pande

Department of Environmental Science, Dr. Babasaheb Ambedkar Marathwada University, Aurangabad – 431 004 MS, India

ABSTRACT

Environmental legislation is evolved to protect the environment health and the resources. The importance of environmental education, awareness about problems are linked with sustainable development. The degradation of our environment is linked to continuing problems of pollution, loss of forests, solid waste disposal, and issues related to economic productivity. The increasing levels of global warming, the depletion of ozone layer and a serious loss of biodiversity have also pointed out in the United Nation Conference on Environment and Development held in Rio de Juneiro in 1992 and the World Summit on Sustainable Development at Zoharbex in 2002. In view of the above this article stated that environmental management has become a part of the human life therefore managing environmental hazards at global level and preventing possible disasters has become an urgent need.

Keywords: *Population explosion, Land degradation, Environmental pollution, Environmental education, Public awareness.*

Introduction

The environment is a physical and social science that integrates knowledge from wide range of disciplines. Many environmental problems we are presently facing are mainly due to (a) over population, (b) wasteful use of resources, (c) destruction and degradation of wild life habitats, (d) depletion and contamination of surface and ground water, (e) depletion of renewable fuels or minerals, (f) deforestation, (g) loss of biodiversity due to species extinction etc. which are interconnected. Along with the rapid population growth the ecological demands are also growing exponentially. As the population is increasing its resource is also increasing exponentially. There is growing evidence that we are decreasing Earth's ability to sustain life, including our own species. Each year more forests, grasslands, and wetlands disappear and some deserts grow larger. Vital topsoil is washed or blown away from farmland and cleared forests, clogging streams, lakes and reservoirs with sediment. Many types of grassland have been overgrazed and fisheries over-harvested.

Present Environmental Issues

The main environmental issues today are wide-ranging and all encompassing: deforestation, biodiversity, soil erosion, and climate change, pesticide build up, industrial and municipal pollution. All have been caused by anthropogenic interventions in the natural and self-sustaining cycles. The human ambition for a higher living standard has hammered 'consumerism', ignoring the lasting and adverse impact of this hammer on earth's limited resources. All these problems could be categorized as under:

Population Explosion

The population of the world is growing by 92 million people adding annually, roughly equal to another Mexico every year. Of this total, 88 million are being added in the developing world due to which ecological demands are exponentially growing resulting in the greater degradation of natural resources in this region.

However, the population increases in poor countries and gluttony patterns of consumption and mass scale production in industrialized countries has upset the balance between population and resources leading to deterioration of the environment. The economic feature of developed and developing countries is not dependent only on the laws of trade and finance, but on environmentally sound development activities. To be beneficial, such development should deal not only with the relations between man and man, but also with the relations between man and nature. And incidentally this formed the basis of UN programmes such as IBP, MAB and UNEP.

Land Degradation

The most urgent environmental problems facing the developing countries are those relating to the use of land. Of particular concern are deforestation, devegetation and desertification. All these are outcome results of many interrelated activities. Overgrazing and land clearing and commercial logging rob the soil of its cover and fertility and reduce agricultural yields. They also increasing flooding, sedimentation and silting up of dams and reservoirs. The problems are worse where the inadequacy of fuel wood compels rural families to burn agricultural residues and dung, thus depriving the source of soil fertility and accelerating erosion. Such deforestation deprives people of other essential products for their survival and threatens wild life reserves and fragile ecosystems. Resources that are in principle renewable are rapidly disappearing as a result of over exploitation. Every year nearly 6,000 million tonnes of topsoil is lost by erosion, which contains over 5 million tonnes of critical plant nutrients, whose replacement value via fertilizer has been calculated to be 2,500 crores of Rupees. Nearly one third *i.e.* over a hundred million hectares of India's land is eroded, salined or water logged.

Environmental Pollution

1. Environmental pollution is a great and growing concern in developing countries, and is typically associated with industrialization. Air and water pollution in metropolitan cities like Kolkota, Mumbai, Delhi etc. are growing to critical level unsanitary living conditions due to inadequate water supply and improper waste disposal. In world, underdeveloped and developing countries, it is observed that water polluted by pesticides, by municipal and industrial wastes, which subsequently cause diseases, malnutrition and death.

2. Agriculture also contributes to pollution wherein the use of chemical fertilizers, pesticides etc. for example, pollute the environment in various ways.
3. *Air pollution*: Air pollution in some Indian cities is rising alarmingly. No major Indian city is free from the threat of chemical accidents or pollution hazards. In Delhi and Mumbai, 1,000 hazardous units and industries have been identified. Nuclear hazards may pose serious threats to people in the vicinity of Nuclear power plants.
4. *Acid rain*: Over the past few years' air pollution has been seen increasing as a regional or global problem, not a local. Incidence of acid rain may occur on earth thousands of miles away from the places of emission of sulphur dioxide and nitrogen oxide. Thus the cloud generated in the developed world may rain in the territory of the developing part of the world. The destruction of forests in Northern Europe is perhaps the clearest consequence.
5. *Ozone depletion*: Ozone is an important chemical species present in stratosphere at an altitude of about 30 km. This ozone layer present in stratosphere acts as a protective shield for the life on the earth. It strongly adsorbs UV rays from the sun in the region 220-330 nm and thereby protects the life on the earth from severe radiation damage, such as DNA mutation and skin cancer. The detection of the "ozone hole" over Antarctica in 1985 attracted the attention of scientific community in the world. Depletion of ozone layer is found to be due to man made chlorofluorocarbons (CFC's), which are used as coolants in refrigerators, air conditioners, and propellants in aerosol sprays and in plastic forms, such as "Thermocole or Styrofoam".
6. *Green House Effect*: the term J. Fourier first coined "Green House Effect" in 1985. Green House Effect means "the excessive presence of those gases (CO_2, CH_4, N_2O, CFCs) blocked in the infra-red radiation from the earth's surface to the atmosphere leading to an increase in temperature, which in turn would make life difficult on earth". The CO_2 is the most common and important "Green House Gas". Volcanoes, oceans, decaying plants as well as human activities, such as deforestation, combustion of fossil fuels and automobile exhausts, release CO_2.
7. *Water pollution*: It is estimated that 70 per cent of the surface water resources are polluted and that in large stretches of major rivers, water is not even fit for bathing. Polluted water is the source of most of the diseases that affects and even kills millions of Indians every year. Reservoirs of most dams in India are silting up at rates much higher than those anticipated, thereby reducing their expected life. The area prone to flooding has tripped since 1960, from 19 to 50 million hectares. Over half of the districts in India have suffered from other floods and droughts, between 1983 and 1986.
8. *Deforestation*: Forests are vanishing at a rate of some 17 million hectares/year. A minimum of 140 plant and animal species are condemned to extinction each day. 58.8 per cent of Indian forests have no or inadequate regeneration, mainly because of degradation. Consequently the productivity of India's forests is very low averaging 0.7 cu.m/hectare/year. Livestock currently requires 932 million tonnes green and 780 million tonnes dry fodder annually. Over 90 million livestock graze in forests in India. There is an urgent need to grow fodder to protect the cattle wealth, the backbone of rural economy. Fuel wood consumption is estimated at 235 million tonnes from forests. While annual production is only 90 million tonnes *i.e.* 40 million tonnes from forests and the remaining from agro wastes. The balance comes from forest capital stock.

Initiative of Government in Environmental Management

Ministry of Environment and Forests, Government of India and each state government have been deeply engaged in the protection, preservation and conservation of environment. Several possible policies, programmes and implementation strategies have been evolved from time to time by these governmental organizations for better environmental management. The programmes and activities are directed towards all aspects of environment and ecology with the purpose of protection and conservation of environment and ecology, such as conservation and restoration of ecological heritage, propagation of ecofriendly technologies for sustainable development synthesizing ecotechnology and biotechnology, involving NGO's and women in all environmental related activities and helping government to formulate laws and regulations.

In addition to "Water Prevention and Control of Pollution Act, 1974, and National Environmental Appellate Authority Act, 1997", about two dozen acts have been enacted in India exclusively for environmental pollution control. Nevertheless there are still some loopholes in these laws of environment, which make for the lawyer to support the cases of polluter. However, these laws if implemented properly can contribute a large extent to protect, conserve the nature and natural resources and take care of the environment.

Environmental Education

Environmental education is an integral process, which deals with man's interrelationship with his natural and man made surroundings, including the relation of population growth, pollution, resource allocation and depletion, conservation, technology and urban and rural planning to the total human environment. Environmental education is a study of the factors influencing ecosystems, mental and physical health, living and working conditions, decaying cities and population pressures.

The Hon'ble Supreme Court heard a public interest litigation case on the urgent need to impart environmental education for all the students–the future citizens of the country–to enable them to protect the public health, prevent pollution and save all forms of life as envisaged under article 51-A(g) of the constitution which has directed all national organizations like UGC which regulates higher education in India and state governments to offer environmental education in the curricula of various degree courses.

Need for Public Awareness?

About 50 per cent of our natural water resources such as mighty and holy rivers *e.g.* Ganga, Yamuna, Cauvery, Godavari, Krishna and their tributaries are highly polluted. Industrial pollution in almost all big industrial cities in India has seriously affected public health. Air pollution from industries and automobiles is posing a serious threat to the health and welfare of the man. Automobile pollution forms 60 per cent of the air pollution in urban areas. Delhi, Kolkota, Ahmadabad, Mumbai, Kanpur etc. are some of the highly polluted cities in India. Around three-fourth of the traffic police in our country are reported to face serious problems of respiratory disorders and cancer due to continued exposure to automobile and industrial pollution. The pollution of air and water due to industries, automobiles, chemical pesticides, fertilizers enormously used in agriculture have resulted in food adulteration and the consequential reduction in natural powers of immunity. The fast growing incidence of cancer, AIDS and other debilitating sicknesses are posing threats to the survival of man.

At many levels in many of the developing countries including India, the problem of pollution is increasing day by day. Environmental deterioration is also fastly increasing due to developmental activities, which are promoted without considering the environmental guidelines and principles of sustainable development. Concern for environmental protection is again stressed in the policies of economic development. Successive plan documents have also given recognition to this important aspect. Inspite of all these aspects, the environmental problems in India have been rising almost in direct proportion to industrialization, urbanization and population growth.

It has been realized that although there are number of acts and legislations which relate directly or indirectly to the environmental protection, but the implementation of the laws will be more effective only with the public support and participation for which an awareness and clear understanding of key environmental problems is highly essential. One has to learn how to live with nature.

Sustainable Environmental Management *Pages* **208–215**
Edited by: **Dr. L.V. Gangawane & Dr. V.C. Khilare**
Published by: **DAYA PUBLISHING HOUSE**

Chapter 23

Nutrient Reduction through Utilization of Bio-indicators: A Case Study of a Tropical Lake

S. Pani, A. Dubey and S.M. Misra

Lake Conservation Authority of Madhya Pradesh, E-5 Arera Colony, Kachner, Paryavaran Parishar, Bhopal, MP, India

ABSTRACT

Selective macrophytes and mollusks were analyzed to evaluate their role in nutrient enrichment in a degraded wetland ecosystem. Based on the analysis of macro and micronutrients, remedial measures for conservation and management of the water body have been suggested.

Keywords: *Heavy metals, Toxicity, Bio-indicators, Conservation*

Introduction

In last few decades most of the tropical water bodies have become polluted due to increase input of nutrients from the surrounding areas through various anthropogenic activities. More over the water bodies particularly in the urban areas usually receive huge quantities of sewage, which increases concentration of certain groups of heavy metal in water. Most of the available nutrients when introduced in the aquatic environment undergo further interactions, thereby accelerating the process of eutrophication. The fate of most of the nutrients is mainly influenced by complex relationship between various abiotic and biotic components. Among the biotic community macrophytes and macro benthos together forms an important component of the lake ecosystem. Macrophyte besides being a significant contributor of primary productivity also acts as a habitat for number of organisms, gastropods being one of them. These two groups not only form important components of the lake biota but also play significant role in nutrients recycling.Thus in order to determine the concentration of some major nutrients including heavy metals, selective groups of macrophytes and mollusca (Lemnae sp) were collected from Upper Lake and analyzed in the laboratory to evaluate their contribution in terms of nutrient enrichment in the lake.

Study Area

The Upper Lake, Bhopal, which is located between 23°12′-23°16′-N, latitude and 77°18′-77°23′ E longitude, is one of the major sources of drinking water for the people of Bhopal and contributes 40 per cent of the total potable water supply. However the rapid growth of the town and human settlements around the lake in last few decades has significantly affected the water quality of the lake. On account of anthropogenic activities, inflow of untreated sewage and agriculture waste, the quality of lake water deteriorated to a considerable extent (Pani *et al.*, 2002). The washing of clothes, bathing and Idol immersion activities are other contributing factors for enhancing nutrient level leading towards the pollution of the lake.

Material and Methods

During the period of investigation *i.e.* 2003-2004 macrophytes were collected from identified sampling points (Figure 22.1) of Upper Lake through quadrate methods. The samples were then collected in Polythene bags and brought to the laboratory. After sun drying, samples were processed for acid digestion. Digested samples were then analyzed through Ion Chromatograph (Dionex 500)

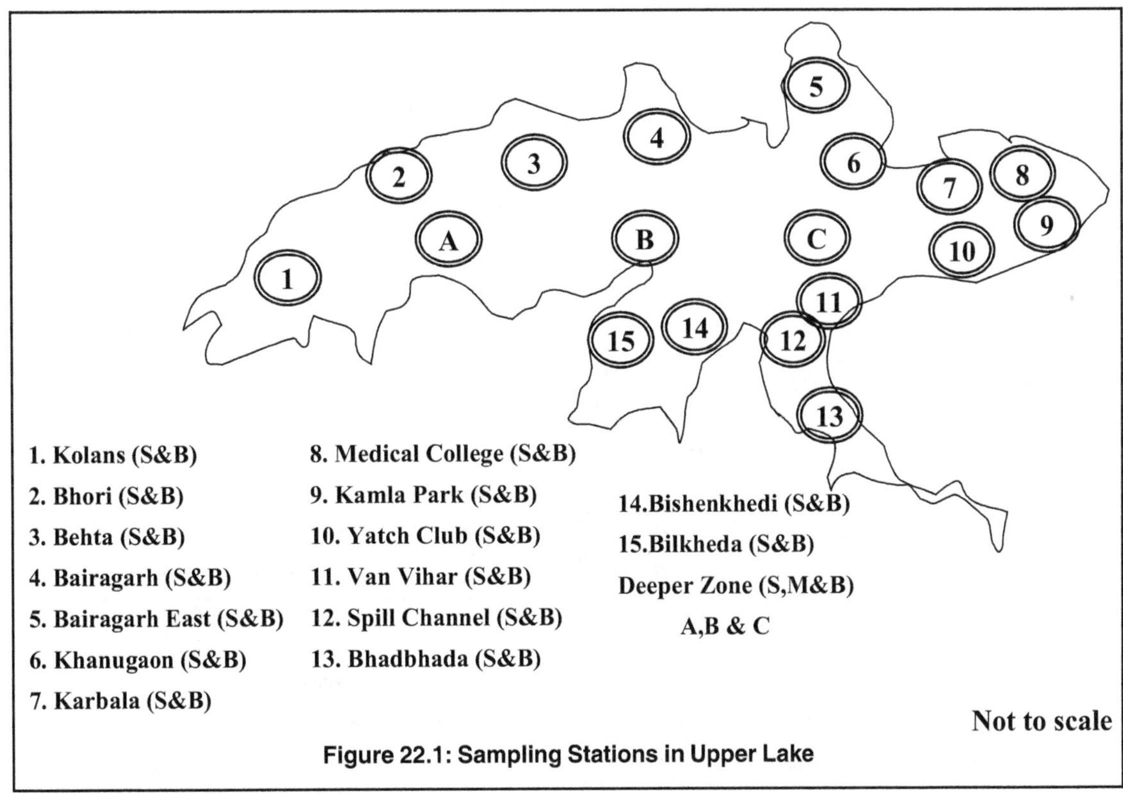

Figure 22.1: Sampling Stations in Upper Lake

for analysis of different cations and anions. For heavy metal determination, sub-samples of different macrophytes were analyzed through AAS (Parkin Elmer Analyst 100). Results were expressed in gm/kg of weight.

For analysis of heavy metal in mollusca, sediment samples from the bottom at different stations were collected by using Peterson grab mud sampler (area 270 cm^2) and were transferred to Polythene bags. The samples were then washed thoroughly in distilled water and dried in oven at 65°C until constant weight was achieved. Dried tissue samples were liquefied for analysis in Atomic Absorption Spectrophotometer (AAS) by nitric acid digestion in closed vessels. The digested samples were aspirated in the AAS (Parkin Elmer AAnalyst 100 with HGA 800) to determine the concentration for Mn, Pb, Cu, Cr, Hg, Cd, Ni and Zn. Background corrections were applied to all the analysis.

Results

Nutrient Concentration in Various Macrophytes

Concentration of selective nutrients in different macrophytes is shown in Figure 22.2. Total seven elements *viz.* NO_3, PO_4, K, Li, Ca, Mg, Na were analyzed in different macrophytes collected from the

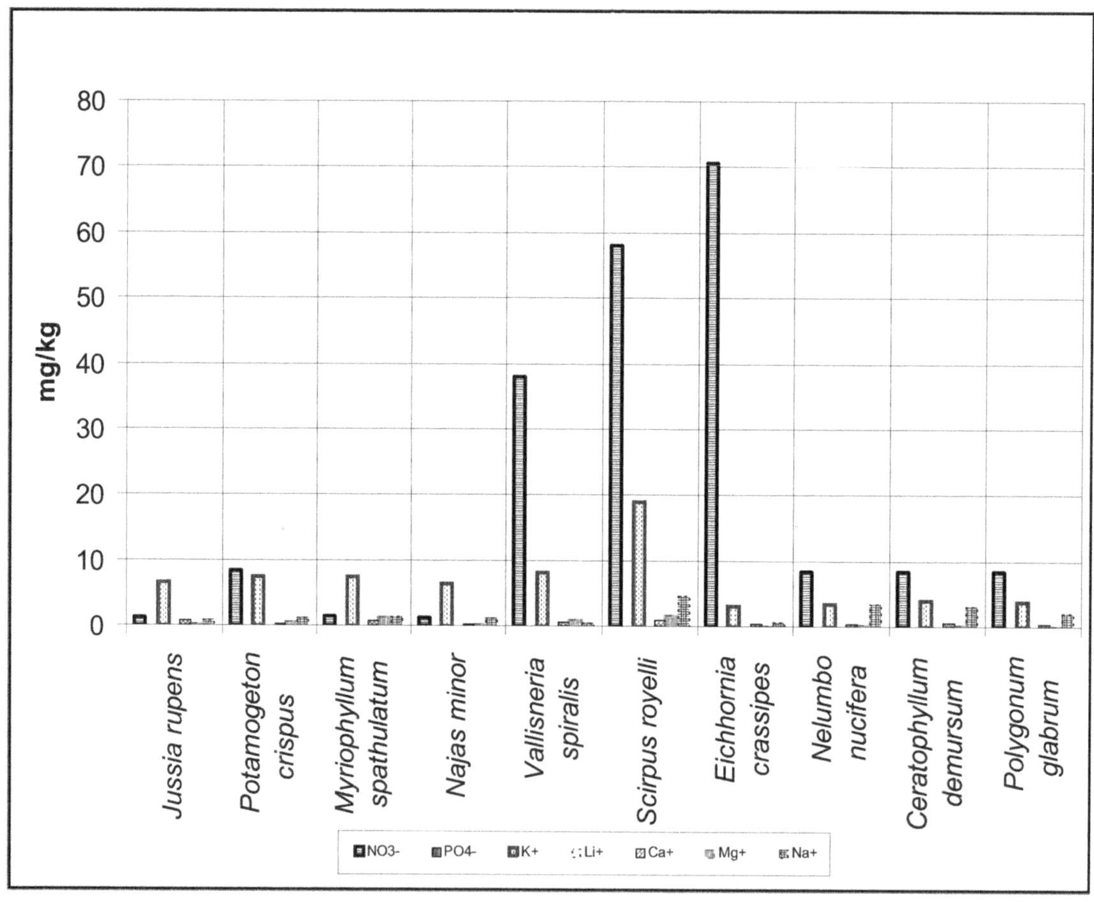

Figure 22.2: Nutrient Concentration in Different Macrophytes of Upper Lake

Upper lake during 2001-2002. Species wise concentration of various anions and cations are depicted below:

Jussia rupens

In *Jussia rupens*, potassium depicted maximum concentration than remaining nutrients. Potassium was followed by nitrate, which is the next most important nutrient available in this plant.

Potamogeton crispus

In *Potamogeton crispus* nitrate was the most available nutrient (8.28mg/kg) followed by potassium.

Myriophyllum spathulatum

In *Myriophyllum* concentration of potassium recorded more than other elements. 4.*Najas minor:* In *Najas minor* also potassium showed higher concentration than other elements.

Vallisneria spiralis

In this species concentration of nitrate was found to be very high (38mg/kg) than other nutrients. In general most of the nutrients of present investigation depicted higher values in this plant.

Scirpus royeli

In *Scirpus royeli* concentration of nitrate was again found to be very high (58mg/kg) than rest of the elements. Potassium was the second most important contributor of nutrients.

Eichhornia crassipes

In *Eichhornia crassipes* the availability of nitrate was maximum (70.6mg/kg) when compared with that of other plants.

Nelumbo nucifera

In *Nelumbo nucifera* also higher values of nitrate were observed than rest of the elements.

Ceratophyllum demersum

In *Ceratophyllum demersum* nitrate was the most important nutrient followed by calcium.

Polygonum glabrum

In *Polygonum* also higher values of nitrate were observed than rest of the elements.

Heavy Metal Concentration in Various Macrophytes

Concentration of selective heavy metals in different macrophytes is shown in Figure 22.3. Total six heavy metals *viz.* Cu, Cr, Mn, Pb, Cd, Zn were analyzed in different macrophytes collected from the lake. Concentrations of these heavy metals in different species are discussed below;

Jussia rupens

In *Jussia rupens*, manganese depicted maximum concentration than rest of the heavy metals.

Potamogeton crispus

In *Potamogeton crispus* availability of Zn was maximum (61.7mg/kg). Zn was followed by copper (25.3 mg/kg).

Myriophyllum spathulatum

In *Myriophyllum* concentration, zinc recorded comparatively higher values than other heavy metals.

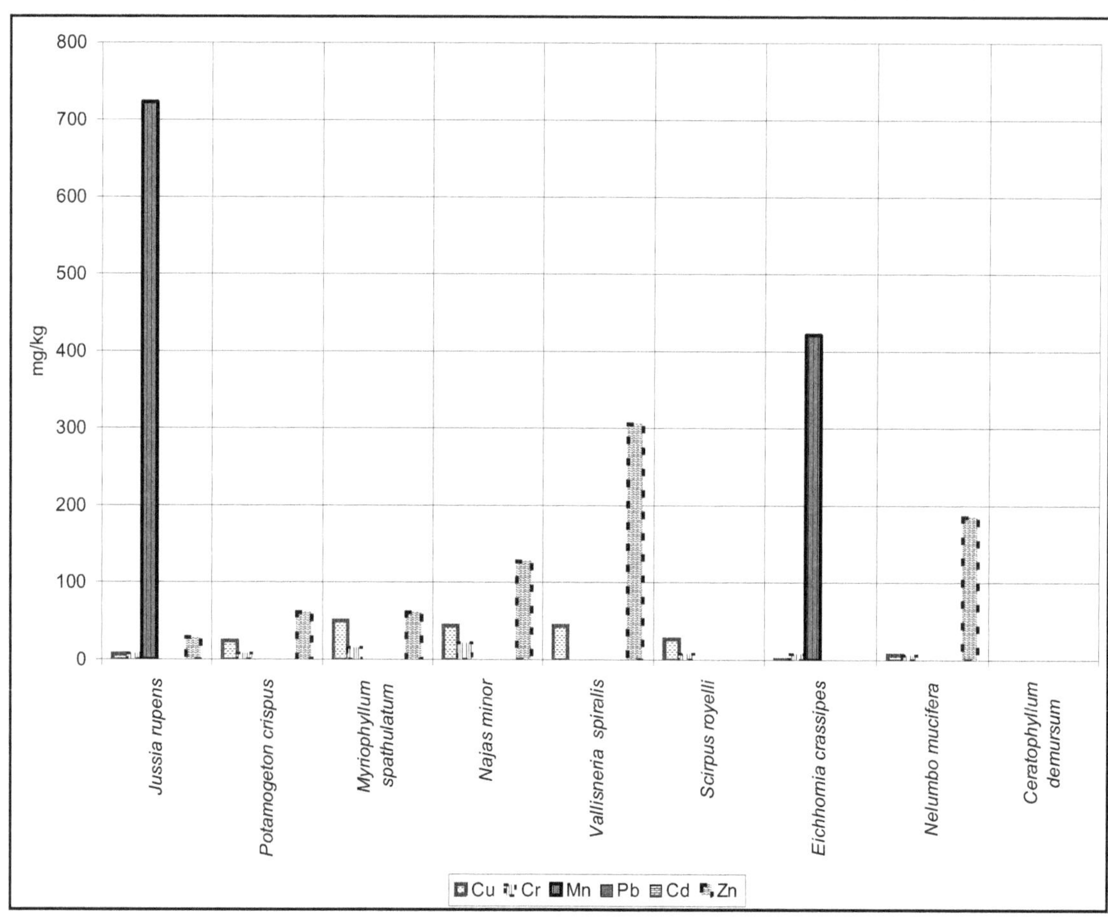

Figure 22.3: Heavy Metal Concentration in Various Macrophytes of Upper Lake

Najas minor

In *Najas*, concentration of Zn was found to be very high, followed by copper (45mg/kg) and chromium (21.95mg/kg)

Vallisneria spiralis

In *Vallisneria* concentration of Zinc was found to be very high (306.4 mg/kg) than other nutrients.

Scirpus royeli

In *Scirpus royeli* concentration of copper (28.2mg/kg) was again found to be very high than rest of the metals.

Eichhornia crassipes

In *Eichhornia crassipes* the availability of manganese was maximum (421.1mg/kg) than rest of the heavy metals.

Nelumbo nucifera

Nelumbo nucifera also depicted higher values of zinc (186.1) than rest of the heavy metals.

Ceratophyllum demersum

In *Ceratophyllum demersum* during present investigation zinc was found in higher concentration.

Polygonum glabrum

In Polygonum also higher values of zinc were observed than rest of the heavy metals.

Heavy Metal Concentration in Mollusca

Concentration of various heavy metals in Mollusca is shown in Figure 22.4. Observations recorded for various heavy metals are as follows:

Manganese

The range of manganese concentrations in different samples of mollusca varied from 9.8 mg/kg to 79.89mg/kg.

Lead

In all the samples lead was found below detectable limit.

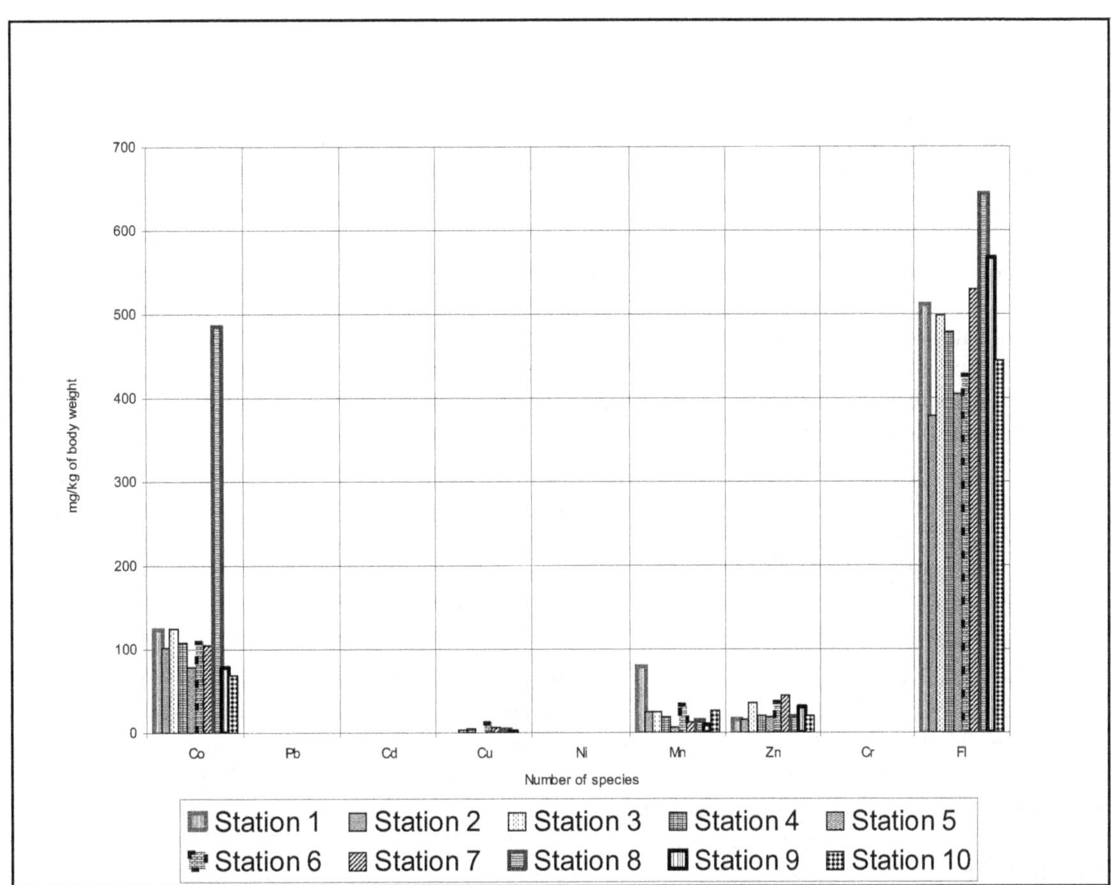

Figure 22.4: Variation in Some Microelements in Molluscan (*Lemnae* sp.) Samples

Copper

Concentration of copper in different samples of mollusca varied from 69.91 mg/kg to 486.57mg/kg.

Chromium

Chromium concentrations were also found to be below detectable limit in almost all the samples.

Cadmium

Cadmium concentrations like chromium were also observed to be absent in all the samples.

Nickel

Nickel concentration was also found to be absent in all the samples.

Zinc

Concentration of zinc in different samples of mollusca varied from 16.26 mg/kg to 44.97 mg/kg.

Cobalt

Cobalt concentration varied from 69.91 mg/kg to 486.57 kg/kg

In general during present observation concentration of cobalt was found very high followed by manganese and zinc.

Discussion

Aquatic plant crops (Macrophytes) frequently increase when nutrient inputs (Phosphorus and Nitrogen) from the catchment increase. The nutrient inputs in general are more due to allochthonus input of intensified agricultural residues and discharge of sewage effluent. The available nutrients in turn absorbed by different plants from the sediments and water. The absorption capacity and requirement of nutrients varies from species to species. From management perspective if information of different nutrient composition of the macrophytes or weeds are available then it becomes easier to formulate strategies for controlling the growth of the nuisance causing weeds. It also helps in deciding the types of species that need to be removed when present in excessive quantities. During present investigation *Eichhornia crassipes* has been observed to absorb maximum concentration of Nitrate and Phosphate along with *Potamogeton crispus* and *Ceratophyllum demersum*.

In aquatic ecosystem nitrogen is required in large quantities for the synthesis of proteins, chlorophyll and enzymes and are available in three forms, ammonical nitrogen, nitrate nitrogen and nitrite nitrogen. Of these three forms, nitrate nitrogen is most stable one, which is readily taken up by the plants for its physiological process. The more rapidly and actively a plant is growing, the more nitrogen it needs. This has been reflected in some of the plants like *Eichhornia, Vallisneria* during present period of investigation.

Mollusca, which are often attached with various macrophytes, are long-term indicators of environmental quality. Many workers have also used certain other species of gastropods as indicators of pollution. (Hellawell, 1986). Gastropods species have sensitive life stages that respond to stress and integrate effects on both short-term and long-term environmental stresses. The structure of the gastropods community reflects the state of the entire ecosystem. Many recent studies have demonstrated changes in the nutrient concentration in mollusca as a response to the pollution level of overlying water ((Hellawell, 1986). In last few decades increase human activities has resulted in elevation of lead, cadmium, mercury and zinc levels in fresh water (Meybeck *et al.*, 1989).

Domestic waste water is the largest single source of heavy metals discharge in the environment through effluents or sewage sludge which generally contain elevated levels of copper, lead, zinc, cadmium and silver (Fostner and Wittmann, 1983). Although in Bhoj wetland the availability of various heavy metals is still below alarming level but if the present trend continues, the level might get elevated. Therefore appropriate strategies are needed for reduction of toxic matter along with the nutrients responsible for degradation of the water body.

The present study reveals that particular plant species has affinity with major nutrients. The plants could be better option for reducing the concentration of nitrogen and phosphorus availablein different forms. The identified plants are as follows;

Eichhornia crassipes: can be used in removal of nitrate, phosphate, potassium and manganese

Scirpus royeli: can effectively used in reducing nitrate and potassium.

Vallisneria spiralis: can be used in reducing nitrate, potassium, calcium and zinc

Myriophyllum spathulatum: can be effectively used in removing calcium, copper

Polygonum glabrum: is effective in reducing nitrate from the system.

Jussia rupens: can be very effective in reducing Manganese from the system

Potamogeton crispus: can be very effective in removing copper

Molluscan species (*Lemnae sp*): can be very effective in removing fluoride concentration from the system.

Thus the above study concludes that all the above species could be effectively used in controlling/ reducing the nutrient load and various pollutants without involving much expenditure and having no side effects, as these species are part and parcel of the limnetic ecosystem.

References

Forstner, U. and Wittmann, G.T.W., 1983. *Metal Pollution in the Aquatic Environment*, 2nd edition. *SpringerVerlag*, Berlin.

Govindasamy, C., Azariah, J.A. and Kannan, L., 1999. Relationship between trace metal concentration in the mussel *Modiolus metcalfi* and sediment. *Poll. Res.*, 18(4): 435–458.

Hellawell, J.M., 1980. *Biological Indicators of Freshwater Pollution and Environmental Management.* Elsevier, New York.

Meybeck, M., Chapman, D.V. and Helmer, R., 1989. *Global Freshwater Quality: A First Assessment.* WHO/ UNEP. Blackwell Reference, Oxford.

Nayak, L., 1999, Heavy metal concentration in two important penaeid prawns from Chilka Logoon *Poll. Res.*, 18(4): 373–376.

Pani, S., Bajpai, A. and S.M.Misra, 2002. Studies on bioaccumulation of selective heavy metals in a tropical ecosystem. *Res. J. Chem. Environ.*, 6 (1).

Shrinivas, V. Sesha and Rao Balaparameswara, 1999. Chromium induced alterations in the oxygen consumption of the Freshwater fish, *Labeo rohita* (Hamilton). *Poll. Res.*, 18(4): 377–380.

Sustainable Environmental Management *Pages* **216–227**
Edited by: **Dr. L.V. Gangawane & Dr. V.C. Khilare**
Published by: **DAYA PUBLISHING HOUSE**

Chapter 24

Biocontrol Potential of *Trichoderma*: A Review

S.S. Patale and D.S. Mukadam

Department of Botany, Dr.Babasaheb Ambedkar Marathwada University, Aurangabad – 431 004, MS, India

ABSTRACT

Trichoderma spp. are well documented as effective biological control agents of plant diseases caused by both soil borne and air borne plant pathogenic fungi. Biocontrol with antagonistic microbes such as the fungus *Trichoderma* is one area of research. Biocontrol mechanism involving *Trichoderma* species operates by way of mycoparasitism, antibiosis, induced resistance and inactivation of host enzymes. *Trichoderma harzianum Rifai* is an effective Biocontrol agent that is commercially produced to prevent development of several soil pathogenic fungi. An additional advantage for *Trichoderma* is that it increases growth in various plants. This review discussed biological control of plant pathogenic fungi by using *Trichoderma* species.

Keywords: *Trichoderma, Mass multiplication, Biocontrol, Viability, Mycoparisitism, Antibiosis, Competition, Mutations.*

Introduction

The genus *Trichoderma* is a well-known bio-control agent used in the present day agriculture for controlling a of plant diseases. Persoon (1801) introduced *Trichoderma*, cosmopolitan fungus with *T. viride* as type species. The fungus is found on decaying plant debris and very commonly can be isolated from the soil. The greenish tufty spreading colonies with tinge are the characteristics feature of *Trichoderma* growth, the mycelium superficial, septate and branched. Conidiophores are macronematous, erect and mononematously branched. Conidiogenous cells terminal, lateral, enteroblastic, phialidic. Conidia one-celled dry or slimy, smooth or encrusted with wart like protuberances hyaline or green. Hypocrea, Podostroma and Thuemenella belonging to hypocreales order of Ascomycotina. It is known by its 2 species distributed in all parts of the world. It seems that it is the most widely studied fungus in taxonomy and industrial microbiology, because of its antibiotic properties. *Trichoderma* spp. is found in all parts of the world like U.S.A., Canada, S. America, African countries, India, Sri Lanka, Japan and other countries. In India alone there are 12 species known which are isolated from *Cannabinus, Cassia,* cotton, *Aesculus, Corchorus, Eucalyptus, Pennisetum,* potato, Nicotiana, Sorghum seeds, *Arachis hypogea,* pods, Kernals etc. Zea, *Cichoriu, Cajanus, Thuja, Cuprassus* and *Quarcus* like host of different substrates in various stages (Vasant Rao *et al.*, 2004). The species of *Trichoderma* have been used in applied sciences for cladistic analysis of isoenzymes polymorphism. Mayer *et al.* (1991), studied DNA fingerprinting and differentiation of strains. It is interesting to note that Gullino (Editor) arranged an international workshop on *Trichoderma* and *Gliocladium* and published the deliberation in Petria (1991), Leong and Berks (Eds.) considered the significance of *Trichoderma* in Molecular industrial mycology. The crude usage of *Trichoderma* cultures beyond doubt has controlled several seed borne, and soil borne pathogens. Its cultures have been sprayed on standing crops to control air-borne diseases. *Trichoderma* has enormous biochemical potentials to be used indifferent fields.

Mass Multiplication of *Trichoderma* spp.

Commercial success of a bio-control agent depends on its bio-efficacy, self-life and more on its mass multiplication ability on suitable and easily available substrates. Backman and Kabana (1975) found that clay impregnated with 10 per cent molasses maximum production of *Trichoderma* spores, similarly molasses containing yeast medium were tried for the growth of *Trichoderma* multiplication and proved a very good substrate for mass multiplication of *Trichoderma* by Papavizas *et al.* (1984), Prasad and Rangeshwaran (2000) and Prasad *et al.* (2002). Sawant and Sawant (1996) and Prasad *et al.* (2002) found that similar to molasses jaggery broth also supported mass multiplication of *Trichoderma.* Henis *et al.* (1978) found significant mass multiplication of *T. harzianum* on wheat bran, while Elad *et al.* (1980) also found that wheat bran in combination with sawdust a good substrate for *Trichoderma* mass multiplication, while Kapoor and Kumar (2004) studied comparative effect of different substrates for *Trichoderma* mass multiplication and observed that wheat bran mixed with maize bran proved to be good substrate than the individual wheat bran or maize bran. Dubey and Patel (2002) while using different agro-waste materials for *Trichoderma* mass multiplication found that pulse bran + saw dust was found to be superior to wheat bran. Many people have worked on the mass multiplication of *Trichoderma* on corn, (Lewis and Papavizas, 1980), barley (A bd E1 Moity and Shatala, 1981), sorghum (Padmanabhan and Alexander,1984, Upadhyay and Mukhopadhyay, 1986). Similarly many people have worked on utilization of animal dung as the best and cheaply available substrate with the farmers for the mass production of *Trichoderma* (Zaidi and Singh, 2004, Sangle *et al.*, 2002 and Dubey and Patel, 2002) who have recorded that among various agro-waste substrates, animal dung and gobargas slurry proved to be the good substrates for mass multiplication of *T. harzianum.* Among the agricultural wastes it has been reported that various workers, have employed sugarcane baggase, rice straw and groundnut shell, Dubey and Patel (2002), used wheat straw, groundnut shells, mushroom bed straw while, Sangle *et al.* (2002), made use of decomposed coffee pulp, Saju *et al.* (2002), used coffee waste, Sawant *et al.* (1995), Tewari and Bhanu (2004) used different agro-industrial cellulosic waste, Mev and Meena (2003) used by-product of sugar industry as the best media for mass multiplication of *T. harzianum.*

The role of physical factors for multiplication of *Trichoderma* has been equally found to be an important condition similar to that of the substrates. Bhatnagar (1996) observed that species of

Trichoderma grow maximum at 25 ± 2°C while poor growth at 10 ± 2°C and 35 ± 2°C,while Saha and Pan (1996) recorded that 30°C was the optimum temperature while 15°C as a minimum and 40°C as maximum. Monacco and Rollan (1999) found that *Trichoderma koningii* preferred 20 ± 2°C as optimum temperature for mass multiplication. Recently Sangle and Bambawale (2004), Tewari and Bhanu (2004) reported on optimum temperature 27 ± 2°C for mass multiplication in *T. harzianum*. The moisture requirement for mass multiplication of *Trichoderma* have been studied by different workers. Papavizas *et al.* (1984) studied in case of *T. harzianum* and *T. viride* and found that moisture level between 40 to 60 per cent proved to be good for multiplication.

It was reported by Saha and Pan (1998) and Zaidi and Singh (2004) that *Trichoderma* population decreased at 75 per cent and above moisture. Similar view was put forth by Jayaraj and Ramabadran (1999) that higher levels of moisture were not found supportive for *Trichoderma* growth. Saju *et al.* (2002) and Mev and Meena (2003) found that substrate with 30–40 per cent moisture supported maximum mass multiplication of *Trichoderma*.

Carbohydrates Requirement of *Trichoderma* spp.

The fungi utilize carbohydrates for their growth but the choice for carbohydrates has been found variable with different fungi. Aube and Gagnon (1969) and Sierota (1977) found that *T. viride* growth was stimulated in the presence of sucrose, Maltose, Fructose etc. Jackson *et al.* (1991) recorded variation in the response of different carbohydrates by *Trichoderma* species, while Monga (2001) stated that *Trichoderma viride, T. harzianum* and *T. koningii* preferred glucose, fructose, Maltose and Sucrose for their maximum growth.

Nitrogen Requirement of *Trichoderma* spp.

Similar to carbon sources fungi require good source of nitrogen for their growth. Aube and Gagnan (1969) found in case of *Trichoderma harzianum* that nitrate and glycine as the best sources of nitrogen for maximum growth, while Danielson and Davey (1973a) stated that similar to glycine, potassium nitrate and Alanine as best nitrogen sources. Jayaraj and Ramabadran (1998) found that ammonium nitrate proved to be the best nitrogen source for maximum mycelial growth, while calcium nitrate as poorest nitrogen source in case of *Trichoderma*. It was interesting to record the results of the experiment conducted by Monga (2001) that *Trichoderma viride* did not like alanine, glycine, potassium nitrate and ammonium chloride as nitrogen sources for its growth and sporulation but on the contrary all these nitrogen sources proved to be superior for growth and sporulation in *Trichoderma harzianum*

Viability

The success of a hyper parasite depends upon those conditions that allow its spores to germinate freely and permit its survival through stress situation in soil. It was studied by Nakkeeran *et al.* (1997) and found that *Trichoderma viride* showed increase in self-life when the spores were stored with talc based and vermiculite formulation. While Monaco and Rollan (1999) found that the spores of *Trichoderma koningii* stored in the form of prills made with the help of sodium alginate, calcium chloride solution and wheat bran. The viability of spore was found to be more than a period of two years at zero degree temperature. Viability of the bioagent in relation to different temperature conditions was studied in case of *Trichoderma harzianum* (Prasad and Rangeshwaram, 2000). It was found that the fungus showed its viability and activity upto 30 days at room temperature while Mev and Meena (2003) stated that *T. harzianum* population showed shelf life in between 7 to 14 days only. The studies carried out by Manav and Singh (2003) showed that *Trichoderma harzianum* formulation irrespective of wet and dry formulation survived better at low temperature at (8°C) than the other temperatures. The studies carried out by Kapoor and Kumar (2004) are that spores of *T. harzianum* embedded with sodium alginate pellets lost its viability within 30 days when stored under water at room temperature but it remained viable for 180 days in refrigerator and for 120 days at room temperature but without water.

Mycoparasitism

The antagonist produce enzymes which are found to be responsible to cause lyses in cell wall components of the pathogenic fungus. Similarly hyphae of the antagonist penetrate the host hyphae and grow in it or produce haustoria within the host cell to function as mycoparasite. Chet and Baker (1981) found successful mycoparasitism of *Trichoderma* on *Pythium* and *Rhizoctonia solani* resulting death of the host fungi Sharma (1986) recorded similar type of mycoparasitism due to *T. harzianum* and *T. viride* on *Fusarium solani, Myrothecium roridum* and *Phoma medicagins. Trichoderma harzianum* showed mycoparasitism on *Sclerotium rolfsii* (Benhamou and Chet, 1996), while *Trichoderma hamatum* and *T. reesei* penetrated to *Drechslera sorokiniana* (Mandel *et al.*, 1999) and *T. viride* on *Crinipellis perniciosa* (De Marco *et al.*, 2000).

Antagonistic Study

Trichoderma has been reported as a superior competitor in the soil microflora due to which the other fungi are either inhibited or suppressed. Singh (1983) found similar thing against *Colletotrichum falcatum.* Panneerselvam and Saravanamuthu (1994) recorded this due to, *Trichoderma viride* against *Fusarium moniliforme* and Sundar *et al.* (1995) with *Rhizoctonia bataticola,* Bhatnagar (1996) with *Fusarium udum,* Jayalakshmi *et al.* (1998) against *Colletotrichum capsici,* Rajeswari *et al.* (1998), against *Macrophomina phaseolina*, Sharma *et al.* (2003) against *Fusarium oxysporum* and recently by Sangle and Bambawale (2004) found against *Fusarium oxysporum* and Singh *et al.* (2004) against *Sclerotinia sclerotiarum.*

Most of the workers have reported that *Trichoderma* treatment has been found successful for the control of soil borne diseases in crops. The tomato wilt due to *Fusarium* was found to be controlled with *Trichoderma* as reported by Padmodaya and Reddy (1998) while significant reduction in stem rot of groundnut was found due to *Trichoderma harzianum* (Biswas and Sen, 2000). Recently Sharma and Gupta (2003) found that *T. longibrachiatum* and *T. harzianum* treatment successfully controlled post emergence fruit rot in French bean caused by *Rhizoctonia solani.*

Similar to soil treatment *Trichoderma* applied to seeds also showed significant results in crop disease control. Pande (1985) found that in case of horse gram seed treatment with *T. viride* successfully controlled *Macrophomina phaseolina.* Sankar and Jayarajan (1996) found in case of sesamum seeds treated with *T. viride, T. harzianum, T. longibrachiatum* and *Gliocladium virens* showed maximum seed germination than in seeds without treatment. Padmodaya and Reddy (1998) reported that seeds treated with *T. viride* in case of tomato successfully controlled wilt caused by *Fusarium oxysporum* f. sp. *lycopersici.* Similar results were obtained by Ram *et al.* (2000), Biswas and Sen (2000) in case of groundnut that *T. harzianum* seed treatment reduced stem rot of groundnut caused by *Sclerotium rolfsii,* while Bunkar and Mathur (2001) tested *T. harzianum* in case of chilli to control chilli rot.

Mode of Action of Antagonists

The mechanisms proposed during last ten years or so to explain the bio-control of plant pathogens by *Trichoderma* or *Gliocladium* are presumptive and are not always clear. The suggested mechanism by which microorganisms affect pathogen are mainly based on the principle of antagonism which is manifested in one or few of the following four ways, (1) direct parasitism or lyses and death of the pathogen (mycoparasitism), (2) competition with the pathogen for food (competition), (3) direct toxic effects on the pathogen by antibiotic substances released by the antagonist (antibiosis) and 4) indirect toxic effects on the pathogen by volatile substances (Cook and Baker, 1983, Hadar *et al.*, 1984).

Antibiosis

Antibiosis is mediated by specific or non-specific metabolites of microbial origin, antibiotics, lytic enzymes, volatile compounds or other toxic substances (Jackson 1965). Several toxic metabolites were produced *in vitro* by *Trichoderma* spp. and there is some evidence that, such metabolites were found to be produced in bits of organic matter in soil (Wright, 1956). Panteleev (1975) observed that a strain of *Pseudomonas* reduces the wilt incidence by producing antibiotics. *P. flurosence* was suppressive to *aeumanomyces graminis* var. *tritici* (Sacc.) Walker, the causal agent of take-all disease of wheat. Production of an antibiotic, phenazine has been implicated in bio-control (Gurusiddaiah *et al.*, 1986).

Elad *et al.* (1983) observed that, degradation and lyses of fungal cell and sclerotial wall were mainly due to key enzymes like b–1, 3 glucanase and chitinase. The relationship between mycolytic enzymes produced by *Trichoderma* and their significance in host cell degradation has been studied by Cook and Baker (1983).

Diffusible (Non-volatile) Toxic Substances

Dennis and Webstar (1971a) studied the production of non-volatile (diffusible) antibiotic substances in *Trichoderma* spp. by an agar layer technique. They noticed that, many isolates produced the non-volatile antibiotics which are found to be active against a range of fungi. The ability to produce such substances varied between the isolates. The susceptibility of pathogenic fungi also varied widely.

Volatile Inhibitory Substances

Antibiosis mediated by volatile substances has received less attention than antibiosis through the production of non-volatile antibiotics, bacteriocins or siderophores (Fravel, 1988). Dennis and Webster (1971b) studied in detail the production of volatile compounds by *Trichoderma* isolates and their inhibitory effect on other fungi. They observed that two isolates of *T. viride* and one isolate of *T. koningii* inhibited the growth of *R. solani*. There was no relation between sporulation by different isolates and their ability to produce volatile antibiotics. The effective isolates especially members of the *T. viride* presented a characteristic coconut kernel smell. The active metabolites had the same morphological effect on all the susceptible fungi resulting in stunted mycelial growth. The hyphae were more highly branched (fasciation) than the normal hyphae.

Upadhyay and Mukhopadhyay (1983) studied the effect of volatile compounds of *T. harzianum* on the growth of *S. rolfsii*, when *Trichoderma* isolates were grown for three days and *S.rolfsii* was exposed for 48 h. vapour action, the growth of *S. rolfsii* was reduced in case of isolates 3 and 4, stimulated in isolates of 1 and 2. The vapour action of *T. harzianum* on 6th day was most inhibitory and thereafter it declined and reached to zero on 15th day. The volatile antibiotic activity of *Aspergillus flavus* was seen against *S. rolfsii* (Deb, 1990).

Synergistic effect of VA mycorrhizae and *Trichoderma* on root pathogens

Harlapur (1988) worked on interactions between VA mycorrhizal fungus (*Glomus fasciculatum*), *Trichoderma harzianum* and a root pathogen, *S. rolfsii* sacc, in pot culture experiments. The results show that the combined inoculation of VA mycorrhizal fungus and *Trichoderma* reduced the severity of foot rot disease of wheat due to *S. rolfsii*. Kulkarni (1992) also studied the biological control of *S. rolfsii* in groundnut with the inoculations of *Glomus fasciculatum* and *Trichoderma harzianum* in pot experiment and showed that inoculations of *Glomus fasciculatum* and *T. harzianum* was most effective in reducing the severity of collar rot of groundnut. Sreenivasa (1994b) worked on the biological control of *S. rolfsii* disease of chilli using combined inoculation of *Glomus macrocarpum* and *T. harzianum* and found to be the most effective in suppressing *S. rolfsii*. Sreenivasa (1997) later studied the different biological activities of VA mycorrhizal fungus, *G. macrocarpum* and *T. harzianum* on *S. rolfsii* at two Phospharous levels *i.e.* 75 and 100 percent recommended dose (75 kg/ha) in three varieties of chilli (Byadagi, Guntur and Sankeswar) and showed that the percentage production of sclerotial bodies was significantly less in the plants inoculated with both *G. macrocarpum* and *T. harzianum*.

Hydrolytic Enzymes Produced by *Trichoderma* spp.

Role of extracellular hydrolytic enzymes such as Cellulases, Pectinases, Amylases, Lipases and Chitinases, produced by species of *Trichoderma* has been found important for biocontrol of pathogens.

Trichoderma viride mutant QM 9123 (pathogenic strain) produced twice as much cellulase, when grown on 1 per cent CMC, 1 per cent Ball-milled cellulose, 5 per cent filter paper, 5 per cent Alpha cellulose or 5 per cent absorbant cotton whereas another isolate QM 9136 (non pathogenic) did not produce cellulase (Mandels *et. al.*, 1971). Mandels and Reese (1957) studied the effect of different carbon sources on cellulase production in *Trichoderma viride* and found that cellulase was produced on glucose, cellobiose, lactose and cellulose. Mandels and Weber (1969) reported the optimal incubation period of 7-12 days for cellulase production in *Trichoderma viride*. Toyama (1952) suggest that an optimum temperature of 40°C for the production of cellulase in *Trichoderma koningii*.

Lora *et. al.* (1995) reported that *Trichoderma harzianum* produced hydrolytic enzymes, β-1,6–endoglucanase. Similarly, *T. viride* produce β-1, 4-endoglucanase and β-1,3-glucanase (Kumar and Gupta, 1999). *T. harzianum* CEET 2413, a mutant strain produced β-1,3- and β-1, 6- glucanase (Rey *et. al.*, 2001).

Margolles *et. al.* (1996) reported that the extra cellular endochitinase was found to be increased five times in *T. harzianum* by using promoter of the (gene for cellulase) chh1 gene from *T. reesei* and the total endochitniase activity increased ten times. De Marco *et. al.* (2000) reported that *Trichoderma harzianum* was able to antagonize in field the phytopathogen *Cirnipellis perniciosa*, the causal agent of witches broom disease of cocoa. A chitinase with molecular mass about 37 KDA, was secreted by the *Trichoderma harzianum* in the culture medium containing chitin. *Trichoderma viride* also produced chitinase (Jebakumar *et. al.*, 2000). Arumugam *et. al.* (2003) reported that *Trichoderma harzianum* produced chitosanase at the optimum pH 6.0, soluble chitosan was used as a substrate

Mutations

Mutations are the ultimate source of all variability in organism. Mukherjee *et al.* (1997) reported that Benomyl–tolerant mutants of *Trichoderma viride* isolate T–15 for use in integrated bio-control of *Botrytis* grey mould of chickpea. Some of the mutants are hyper producers of antifungal substances in culture filtrate, and are equally effective as the wild type strain in disease control potential. Kumar and Gupta (1999) reported that UV and g–rays treatment of *T. viride* tolerant to 50–100 ppm of tebuconazoic. Selvakumar *et al.* (2000) reported that carboxin tolerance in *T. viride* was induced by exposing the culture to UV light and EMS (Ethyl Methane sulphonate). Three stable mutants showed different colony characters, fungicide tolerance and antagonistic potential against *Ustilago segatum*. Manav and Singh (2003) reported that shelf life of different formulations of mutant and parent strain of

Trichoderma harizianum at variable temperature. Shelf life of UV induced mutants and parent strains of *Trichoderma harzianum* stored at room temperature (18–25°C), low temperature (8 + 2°C) and high temperature (25 + 2°C) was estimated after different stage period *i.e.* 0, 30, 60, 90 and 120 days. The cfu/g of mutant strain was constantly higher in all observations as compare to wild type strain. Roy and Pan (2004) reported that g–ray induced mutant of *T. harzianum* and *T. virens* both wild type and mutants in non-sterilized soil showed that population increased as compared to the wild type.

Biological Control

Biological control mainly consists of using a micro-organisms to control harmful microorganisms (by hyperparasitizing) causing plant disease without disturbing the ecological balance. Weindling (1932) suggested the potential use of *Trichoderma* spp. as a bio-control agent against the soil borne plant pathogens like *Rhizoctonia solani.* The biological control of root diseases of crop plants by introduction of antagonistic microorganisms has been suggested as an environmentally safer alternative to the use of fungitoxic chemicals (Baker and Cook, 1974).

Gaur and Sharma (1991) reported some microorganisms like *Trichoderma* spp., *Aspergillus* spp., *Streptomyces* spp., *Penicillium* spp. and *Bacillus* spp. antagonistic to *Fusarium udum* Butler., from rhizosphere soil of pigeon pea. *Trichoderma viride* was the most effective in controlling the disease. Sumitha and Gaikwad (1995) studied 25 isolates from rhizosphere soil and healthy seeds. They reported that *Trichoderma harzianum* and *Bacillus subtilis* produced a wide zone of inhibition against *Fusarium udum* and inhibited spore germination completely. Further these antagonistic organisms showed no adverse effects on pigeon pea seed germination. The antagonistic activity of three *Trichoderma* spp. *Viz.*, *T. viride, T. harzianum* and *T. koningii* against *Fusarium udum,* the causal organism of pigeon pea wilt, was studied at different temperatures, pH and C/N ratio, by Bhatnagar (1996). He concluded that the antagonistic activity was not much altered by changing the Environmental conditions. Siddiqui and Mahmood (1996) reported the use of *Trichoderma harzianum, Glomus mosseae* and *Verticillium chlamydosporium,* alone and in combination, in the management of wilt disease complex of pigeon pea caused by *Fusarium udum* and *Heterodera cajani.* Simultaneous use of bio-control agents against pathogens gave better control than their individual application. Somasekhara *et al.* (1996) studied under green house conditions the biological control of *Fusarium udum,* the pigeon pea wilt pathogen, by six isolates of *Trichoderma* spp. A two system of application was adopted *viz.*, seed and soil application. *T. viride* isolate–H significantly reduced the number of *F. udum* propagules and wilt incidence and it was suggested that this isolate would be useful in the management of pigeon pea wilt. The use of *Trichoderma viride* as a seed treatment against *Fusarium udum* was evaluated by Bidari and Gundappagol (1997) in a wilt sick garden at Gulbarga. A comparable reduction in wilt was obtained when seeds were treated with *T. viride* over control.

Nakkeeran and Devi (1997) reported *Aspergillus flavus, A. niger, Penicillium* sp., *Cladosporium herbarum, Rhizopus stolonifer, Alternaria alternata, Macrophomina phaseolina* and *Fusarium udum* as seed-borne fungi of Pigeon pea. For control of these microorganisms, talc based formulation of *T. viride, T. harzianum,* etc. were used. Results indicated that all the seed borne fungi detected were most effectively controlled/reduced by seed dressing with *T. harzianum* and *Bacillus subtilis.* The efficacy of *Trichoderma* spp. as a bio-control agent against the wilt pathogen *Fusarium udum* of pigeon pea, was tested by Biswas and Das (1999), using five species of *Trichoderma in vitro* dual culture. The conclusion was that *T. harzianum* was the most effective antagonist followed by *T. longicornis* while *T. koningii* and *T. viride* were ineffective. Chakraborty and Sengupta (1998) studied the management of Fusarial wilt of pigeonpea by inoculation of plants with non-pathogenic strains of *Fusarium.* The susceptible variety

ICP 2376 seedlings were inoculated with soil-borne *Fusarium* spp. pathogenic to hosts other than pigeon pea. This inoculation resulted in production of some substances that were antifungal to *F. udum.* After inoculation with *F. udum,* an appreciable reduction in the germination of conidia in the stem extracts and xylem fluids of seedlings was observed. Manoranjithan *et al.* (2000) studied that using talc based formulations of *Trichoderma viride* and *Pseudomonas* sp.on damping-off disease, growth of chilli seedlings and population of *Pythium aphanidermatum* as studied under pot culture conditions. Seed treatment with *T. viride* (4g kg^{-1}) + *Pseudomonas* sp. (5g kg^{-1}) showed 7.00 and 12.50 percent pre and post-emergence damping-off, respectively against 27.50 and 54.75 percent in control. The treatment also increased the shoot length, root length and dry matter production of chilli seedlings, and reduced the population of *P. aphanidermatum.*

Chattopadhyay and Varaprasad (2001) studied that using bioagents in castor wilt management, ten effective bioagents along with neem (*Azadirachta indica*) leaf extract, carbendazim as seed treatment and carbofuran @ 3 kg a.i. ha^{-1} as pre-sowing soil application were tested for their effect on castor wilt incidence in two sets of field experiments in two separate wilt–sick plots (E1 and G1). Highest reduction of wilt over control was recorded in G1 by *Trichoderma viride* Pers isolated from healthy safflower plant root (T6), which was the best among the bio-agents in E1 and second to carbendazim among all treatments of the plot. Some of the bioagents showed reduction in population of *Fusarium oxysporum* and *Rotylenchulus reniformis* in soil. Significantly high yields were recorded in all the treatments compared to control. Arya and Kaushik (2001) studied that efficacy of six species of *Trichoderma, Gliocladium virens* and *Chaetomium globosum* was studied against the forest tree–nursery damping–off, namely caused *Fusarium oxysporum, Pythium aphanidermatum* and *Rhizoctonia solani* both *in vitro* and *in vivo.* Evaluation of the fungal antagonists by dual culture method revealed that *T. viride* caused maximum inhibition of all the three pathogens, followed by *G. virens, T. harzianum, T. hamatum* and *T. longibrachiatum.*

Mani and Hepziba (2003) reported that effect of organic amendments and bio-control agents *viz., Trichoderma viride* and *T. harzianum* was tested against root rot of sunflower caused by *Macrophomina phaseolina* for two consecutive seasons. The application of organic amendments such as FYM (12.5 t/ ha) showed disease incidence of 35.29 percent as compared 55.53 percent in the untreated control. Seed treatment with *T. viride* combined with application of neem cake showed 19.18 percent of disease incidence followed by seed treatment with *T. harzianum* in combination with soil application of neem cake (29.68 percent). The treatment with *T. viride* and neem cake combination resulted in highest yield (822.5 kg/ha) as compared to control (537.50 kg/ha). Dubey (2003) reported that seed treatment with slurry or water mixed spores of *Trichoderma viride* and *Gliocladium virens* gave best protection to germinating seeds of urd/mung bean against *Rhizoctonia solani.*

Trichoderma as a Decomposer

Trichoderma spp. has been found to work as decomposers of organic matter. Composting process is fastened, if it is added to the raw material of compost. Generally farmers destroy the straw of paddy and wheat by burning, in the field or use as fodder. Virginia (1993) carried rapid rice straw composting with the help of activator *T. harzianum.* The activator complements soil microbes as a source of waste cellulose decomposers, thereby increasing the number of decomposers and the rate of decomposition so that farmers can use the compost sooner. Srivastava (2004) reported that carbon containing rice straw is mixed with nitrogen containing material like animal dung etc. The rice straw fully saturated with water is uniformly put over harvested field. Thickness of the rice straw is maintained to 10–15 cm, 25 kg of *Trichoderma* culture is broadcasted over rice straw and covered with banana leaf or polythene sheet. Water content is maintained upto 60–65 per cent. Manure is prepared within 30 days. Total quantity of the manure is found up to30 per cent of the original material.

Trichoderma Toxins

Aspite *et al.* (1996) extracted trichodermin from submerged cultures of *T. viride* and *T. lignorum.* The two antagonists checked barley root rot. Trichodermin treatment not only checked the disease and increased barley yield but improved the biological activity of soil, Trichonitrin was obtained from *T. harzianum* strain B1. It is used for seed treatment and controlled *Rhynchosporium secalis* in spring barley (Kulisler, 1997).

Trichoderma as a Disease Causing Agent

Trichoderma spp. can cause rot in fruits of apple in storage condition. Rot caused by an unidentified species of *Trichoderma* reported in Washington by English (1944). It was found on Jonathan apples (*Malus sylvestris* Mill.) held in storage for 2.5 months The rot in this case had its inception at the calyx or core of the fruit, and in some cases, almost half of the apple was rotted. *Trichoderma harzianum,* a fungus, not usually associated with apple maladies, is found to cause a brown firm rot in apples (Conway, 1983). A post harvest decay of citrus fruits caused by *Trichoderma lignorum* reported by (Knosel and Schickedanz, 1976).

References

Abd-El Moity, T.H., Shatala, M.N., 1981. *Phytopath. Z.* 100: 29–39.

Apsite, A., Viesturs, U., Berzina, G., Shteinberga, V., 1996. *Latvijas Lauksaimnieciboy Universitates–Raksti,* 283(6): 3–16.

Arumugam, P., Palani, P. and Balasubramanian, R., 2003. *J. Mycol. Pl. Pathol.*, Vol. 33, No. 3:414–420.

Arya Sanjay and Kaushik, J.C., 2001. *Pl. Dis. Res.* 16(1): 46–51.

Aube, C. and Gagnon, C., 1969. *Canadian Journal of Microbiology.* 15: 703–706.

Backman, P.A., Kabana, R.R., 1975. *Phytopathology*, 65: 819–821.

Baker, K.K. and Cook, R.J., 1974. *San. Franscisco: Freeman.* pp. 433.

Benhamou, N. and Chet, I., 1996. *Phytopathology,* 86: 405–416.

Bhatnagar, H., 1996. *Indian J. Mycol. Pl. Pathol.*, Vol. 26, No. 1: 58–63.

Bidari, V.B. and Gundappagol, K.C., 1997. *Adv. Agric. Res. In Indian.* 7: 65–69.

Biswas, K.K. and Das, N.D., 1999. *Ann. Pl. Protec. Sci.* 7(1): 46–50.

Biswas, K.K. and Sen, C., 2000. *Indian Phytopath.* 53(3): 290–295.

Bunker, R.N. and Mathur, K., 2001. *J. Mycol. Pl. Pathol.*, Vol. 31, No. 3: 330–334.

Chakraborty, A. and Sengupta, P.K., 1998. Fusaria. *Annl. Pl. Protection Sci.* 6: 2, 121–126.

Chattopadhyay, C. and Varaprasad, K.S. (001) *Indian J. Plant Prot.* 29 (1 and 2): 1–7.

Chet, I. and Baker, R., 1981. *Phytopathol.* 71: 286–290.

Conway, W.S., 1983. *Plant Disease* 67 No. 8: 916–917.

Cook, R.J. and Baker, K.F., 1983. *American Phytopathological Soc. St. Paul. Minnsota*, p. 539.

Danielson, R.M. and Davey, C.B., 1973a. *Soil Biology and Biochemistry*, 5: 505–515.

De Marco, J.L., C. Lima, L.H. de Sousa, M.V. and Felix, C.R., 2000. *World Journal of Microbiology and Biotechnology*, 16: 383–386.

Deb, P.R., 1990. *Acta Botanica Indica*, 18: 159–162.

Dennis, C. and Webster, J., 1971a. *Transactions of British Mycological Society*, 57: 25–29.

Dennis, C. and Webster, J., 1971b. *Transactions of British Mycological Society*, 57: 41–48.

Dubey, S.C. and Patel, B., 2002. *Indian Phytopath.* 55(3): 338–341.

Elad, Y., Chet, I. And Katan, J., 1980. *Phytopathology*, 70: 119–121.

Elad, Y., Chet, I., Boyle, P. and Henis, Y., 1983. *Phytopathology*, 73: 85–88.

English, H., 1944. *Plant Dis. Res.* 28: 610–622.

Fravel, D.R., 1988. *Annual Review of Phytopathology*, 26: 75–91.

Gaur, V.K. and Sharma, L.C., 1991. *Ind. National Sci. Acad. Part–B: Biol. Sci.* 57(1): 85–88.

Gurusiddaiah, S., Weller, D.M., Sarkar, A. and Cook, R.J., 1986. *Antimicrobial Agents Cheomotheraphy*, 29: 488–495.

Hadar, Y., Harman, G.E. and Tailor, A.G., 1984. *Phytopathology*, 74: 106–110.

Harlapur, S.I., 1988. Studies on some aspects of foot rot of wheat caused by *Sclerotium rolfsii* Sacc. M.Sc. (Agri.) Thesis, University of Agricultural Sciences, Dharwad, p. 98.

Henis, Y., Gaffar, A., Baker, R., 1978. *Phytopathology*, 68: 900–907.

Jackson, A.M., Whipps, J.M., Lynch, J.M. and Bazin, M.S., 1991 *Biocontrol Science and technology*, 1: 43–51.

Jackson, R.M., 1965. Antibiosis and fungistasis of soil micro–organisms. In: Ecology of soil Borne plant pathogens (eds. Baker, K.F. and Snyder, W.C.), University of California Press, Berkeley, pp. 363–369.

Jayaraj, J. and Ramabadran, R., 1998. *J. Mycol. Pl. Pathol.*, Vol. 28, No. 1: 23–25.

Jayaraj, J. and Ramabadran, R., 1999. *Indian Phytopath*. 52(2): 188–189.

Jebakumar, R.S., George, G.L., Babu, S., Raghuchander, T., Gopalswamy, G., Vidhyasekaran, P., Samiyappan, R., Raja, J.A.J. and Balsubramanian, P., 2000. *Indian Phytopath*. 53 (2): 151–161.

Kapoor, A.S. and Kumar, P., 2004. *Pl. Dis. Res.*, 19(1): 55–56.

Knosel, D. and Schickedanz, F., 1976. *Phytopathol. Z*. 85: 217–226.

Kulisler, R., 1997. *Ochrana Rostin–UZPI* 33: 213–219.

Kulkarni, S.A., 1992. Studies on collar rot of groundnut caused by *Sclerotium rolfsii.* M.Sc. Thesis, University of Agricultural Science, Dharwad.

Kumar, A. and Gupta, J.P., 1999. *Indian Phytopath.* 52(3): 263–266.

Lewis, J.A., Papavizas, G.C., 1980. *Phytopathology*, 70: 85–89.

Lora, J.M., Cruz, J. de La, Llobell, A. Renitez, T. Pintor Toro, J.A., 1995. *Molecular and General Genetics*, 247: 639–645.

Manav, M. and Singh, R.J., 2003. *Pl. Dis. Res.* 18(2): 144–146.

Mandal, S., Srivastava, K.D., Aggarwal, R. and Singh, D.V., 1999. *Phytopath*.52 (1): 39–43.

Mandels, M. and Weber, J., 1969. The production of cellulases. Cellulases and their applications, Adv. Chem. Ser. 95 (Ed. Gould, R.F.), pp. 391–414.

Mandels, M., Weber, J. and Parizek, R., 1971. *Appl. Microbiol.* 21: 152–154.

Mandels, M. and Reese, E.T., 1957. *J. Bacteriol.* 73: 269–278.

Mani, M.T. and Hepziba, S.J., 2003. *Pl. Dis. Res.* 18(2): 124–126.

Manoranjitham, S.K., Prakasam, V., Rajappan, K. and Amutha, G., 2000. *Indian Phytopath.* 53(4): 441–443.

Manoranjitham, S.K., Prakasam, V., Rajappan, K. and Amutha, G., 2000. *J. Mycol. Pl. Pathol.*, Vol. 30, No. 2: 225–228.

Margolles–Clark, E. Harman, G.E., Pentilla, M., 1996. *Applied and Environmental Microbiology,* 62: 2152–2155.

Mev, A.K. and Meena, R.L., 2003. *J. Phytol. Eres.* 16(1): 89–92.

Meyer, R., 1991. *Appl Environ.Microbiol.*, 57: 2269–2276.

Monaco, C. and Rollan, M.C., 1999. *World J. of Microbiol. and Biotechnology,* 15: 123–125.

Monga, D., 2001. *Indian Phytopath*. 54(4): 435–437.

Mukherjee, P.K., Haware, M.P. and Raghu, K., 1997*Indian Phytopath.* 50(4): 485–489.

Nakkeeran, S. and Devi, P.R. (1997). Seed–borne microflora of pigeonpea and their management. *Pl. Dis. Res.* 12(2): 197–200.

Nakkeeran, S., Sankar, P. and Jeyarajan, R., 1997. *Indian J. Mycol. Pl. Pathol*. 27: 60–63.

Padmanabhan, P. and Alexander, K.C., 1984. *Sugarcane*. 4: 11–12.

Padmodaya, B. and Reddy, H.R., 1998. *J. Mycol. Pl. Pathol*. Vol. 28, No.3: 339–341.

Pande (1985). Biocontrol characteristics of some moulds. *Biovigyanam,* 1(1): 14–18.

Panneerselvam, A. and Saravanamuthu, R., 1994. *J. Indian Bot. Soc. Vol.* 73: 265–267.

Panteleev, A.A., 1975. *Rastenii*, 2: 51.

Papavizas, G.C., M.T. Dunn, J.A. Lewis and J. Beagle–Restaino (1984). *Phytopathology*, 74: 1171–1175.

Persoon, C.H. (1801). Synopsis methodical Fungorum, pp. 706, Gottingen.

Prasad, R.D. and Rangeshwaran, R., 2000. *J. Mycol. Pl. Pathol., Vol.* 30, No. 2: 233–235.

Prasad, R.D., Rangeshwaran, R. and Sunanda, C.R., 2002. *Pl. Dis. Res.* 17(2): 363–365.

Ram, D., Mathur Kusum, Lodha, B.C. and Webster, J., 2000. *Ind. Phytopath*. 53(4): 450–454.

Rey, M., Delgado–Jarana, J. and Benitez, T., 2001. *Appl. Microbiol. Biotechnol,* 55: 604–608.

Roy, A. and Pan, S., 2004. *Indian Phytopath.* 57(3): 255–259.

Saha, D.K. and Pan, S., 1998. *Indian Phytopath*. 51(1): 51–56.

Saju, K.A., Anandraj, M. and Sarma, Y.R., 2002. *Indian Phytopath.* 55(3): 277–281.

Sangale, U.R. and Bambawale, O.M., 2004. *J. Mycol. Pl. Pathol., Vol.* 34, No. 1: 107–109.

Sangale, U.R., Wuike, R.V. and Gade, R.M., 2002. *Pestology, Vol.* XXVI, No. 9: 46–48.

Sankar, P. and Jeyarajan, R., 1996. *Indian Phytopathology*, 49 (2): 148–151.

Sawant, I.S. and Sawant, S.D., 1996. *Indian Phytopathology*, 49(2): 185–187.

Sawant, Indu, S., Sawant, S.D. and Nanaya, K.A., 1995. *Indian J. Agric. Sci.*, 65: 842–846.

Selvakumar, R., Srivastava, K.D., Agarwal, R., Singh, D.V. and Prem Dureja 2000. *Indian Phytopath.* 53 (2): 185–189.

Sharma, M. and Gupta, S.K., 2003. *J. Mycol. Pl. Pathol. Vol.* 33: 345–361.

Sharma, R.K., 1986. Seed borne pathogens of Soybean and their biological control. *M.Sc. Ag. Thesis. J.N.K.V.V. Jabalpur*, pp. 90.

Sharma, R.L., Singh, B.P., Thakur, M.P. and Trimurthy, V.S., 2003. *J. Mycol. Pl. Pathol., Vol.* 33, No. 3: 411–414.

Siddiqui, Z.A. and Mahmood, I., 1996. *Israel J. Pl. Sci.* 44(1): 49–56.

Sierota, Z.H., 1977. *European Journal of Forest Pathology*, 7: 65–76.

Singh, N., 1983. *Indian Bot. Reptr.* 2(2): 142–144.

Singh, R.S., Mann, S.S., Kaur Jaspal, Astha and Kaur Ramandeep 2004. Variation in antagonistic potentiality of *Trichoderma harzianum* isolates against *Sclerotinia sclerotiorum* causing head rot of sunflower. *Indian Phytopath.* 57(2): 185–188.

Somasekhara, Y.M., Anilkumar, T.B. and Siddaramaiah, A.L., 1996. *Mysore J. Agric. Sci.* 30(2): 163–165.

Souza, A.D., Roy, J.K., Mahanty Bibekananda and Dasgupta, D., 2001. *Indian Phytopath.* 54(3): 340–345.

Sreenivasa, M.N., 1994b. *Environment and Ecology*, 12: 319–321.

Sreenivasa, M.N., 1997. *Environment and Ecology*, 15: 343–345.

Srivastava, V.K., 2004. *Pestology*, Vol. XXVIIII No. 8: 41–45.

Sumitha, R. and Gaikwad, S.R., 1995. *J. Soils and Crops.*5(2): 163–165.

Sundar, A.R., Das, N.D. and Krishnaveni, D., 1995. *Indian J. Plant Prot.* 23: 152–155.

Tewari, L. and Bhanu, C., 2004. *J.l of Scientific and Industrial Research, Vol.* 63: 807–812.

Toyama, N., 1952. *J. Fermentation Technol.* 30: 89–93.

Upadhyay, J.P. and Mukhopadhyaya, A.N., 1983. *Ind. J. of Mycol.and Pl. Pathol.*, 13: 232–233.

Upadhyay, J.P., Mukhopadhyay, A.N., 1986. *Tropical pest management*, 32: 215–220.

Vasant Rao, Manoharachary, C., Sureshkumar, G. and Subodh, K., 2004. *Fungi around some aquatic bodies in Andhra Pradesh, India*, pp. 110–154.

Virginia, C., 1993. Rapid composting fits rice farmers, ILELA, 9(2): 11–12.

Weindling, R., 1932*Phytopathology*, 22: 837.

Wright, J.M., 1956. *Annals of Applied Biology*, 44: 461–466.

Zaidi, N.W. and Singh, N.S., 2004. *Indian Phytopath.* 57(2): 189–192.

Sustainable Environmental Management *Pages 228–237*
Edited by: **Dr. L.V. Gangawane & Dr. V.C. Khilare**
Published by: **DAYA PUBLISHING HOUSE**

Chapter 25

Oceanography as Environment

S.K. Patel

Department of Life Sciences, Bhavnagar University, Bhavnagar – 364 002, Gujarat, India

ABSTRACT

Oceanography deals with the study of Ocean in scientific way. It is divided to chemical, biological physical and geological types. Ocean is very immense and productive. As per data Ocean has volume of 1350 million km^3.It offers food with less space and efforts, then also provides shells, pearls, chemicals *viz.* iodine, bromine, K, Mg, Mn. etc. It is a source to get edible salt. Accumulation of oil and gas below sea floor becomes natural resources. It affects the weather and climate patterns of earth. It becomes media between various countries for trade through ships. It is a good source to obtain energy. In this article different aspect of ocean environment is discussed.

Keywords: *Ocean science, Utilities.*

Introduction

Oceans are very immense and productive which covers 72 per cent of world' area with average depth is 10,000 meter. It offers food with less space and efforts and also provides shells, pearls, and different chemicals. Accumulation of oil and gas below sea floor becomes natural resources, solar energy is also obtained from the Ocean. It affects the weather and climate patterns of earth. It becomes media between various countries for trade through ships. There are about more than 20 Universities and research laboratories in world, established for advanced training in this stream. There are 150 colleges giving training of marine science. Some important of them are Woods Hole Oceanographic Institute, School of Marine and Atmospheric Science, University of Miami, and University of Washington offers marine biology courses at USA. Oceanography modified after 1980's due to laws of sea implemented and more applied applications to use it in better way.

History

Benjamin Franklin's Map of Gulf stream established early at U. S. A. Maury's study of Ocean currents. European contribution by Charles Darwin on Beagle, Thomas Huxley did World Wide Voyage with the help of Challenger Ship (1872-1876), he became one of the pioneer to start oceanography as science. During 1925 modern oceanography started with work of Mete at South Atlantic. During this time Scripps Institute at California and Wood Hole Institute were established and founded well the study of oceanography.

By 1960's more advanced research was established on international level. Projects on deep-sea drilling were sanctioned and more research on applied aspects on climate, energy from sea and other oceanic concerns ware established.

Instruments and Techniques

Ships and instruments are key methods for working in ocean and sampling marine environmental. The research ship can range size from large deep-sea research vessel to small ship for shore work. Satellites buoy systems and fixed or floating plat-forms are important oceanographic instruments. Loran and other electronic aids have led to recent improvements in oceanography. Electronic Echo-sounding is used for knowing water depth. Pingers, cameras and side-scan sonar's are specific tools used for oceanographic research.

Biological Oceanography

Sea is as a biological environment, animals and plants of sea, their interaction with ocean and process of production of organic matter (food) from sea is concerned with Biological Oceanography.

History

Seafood was popular to Egyptians; they were using fishes caught from Red Sea and Nile River. Early marine Biologist was C. G. Ehrengerg (1798-1876). He collected rocks having microscopic organisms as diatoms, radiolarians, sponges etc. living in ocean. Edward Forbes 1844 was farther of marine biology, divided ocean in to 8 zones on basis of marine organisms. Challenger expedition (1872-1876) held that found Bathybius samples of primitive life from ocean. Biological study was done by J. Vaughan, Thompson, Johannes Muller, Sir John Murry and Victor Hensen studied planktons.

Biological Environmental Properties of Sea

Marine animals constantly immersed in seawater. The physical and chemical characteristics of seawater tend to be relatively stable. Organisms are free from effects of desiccation.

Water is a solvent and hold critical gases and minerals for animal and plant life. Because of density provide support to many organisms. Jellyfish etc without skeleton are supported by water. Largest whale exist in the ocean, seawater is buffered solution-slightly alkaline having PH range between 7.5 to 8.4. This initiate mollusc to secrete $CaCo_3$ shells. The carbon is abundant needed by plants to produce organic matter. Water is transparent hence light penetrate to more depth up to 100 m. It has high heat capacity and high latent heat of evaporation, prevent rapid change in temperature. Dissolved elements in seawater are important biologically. Ocean water temperature ranges from – 2°C (28°F) at depth to 40°C (104°F) in Gulf. Salinity very from near zero in estuary to 40 per cent near shore. Ocean depth can reach over 10,000 meter (32810 feet) pressure ranges from 1 at m. (1 kg per square cm or about 14.7 lb per square inch) at surface to over 1000 at m. at depth. There is 1 at m increase with each 10 m 32.8 ft of depth.

Biological Communities of Marine Environment

A community is composed of organisms occurring together that appear to be dependent on each other or on common environment.

Plankton Community

Plankton = wandering micro living plants or animals shows weak locomotion, move by ocean currents. There are mainly phytoplanktons and zooplanktons. Phytoplanktons are primary producer of organic matter in ocean. These show vertical and seasonal variation. Light penetrates 1 meter onwards up to 20 m. depth. These are epiplanktons eg. diatoms, dianoflagellates, brownalgae greenalgae, sea weeds (*Sargassum*) etc.

Planktons by size are nannoplankton (smaller than 60 m), microplanktons (smaller than 1 mm) and macroplanktons (large than 1 mm). By duration of life planktons meroplankton (part of life as planktons), Holoplanktons (all of life as plankton). Zooplanktons are heterotrophic depend for food on others eg. a. Holoplanktons are foraminifers, radiolaria, arrow worms, snails, jelly fishes, sagitta other crustacean larvae.

Meroplanktons (Temporary larval forms) are coelenterata larvae-planula, ephyra, medusa of obelia, cerianthus larval, rhizostoma, ctenophora, bougainvillia, pleurobranchia etc. Larvae of polychaeta (annulida) chaetognathus as sagitta, spadella, cliona, carinaria, Mollusc larvae, Bipinnaria, echinopluteus, auricularia etc. Tornaria of balanoglossus and Lucifer of decapoda. These are important in economy of sea.

Nekton Community

Means swimmers, swim freely includes fishes, mollusca, annelids, coelenterata, and echinoderms. These are able to find food and avoid predators. They can migrate in ocean up to 4000-meter depth. These become food to higher marine carnivores, large fishes and human beings. These feed on planktons and check their population. These are affected by factors like temperature, pressure, O_2 level, food availability etc. these include sepia, octopus, loligo, oysters etc.

Bony fish many with tuna, mackrel, eel, suckur fish, clupeids sardine, polynemids, sea birds, sea turtles and snakes, sharks, cods, rays etc.

Deep-sea Benthoes

Benthic means deep-sea or bottom dwellings are barnachles, oysters, worms, clams, and burrowing forms. Marine Bacteria are of radiolaria, foraminifera. Benthoes are predators and mostly are scavengers of detritus feeders. Deep-sea is rich with food. It has sea cucumber, sea lilies, brittle stars, sea urchins, benthic fishes, crustaceans, clams, sea anemones, molluscs etc. metabolicate is low, anaerobic respiration, blind, bioluminous animals. Prawns, crabs, octopus, loligo, glass sponges, pennatula, sponges, flat fishes, holothurians, gastropods, hermitcrabs are present.

Intertidal Marine Zone

It is rich with faunal diversity as most of animals breed in this area. It has ideally (*i*) Sandy (*ii*) Rocky and (*iii*) Muddy habitats. Animals are adopted well for above habitats.

Sandy Marine Community

It is a changing and shifting substratum. Fauna is protozoa, turbellaria, Annelids, Fishes various types:–seahorse, eel, pomphret, pterois, clupeids, bombay duck, sardine, dara, mackerel etc. Hydrozoa, rotifera, polychaets, nudibranch, isopods, copepods, nereis, hermitcarbs, crustaceans, snails, crabs physalia, vellela, corals, zooanthus etc. Protochordata amphioxus, balangolossus, acidians, shrimps, molluscs–many gastropods, lamellibranches, dentalium etc. present.

Rocky Marine Fauna

Limpets, patella, Haliotis, oyzters, barancles, broyozoa, sponges, obelia, porpita, vallela.

Anthozoans–Sea anemone, zooanthus.

Annelids–Polynoe, sabella, eurythoe, serpulids, sabella.

Mollusca–Gastropods, mytilus, trochus, chiton, nerita turbo, aplysia, oysters, patella, doris, murux.

Echinmoderms–Starfish, sea urchin, sea cucumber.

Tube living animals, sea lily ' Nereis, gyrosoma.

Fishes–Eel, seahorse, pterois, exocoetus.

Crabs–Grapsus, pilummus, galasimus etc. are present.

Muddy Fauna

Onchidium, snail, bivalve, solen, splysia. Fishes and mud-skippers more, crabs; coelenterate: plumularia, sertularia, seapan (Vergularia), cirianthus etc. are present.

Light penetration vary can penetrate 1000 m (3281 feet). Temperature similar to larger part of ocean.

Salinity is constant in surface water of open sea ranging between 33 to 37 per cent. Only in isolated area get higher. Salinity of deeper water is more uniform, range of 34.6 to 35 per cent. Movement of water is important for moving nutrients for growth of plants. Water disperse eggs, larvae and adults in ocean.

The density of life in the ocean can be expressed by a measure called biomass (the amount of organisms in grams per square meter of ocean bottom). Biomass normally shows high value in littoral region and low values in deep-seawater as following:

Area	*Biomass (g/m²)*
Coastal Zone	100–5000
50–200 m. depth	200
4000 m.	about 0.5
Central part of ocean	0.01
6000 m.	1.20
8500 m.	0.30
Tonga Trench, 10,500 m.	0.001

The water currents are higher at tidal zone of ocean, while at deep waters are normally slow. Deep-sea area is complete dark. The benthic faunal 350 species have been identified from deep tidal zone of ocean.

Biological oceanography emphasizes the study of the animals and plants in the ocean and their interaction both with themselves and their environment. Organisms that live in ocean have several advantages over terrestrial ones, but they also have some disadvantages. For plants small portion of sea floor is adequate for attached growth, so most of plants are floating organisms and deal photosynthesis (phytoplanktonns). The production of organic matters by plants is a key process in the ocean as is needed by all other organisms for survival. The zooplanktons being hetrotrophs depend for food on phytoplanktons and these are fed by Nektons. These form a food chain in Ocean.

The marine environment is divided into two major, (*i*) Pelagic (the overlying water) and (*ii*) Benthic (the bottom). Life is more abundant in near shore regions as food supply is abundant. Most marine organisms are fundamentally related to each other via the food cycle, which is initiated by organic matter production, by phytoplanktons. Phytoplanktons fed by zooplanktons (herbivous) higher life nektons. (sardine, other fishes) are eaten by bigger predators as tunafish. Then there is role of scavengers, and decomposers in food cycle to change dead bodies of organisms to simpler elements that part in recycling.

Mangroves

Mangroves are nature's coast guards. Not all tropical coasts are lined with mangroves, but India has been blessed with luxuriant mangroves along almost its entire coastline.

Mangroves are general term applied to plants, which live in muddy, loose, wet soils in tropical and near tropical tidewaters. They are trees or shrubs that grow between the high watermark of spring tides and a little above the mean sea level. They are circumstropical on sheltered shores and often grow along the banks of rivers as far inland as the tide penetrates. The word "mangrove" is derived from a combination of the Portuguese work "mangue" for tree and the English "grove" for a strand of trees. A mangrove forest is also called "mangal", while the word mangrove is used with reference to specific types of trees.

It has been estimated that 60 per cent to 75 per cent of the tropical coastline is lined with mangroves trees. Some favourable factors for their growth are temperature 20 to 25°C, high salinity, soft mud with rich organic matters and wide tidal range of coast. Protected shores of estuaries, creeks and Gulf are very good habitat at Gujarat.

In India, the important mangrove forests are found din the Andaman, Nicobar and nearby islands and the east and west coasts of the country's mainland. Andaman and Nicobar coast of the best quality of mangrove forests, while those of Gujarat and Tamil Nadu are considered generally good.

Gujarat, particularly Saurashtra and Kachchh, have about 20,000 ha of mangroves. Locally the mangrove forests are called "cher forest".

The eastern seaboard has deltaic mangroves, as a result of the alluvial deltas of the great rivers; the western seaboard has estuarine backwater type of mangroves, supported by the estuaries of the major rivers here.

Mangrove Adaptations

Mangroves have a characteristically low and dense forest type of vegetation with a special root system and viviparous seeds. Most of the mangroves have adjusted to their environment through mechanical adaptation for attachment in soft or loose substrate; formation of respiratory roots and aerating devices; evolution of vivipary; use of specialized means of seed dispersal and development of xerophytic structures.

Usually, mangroves are shallow rooted and lack well developed tap roots as a result of high salt concentrations and the water-saturated, organically rich anaerobic substratum. A number of adaptations in root morphology are typical in most mangroves. The roots can be "prop" (stilt like protrusions from the lower part of the stem) or "drop" (from branches and upper part of the stem) type that terminates only a few centimetres below ground (*e.g. Rhizophora*). In other cases, surface or horizontal roots (like cables- 1 to 5 cm below the surface soil) grow out from the stem base and produce negatively geotropic, erect, aerial tips called "pneumatophores" which protrude from the subsoil horizontal roots (*e.g. Avicennia*).

All these varieties of roots act as a ventilation system for the trees. They also help in anchoring and feeding the plants, the latter function being carried out by absorption. Another relatively common feature in a mangrove is 'vivipary', where germination takes place while the seed is on the parent plant. This helps the seed to anchor itself in the soil without being washed away, which might happen if the seed were to fall into water. The stem and leaves also show specialized adaptations to high salinity and fresh water shortage. These plants have salt excreting glands as well as water storage tissues, salt balance being a very important physiological need of a plant.

Environmental and Economic Role

As much as 75 per cent of the low-lying tropical coastline up to 25°N and S latitudes is characterized by reduced wave energy and freshwater drainage and id dominated by a rich and diverse littoral mangrove swamp community. The tree canopies, masses of aerial roots, muddy substrates, as well as associated creeks and embankments (formed by tidal inundation) that characterize the mangrove ecosystem, provide a diverse structural habitat for a complex community of invertebrates, fish and birds.

Mangroves indirectly play an important role as feeding and nursery grounds for commercially important fish and shellfish species. As much as 90 per cent of tropical marine fish species pass at least one stage of their life cycle in mangrove estuaries.

In addition to providing protective and feeding habitat, mangroves function to reduce coastal erosion and to maintain high estuarine water quality by a) Sediment entrapment from surface water runoff and b) Nutrient release in steady-state equilibrium.

The economic benefits from mangroves are numerous. They provide a multitude of products directly and indirectly: fish, firewood, wood condiments, sugar, alcohol, paper, green manure, fodder, glues, tannins, synthetic fibres et.

Being a dynamic living resource, mangroves are self-maintaining and renewable, but only if the ecological processes governing the system are maintained. The internal ecological processes maintaining and renewing the mangrove ecological system are regulated mainly by external processes that depend on: a balance between fresh and slat water; adequate supply of nutrients; and a stable substrate.

Changes or modifications of one or more of these very important factors can seriously impede or eliminate the renewability of the resources. Several human activities can be responsible for this.

Important of Observing the Ocean

The oceans are central to the life system of the earth. They redistribute the heat and freshwater across the geosphere and thus regulate global climate. They sustain significant part of global living and non-living resources. They are also the largest reservoirs for man-made pollutants, with a phenomenal assimilative capacity.

Our abilities to forecast weather and global climate changes, thus pre-empting disastrous effects of events such as cyclones and El Nino, depends on our ability to observe the changes in ocean processes at spatial and temporal scales with high resolution.

Effects of oceanic processes have global reach cutting across national boundaries. This emphasizes the need for internationally coordinated observation of ocean parameters.

While some stocks of living marine resources such as benthic organisms are local or national in distribution, several others such as the straddling stocks of tunas or sharks are regional or even global in distribution. Their trans-national nature becomes even more striking when several nations share common sea areas, as in the Mediterranean or the North Sea.

Oceanic and coastal circulations also actively transport pollutants across geopolitical boundaries, with sources and impacts separated on various space and time scales.

It is impossible, politically, logistically and economically, for any one nation to collect all the information it needs for national prosperity and environmental security. Global Ocean Observing System (GOOS) answers this need. It is an operational, global network that systematically acquires and disseminates data and data products on past, current and future states of the marine environment to meet the needs of users ranging form governments and industries to scientists, educators, non-governmental organizations and public. The observing system comprises of two related and convergent modules:

1. A global ocean module concerned primarily with changes in the ocean-climate system and improving marine services including hazard warnings and
2. A coastal module concerned with the effects of large scale changes in the ocean-climate system and of human activities.

At global level Ocean is the largest sink for CO_2 acting as a buffer for climate regulation. About 10 per cent of the world protein resources are stored in the oceans.

Effects of changes in open oceans impact coastal regions ultimately. The costal module adds a human dimension to GOOS. Some highlights of coastal activities are:

1. 50 per cent of the world's population lives within 150 km from coast and nearly 40 per cent within 100 km.
2. Aquaculture is about to touch 20 per cent of the total living ocean resources.
3. 90 per cent of continental pollution is thrown into the ocean.

A rapid growth in population pressure has altars the value of coastal zone ecosystem services and goods. Some important socio-economic activities in the coasts are: offshore drilling for oil and other mineral resources, fisheries, aquaculture, maritime transport, harbour and coastal management, environmental management, defence activities and tourism. Over fishing is a problem with many marine stocks; physical alterations to coastal habitats increased the susceptibility of coastal populations to flooding and erosion; inputs into the sea of untreated sewage carry nutrients that cause coastal eutrophication and human health risks; introduction of non-native species that affect local flora and fauna and eventually the local fisheries.

The coasts of environmental degradation and the benefits derived from successful mitigation are becoming increasingly clear. Lack of an integrated and sustained observing system to assess and predict critical changes in the coastal environment harms our ability to sustain healthy ecosystems and their resources, including the vast reservoir of bioactive molecules harboured by coastal flora and fauna.

Our abilities to forecast weather, use the marine resources sustainably and maintain the ecosystem values of the marine environment depend on rapid detection and timely prediction of the changes in ocean processes. This requires a cost-effective observing system that responds to the needs of a variety of users. This is possible only through an unprecedented level of regional and global co-operation. Nations are thus seeking appropriate regional cooperation to collect and access required information. This regional approach-share obligations and co-sponsor sophisticated equipment and quality-assured observations-would be the hallmark of GOOS Regional Association (GRAs) from which the truly Global Ocean Observation System would mature.

Major goals GOOS is to accelerate acquisition, processing and analysis of data of known quality so as to deliver environmental data and information to the end-users within time scales on which environmental decisions and predictions are made. A data management and communication system that provides rapid access to data and information will be the 'lifeline' of the observing system. In accordance with GOOS design principles, the GRAs will need to develop a hierarchical, distributed network of local, national, regional organizations feeding eventually into a global framework, that use common standards and protocols for quality control, access to and exchange of data and archival.

State of the Oceanography

Modelling and Assimilation

This involves developing eddy-resolving models and implementing global oceanic models. These require access to larger computing facilities. Meteorological data assimilation has paved the way for oceanography, and consequently oceanic data assimilation is now adequately developed.

Data Flow

The surface of the ocean is adequately measured from space, by theromography, scatterometry and altimetry. These techniques have nearly matured and have been operationalized. Compared to surface of ocean, the interior of the ocean is unfortunately only marginally measured, although progresses recently achieved concerning autonomous sounding vehicles are improving the situation at reasonable cost. Thanks to new technology, researchers have been able to get detailed images of the Earth's surface below the seafloor. According to John Orcutt of the University of California's Scripps Institution of Oceanography only recently hat it been possible to actually build instruments that could last long at the bottom of the ocean.

Predictability

Ongoing programs have shown that oceanic processes may exhibit a large predictability, opening the way to operational oceanography, *i.e.* global and real-time oceanography including forecasting products.

Programmes

Earlier International Programmes (COOS, GCOS, WCRP, IGBP) have provided top down push for development of networks resulting in gigantic plans, with marginal impact at the national levels. Recent trend has been towards a more bottom-up approach, where national programmes, are established and are pushing for the establishment of international co-operation. The latest example of this is Indian Ocean Global Ocean Observing System (IOGOOS).

Oil Pollution

Oil pollution increasingly threatens the aquatic ecosystems of the world is oil spills and oil leakages. The situation needs a close look because our own Gulf of Kachchh and Arabian Sea are about to be critically endangered.

The Gulf of Kachchh is the repository of some of the most diversified and rare marine life. Saurashtra and Kachchh coastal areas have some of the richest fishing grounds supporting commercial fisheries worth hundreds of crores of rupees (which can be converted to value-added products many times over). Fishing activities support thousands of families directly and via ancillary industries. Seafood is the staple of many millions of human beings inside and outside India; and cattle and poultry consume its by-products. Most important, we owe it to mother nature to conserve the incredibly rich biodiversity of life in the ecosystems in these areas. Therefore understanding about oil spills, leakages and their effects is of great importance.

Oil spills can have a serious impact on coastal ecosystems. Recovery can take many years and in some cases damage may not be repairable at all. Oils can also pose hazards to human health, in particular from prolonged skin contact and inhalation of vapours. Some crude oils contain hydrogen sulphide gas, which is highly toxic.

According to the International Maritime Organization (IMO), about 2.0 million tonnes of oil entered the sea in 1975 as a result of maritime transportation losses. Two-third of this is from vessel operational discharges.

Since then, interoceanic oil transportation has increased manifold. Tanker accidents- oil spills and oil leakages–have been increasingly reported from all over the world. Major spills throughout the world shows that the majority (about 75 per cent) occur in ports during ship operations like loading, discharging, bunkering etc.

Oil spills statistics show that other kinds of ships (cargo and passenger ships) also cause numerous oil spills due to the large amount of bunker oil that they carry.

In Gujarat we have all the more reason to be concerned because we have oil terminals and refineries on the coastline. Gujarat has two oil terminals (Kandla and Vadinar) in the Gulf of Kachchh. Two refineries are coming up in the Gulf and a pipeline from Oman to Gujarat is on the anvil. This will have its landing point somewhere near Okha port or on the Kachchh coastline.

Spills and Leakages into the Marine Environment (1975)

Marine Transportation	*Million Tonnes per annum*
Tanker operations	1.08
Dry docking	0.25
Marine terminals	0.003
Bilge and fuel oils	0.5
Tanker accidents	0.2
Non-tanker accidents	0.1
Total	**2.133**

Properties of Oil

Petroleum products are derived by refining crude oils. The refined products are commonly known as gasoline (motor spirit), kerosene, gas oils, fuel oils (light, medium and heavy), and lubricating oils.

Oils are classified either as crude oils and refined products or according to their viscosity. Different oils behave differently in and out of water and by different oil-environment interactions.

When an oil spill occurs at sea, most oil will float and begin to spread. A few, whose density exceeds that of water, sink. An oil slick will not usually stay in the same position but will drift with the waves, tides and currents. Density (sp. grvity) dictates the buoyancy of oil on water and it influences spreading and dispersal. Generally, oils with low density tend to have low viscosity and relatively high proportion of volatility. Boiling point and boiling range decide the rate of evaporation of oil. The lower these are, the faster evaporation will occur.

Viscosity of oil is its resistance to flow. Oils with high viscosity will flow with difficulty and those with low viscosity are highly mobile. Viscosity decreases with increase in temperature, thus seawater temperature and absorption of heat from the sun play an important part. The temperature below, which the oil becomes semi-solid and will not flow, is called the pouring point. Flash point is the lowest temperature at which sufficient vapour exists above the spilled oil to yield a flammable mixture. This is an important factor in relation to safety during clean-up operations. Solubility of oil in water is significant in the context of toxicity to marine life. Oil spilled on sea undergoes a series of processes called weathering, which change its characteristics and behaviour. Spreading of an oil slick on the surface is a rapid and dominant process at the time of release, decreasing steadily until it has essentially stopped within one to ten days. Many factors affect the spreading–viscosity, evaporation, volatility, currents, sea state, wind, solar radiations, density, temperature and others.

It is true that the understanding of the effects of oil spills and leakages has increase considerably over the last two decades, but the results of such research continue to be limited by insufficient knowledge about the total ocean ecosystem. Field research has to supplement results of laboratory experiments. Work needs to be done on various aspects: for instance, changes in behaviour of marine organisms resulting from exposure to oil; effects on ecosystems or populations; and the interaction of oil with other contaminants.

The effect of a particular spill or leakage will depend on many factors, including volume of oil spilled/leakage (like temperature, wind), time of year, the presence of structures or resources in the path of spill, location of spill/leakage in relation to the nature and mixing of sediments, sea bottom topography, and geomorphology of the coast. The variability of these and other factors and their interactions can lead to a wide range of ecological, economic and physical effects.

References

Bames, R. S. K. and Hughes, R. N., 1982. *An Introduction to Marine Ecology.* Blackwall Pub. Comp., N. Y., pp. 1–200.

Barnes, H. and Barnes, M., 1988. *Oceanography and Marine Biology.* Aberdeen Univ. Press, Scotland, pp. 1–580.

Dietrich, G., 1963. *General Oceanography.* Oxford Pub. Comp. N. Y., pp. 1–150.

Fredrich, H., 1969. *Marine Biology,* Sidgwick and Jackson, pp. 1–300.

Johnson, R., 1976. *Marine Pollution,* Academic Press, Pp. 1–208.

May, R.M., 1984. *Exploitation of Marine Communities,* Springer Verlag, pp. 1–250.

Reish, D.J., 1969. *Biology of the Oceans.* Dickenson Pub. Comp., California, pp. 1–235.

Thomes, Mary-Frances, Rachakonda, S. and Rachakonda, N., 1986. *Biology of Benthic Marine Organisms,* Oxford and IBP Pub. Comp., New Delhi, pp. 1–605.

Sustainable Environmental Management *Pages* 238–243
Edited by: **Dr. L.V. Gangawane & Dr. V.C. Khilare**
Published by: **DAYA PUBLISHING HOUSE**

Chapter 26

Ecological Studies of Leaf Litter Fungi of *Citrus aurantium*

Ch Ramesh and K.C. Raju

Mycology Laboratory, Department of Botany, Karnatak University, Dharwad – 03 Karnataka State, India

ABSTRACT

Decomposition of organic matter is important for healthy environment. Comparative evaluation of different methods of isolation of fungi involved in the decomposition of leaf litter of *Citrus aurantium* L. was investigated. The Dilution Plate Method revealed quantitative picture of fungal population. On the other hand Leaf Impression method, besides revealing the association of some common fungi, facilitated detection of slow growing and nutritionally specialized fungi. The Moist chamber method revealed the presence of only a few folicolous fungi.

Keywords: *Leaf litter, Decomposition, Isolation*

Introduction

Fungi may occur as mycelia fructifications or spore. Only mycelia decompose substrates. Fungal ecology is still dominated by process-oriented point of view and it therefore seems desirable to identify those fungi that are present in an active mycelial state. Unfortunately direct microscopic examination reveals mainly sterile hyphae whose identity or activity often cannot be directly established. Indirect methods must therefore be used. The simplest methods the dilution plate technique adopted from Bacteriology yields mostly colonies originating from spores, which do not participate in decomposition processes, and may simply have been blown from elsewhere (Garrett, 1963) an alternative method was introduced by Warcup (1957).

Ecology has been defined by the Daubenmire (1974) as the study of the reciprocal relations between organisms, which cannot manufacture their basic food requirements and so are dependent on food materials produced by other organisms either by saprobes or parasites. They can thus be placed in four of the six categories introduced in Daubenmire (1974) classification of the types of symbiosis, which is defined as embracing all kind of relationships between organisms.

Symbiotic fungi demonstrate antagonism, neutralism or mutualism in their relations with each other or with other organisms, and the relationship may be facultative or obligate as for as nutrition relationship are concerned, fungi may be saprotrophic, deriving nutrients from dead organic matter, necrotrophic, deriving nutrients from living cells.

Most of the fungi have a multicellular thallus or mycelium composed of many fine threads called hyphae. In some groups, the mycelium is not divided into cells but is elongate, branched and coenocytic. In other groups the thallus is unicellular, and each cell is minute single entity. Those portions of the thallus in which the cell wall has not become thickened are in the intimate contact with their environment at all times and on all surface, except where formed into hyphal strands, rhizomorphs or sclerotia. Such cells or hyphae may be compared with the root hairs of higher plants in their youngest stages.

Spore germination, mycelial development, fruiting body production, spore discharge and dissemination and the development of rhizomorphs, sclerotia and resting cells, all fall within the province of fungus ecology to a large extent, fungi comprise a group of ubiquitous and omnivorous organisms. Their existence has been recognized almost since the beginnings of man recorded experiences and impressions of nature (Ansiworth1976). They are found whenever they are searched for, and their spores are found any where that spore traps (Davies, Zajic and Wellman 1974; Krarmer and Pady 1966) are exposed or soil samples are plated (Cooke 1958) parallel with the development of systematic mycology culminating in such universal compendia as those of Saccardo (1882-1931) and Oudemans (1919-1924) was the realizations that certain fungus species are associated with or cause, plant diseases. The description of this development (large, 1940) and that of plant pathology from an important portion of the history of fungus ecology.

Litter can induce fallen leaves dead herbaceous material, part of flowers and fruits, twigs, branches and logs including bark, and for some even the roots, which penetrates shallowly or even deeply into the soil. The fungal populations of the litter can be derived from pervious generations of litter material, the soil populations or the Phyllosphere and related areas of plant surface which have become colonized from their initiations by one fungus or another (Dickinson and Preece 1976, Preece and Dickinson 1971) for Pugh (1974) litter also includes any man made decomposable organic product in place or not. Few studies have been made on the succession of fungi involved or their contributions to the rate of decay.

Dickinson and Pugh (1974) brought together information on the ecology of plant litter decomposition. Here the decompositions of litter generated by such divers organisms as lower plants, herbs, angiosperms trees, coniferous trees, wood roots and digested litter are considered by people who have devoted time to the study of each area. The organisms involved in the decomposition of citrus litter, including aquatic and terrestrial fungi are discussed. Leaf litter fungi were studied by dilution plate technique, leaf print method and moisture chamber method.

As already started in the introduction, a part of the present investigation was directed towards ecological studies of the fungi colonizing leaf litter of *Citrus aurantium*. For this purpose litter sample from three different layers were collected from the sampling area at monthly intervals beginning from December 1994 to November 1996. The nature and distribution of the mycoflora,the periodicity and frequencies of occurrences of different fungi occurring on these litter samples of *Citrus aurantium* were then determined.

Twenty-two species belonging to eighteen genera were recorded from G_1 layer of these two were members of Zygomycetes, one Ascomycete and the remaining nineteen belonging to Mitosporic fungi. Species of fungi recorded in different number of sampling are listed in (Table-1) along with data on the frequency of occurrence of each species.The numbers of species recorded per sampling varied from 5-15 comparatively greater number of species were recorded during cooler months. Of the twenty two species record only one species was most frequent and was recorded in almost all the sampling (ie-16) one species of *Penicillium expansum* was occasional and the rest of the species recorded were rare and these occurrence was confirmed to 1-5 samplings (*i.e.*-2). The percentage frequency of some of the species are presented in (Table 26.1) The frequency of occurrence of species in individual samplings varied from 10-65 per cent. However, only few species have shown more than 10 per cent of average percentage frequency and the maximum being 14.32 per cent.

During the entire sampling period maximum frequency was recorded in Feb and March 1995 for *Aspergillus oryzae* after which it gradually declined. For *Penicillium expansum* and *Drechslera australiensis* the highest was recorded in Feb.1995. Some of the fungi such as *Alternaria* and *Stachybotrys* although occurred in less samplings and their frequency was also less.

Discussion

Some interesting conclusions emerged from the result of this investigation. Firstly, from each little layer a large number of fungal species were recorded. In the present study a some what arbitary classification of the various species of fungi had to be made on the basis of their Periodicity of occurrence. ie- the number of samplings in which each fungus was found to occur as against the totally number of samplings which is 24 during two years of study. Five categories are recognized species occurring in 16-24 samplings were regarded as most frequent, those occurring in 11-15 as common, 6-10 occasional and 1-5 as rare not that this perfect classification, but yet it was felt that this would give some ideas on the Periodicity of occurrence and facilitates descriptions. Although general fungal species were common components of the mycoflora of all the three layers, it was noticed that each layer had its won characteristics flora with regard to Periodicity of occurrence. The species comprising most frequent to common other different layers. Similar difference were observed by Vittal(1973)and Sudha(1978).

It was found that a large number of species colonized the phylloplane of these plants which have been studied and several of these were found to occurs in litter layer layers also It would be therefore appear that the phylloplane is an ideal substrate for colonization. In the present study, frequency of occurrence of a fungus has been taken as an index of its colonizing efficiency in different litter layer

Thus it is clear that in different layer of litter shifts in activity of the various species of the mycoflora occurred indicating peaks and troughs and troughs of activity for some of such the species. An assessment of such activity is based on frequency of occurrence of these fungi different layers of litters computed on the bases of sporulating colonizes on the litter and not on dilution plate counts and the data so obtained may be considered sufficiently reliable.

Table 26.1: Fungi Recorded *Citrus aurantium* L. from Leaf Litter and their Frequency (%) of Occurrence in Different Months of Sampling by Using Dilution Plate Method

	D	*J*	*F*	*M*	*A*	*M*	*J*	*J*	*A*	*S*	*O*	*N*	*D*	*J*	*F*	*M*	*A*	*M*	*J*	*J*	*A*	*S*	*O*	*N*	*D*	*Average Frequency*
Zygomycetes																										
1.*Circinnella mucoroides*			90																							3.75
2. *Rhizopus stolenifer*													15													0.62
Ascomycetes																										
3.*Emericella.sp*																35										1.45
Mitosporic Fungi																										
4. *Alternaria alternata*																						55				2.29
5. *Aspergillus niveus*																								40		1.67
6. *Aspergillus oryzae*				65	65	15	10	5		20	5		15		15	15	15	25		5		35	10	25		14.37
7. *Aspergillusristictus*							20																			0.83
8. *Chaetopsis grisea*																		10								0.41
9. *Curvularia clavata*																						40				1.67
10. *Cylindrocaldium floridanum*																										1.45
11. *Drechslera australiensis*																										1.04
12. *Drechslera biseptata*			115																							4.79
13. *Fusarium moniliformae*					20																					0.83
14. *Penicillium expansum*													5	15		25	5	10	5	5	35					4.37
15. *Polyscytalina grisea*										25																1.04
16. *Rhizoctonia* sp.	35																									1.45
17. *Scolecobasidum echinophlium*									15							20										1.45
18. *Scopulariopsis brevicaulis*																								40		1.67
19. *Stachybotrys kampalensis*																										2.29
20. *Stachybotrys nephrospora*																										1.25
21.*Veronaea coprophlila*																			25							2.50
22. *Zygosporium masonii*																										0.41

References

Ansiworth, G.C., 1976. University of Cambridge press, Cambridge.

Burges, A., 1939. *Serie de ckencias naturaie* VIII, fasc. 11, 64–81.

Cook, W.B., 1958. Continuous sampling of trickling filter populations 1 procedure sewage, Ind. wastes 30:21.

Daubenmire, R.F., 1974. *Plants and Environment,* 3rd ed John Wiley and sons New York.

Davies, T.S., Zajic, J.E. and Wellman, A.M., 1974. *Microbiol.*, 15: 256.

Dickinson, C.H., 1965. *Trans. Br. Mycol. Soc.*, 48: 603–610.

Dickinson, C.H., 1967. *Can. J. Bot.*, 45: 915–927.

Dickinson, C.H. and Preece, T.F., 1976. *Microbiology of Aerial Plant Surface.* Academic Press, London, pp. 659.

Dickinson, C.H. and Pugh, G.J.F., 1974. *Biology of Plant Litter Decomposition.* Academic Press, London, Vol. 1, p. 146, 175.

Eickev, A., 1973. *Biol. Biochem.*, 5: 441–443.

Garrett, S.D., 1963. *Soil Fungi Soil* Fertility. Pergamon Press, Oxoford.

Hering, T.F., 1995. *Trans. Br. Mycol. Soc.*, 48: 391–408.

Hudson, H.J. 1968. *New Phytol.*, 67: 837–874.

Hogg, B.M. and Hudson,H.J., 1966. *Trans. Br. Mycol. Soc.*, 49: 185–192.

Kramer, C.L. and Pady, S.M., 1966. *Phytopathology*, 56: 517.

Large, E.C., 1940. *The Advance of the Fungi.* Henry Holt, New York.

Macauley, B.T. and Thrower, L.B., 1966. *Trans. Br. Mycol. Soc.*, 49: 509.

Oudemans, CATA, 1919–1924. *Enumeratic Systematica Fungorum, 5* Vol. Martinus Nijhoff, The Hagne.

Preece, T.F. and Dickinson, C.H., 1971. *Ecology of Leaf Surface Microorganisms.* International symposium, Press, New York, p. 631.

Pugh, G.J.F., 1974. Fungi in intertidal regions. Marine mykrobiologic II. Veroffentlichen institut meeresfor schung Bremerhaven supplement, 5: 403–418.

Ryvarden, A.J., 1985. *Trans. Br. Mycol. Soc.*, 85: 539–540.

Saccardo, P.A., 1882–1931. Syllog: The sylloge Fungorum omnium hycuoque oognitorum.

Saito, T., 1956. *Ecol. Rev. Sender*, 14: 141–147.

Sharma, K.R., Behera, N. and Mukerji, K.G., 1974. *Trans. Br. Mycol. Soc. Japan*, 15: 223–233.

Subramanian, C.V. and Vittal, B.P.R., 1979. *Zurnova Hedovigia,* pp. 361–369.

Sudha, K., 1978. A study of mycoflora of leaves litter. *Ph.D Thesis,* Madras university.

Visser, S. and Parkinson, D., 1975. *Can. J. Bot.*, 53: 1640–1451.

Vittal, B.P.R., 1973. Studies on mycoflora of leaves litter. *Ph.D Thesis,* Madras University.

Vittal, B.P.R., 1976. Studies on litter fungi I mycoflora of Atlantia and Gymnosporia litter. *Proc. Indian Acad. Sci.*, 83(B): 133–138.

Warcup, T.H., 1957. *Trans. Br. Mycol. Soc.*, 40: 237–262.

Warcup, T.H. and Talbot, P.H.B., 1962. *Trans. Br. Mycol. Soc.*, 45: 495–518.

Watson, E.S., McClurkin, D.C. and Huneycut, M.B., 1974. *Ecology,* 55: 1128–1134.

Webster, J., 1956. *J. Ecol.*, 45: 1–3.

Sustainable Environmental Management *Pages* **244–247**
Edited by: **Dr. L.V. Gangawane & Dr. V.C. Khilare**
Published by: **DAYA PUBLISHING HOUSE**

Chapter 27
Environmental Protection in Relation to Vipassana Meditation

J.S. Sardar

Director, Dr. Babasaheb Ambedkar Research Centre, Milind College of Science, Aurangabad – 431 002, M.S. India

ABSTRACT

It is the mind that pollutes matter. Environmental pollution is due to impure mind. Purification of mind can be achieved by awareness, which is obtained by technique of meditation known as vipassana. This meditation purifies mind. The pure mind or pure consciousness purifies polluted environment.

Keywords: *Environmental pollution, Meditation, Vipassana*

Introduction

The problem of environmental pollution can never be completely understood just by intellect for origin of such pollution lies not out there in space but lies in the human mind. Thus the mind that is polluted first and then the outer environment. Hence, until the thought process is cleaned and purified, the tall talks of environmental conservation will be of no use.

This article, putforth in two parts. First part will deal with environment and the second part with that of purification and conservation of environment through vipassana meditation. It is just not possible to escape from the consequences of our good or bad deeds. "Neither in the sky nor in the mist of the sea. Nor by entering into the clefts of mountains is there known a place on earth where stationing himself a man can escape from his evil deeds"–*Dhammapada*

Now let us deal with the first part, the environment. Environment is a certain set of circumstances surrounding a particular recurrence. The environment of an organism includes a biotic or physical milieu in which we think about geographic location, climatic conditions etc. The organic matter includes non-living organic matter and all other organisms, plants and animals.

The environment of human being includes the abiotic factors of land, water, atmosphere, climate, sound, odour etc. The biotic factors like animals, plants, bacteria and viruses also involve the environment. The social factors of aesthetics. Environment is undergoing extensive degradation these days.

The degradation can be of the following types, which can be summed up in the lines below:

Air air everywhere
No pure oxygen to breath,
Water water everywhere
No pure water to drink,
Food food everywhere,
No pure food to eat.
Tension tension everywhere
Speed speed to peace
Pollution pollution everywhere
Population blast so deep

Mother earth ever suffering with grief of the environment (Chitkar, 2000). When we look around, we see there pollution is everywhere, which is the cause of all suffering. Let us see this reaction of atmospheric oxygen (Manhan, 2000).

The oxygen cycle is critically important in atmosphere chemistry, geochemical transformation and life processes. Oxygen in the troposphere plays a strong role in process that occurs on the earth's surface. Atmospheric oxygen takes part in energy producing reaction such as

$$CH_4 + 2O_2 \rightarrow CO_2 + 2H_2O$$

Some oxidative weathering processes consume oxygen such as:

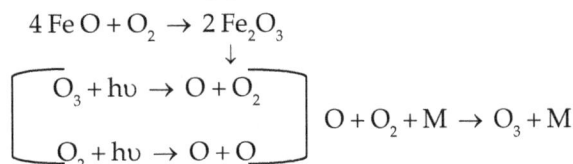

$$4\,FeO + O_2 \rightarrow 2\,Fe_2O_3$$

$$\downarrow$$

$$\left[\begin{array}{l} O_3 + h\upsilon \rightarrow O + O_2 \\ O_2 + h\upsilon \rightarrow O + O \end{array}\right] \quad O + O_2 + M \rightarrow O_3 + M$$

Ozone shield absorption + UV radiation from 220–330 nm.

Where M is N_2 or O_2 which absorbs excess energy given off by reaction and keeps the O_3 molecule together.

$$2CO + O_2 \rightarrow 2CO_2$$

Oxygen is returned to the atmosphere through plant photosynthesis.

$$CO_2 + H_2O + h\upsilon \rightarrow \{CH_2O\} + O_2$$

Molecular oxygen is somewhat unusual in that its ground state is a triplet state with two unpaired electros ($^3O_2$) that can be excited to singlet state ($^1O_2$).

In addition to O_2 the O, O^*_2 and O_3 are the species of upper atmosphere. Atomic oxygen, O, is stable in thermosphere. Atomic oxygen is produced by a photochemical reaction.

$$O_2 + h\upsilon \rightarrow O + O$$

Ionic oxygen O+ is produced by UV radiation

$$O + h\upsilon \rightarrow O^+ + e^-$$

Which further reacts with molecular oxygen on nitrogen in atmosphere.

$$O + O_2 \rightarrow O^*_2 + O$$

$$O^* + N_2 \rightarrow NO^+ + N$$

The diatomic oxygen can also be produced by photochemical reaction of low energy x-ray *i.e.*O_2 + hv $\rightarrow O_2^+ + e^-$ and by the following reaction

$$N_2^+ + O_2 \rightarrow N_2 + O_2^+$$

The formation of O_3 is by photochemical reaction.

$$O_2 + h\upsilon \rightarrow O + O$$

$$O + O_2 + M \rightarrow O_3 + M \text{ (increased energy)}$$

As the function of O_3 is to protect living beings on the planet earth from ultra violet radiation. The region of maximum ozone concentration is found in the range of 25–30 km high in the atmosphere. Some important reactions of atmospheric nitrogen are presented here:

$$N_2 + h\upsilon \rightarrow N + N$$

$$N_2{+} + O \rightarrow NO^+ + N$$

$$NO^+ + e^- \rightarrow N + O$$

$$O{+} + N_2 \rightarrow NO + N$$

$$N_2 + h\upsilon \rightarrow N_2 + e^-$$

$$N_2 + O \rightarrow NO^+ + N$$

$$NO + h\upsilon \rightarrow NO^+ + e^-$$

$$N_2 + h\upsilon \rightarrow N_2^+ + e^-$$

$$NO_2 + h\upsilon \rightarrow NO + O$$

Atmospheric CO_2

CO_2 along with water vapor is responsible for absorption of infrared energy re-emitted by the earth. The current researches suggest that atmospheric carbon dioxide (CO_2) level will alter the earth's climate through the green house effect.

Atmosphere Water

Water vapor absorbs infrared radiation even more strongly than does CO_2. Water vapor in the form of clouds reflects light from the sun and lowers the temperature. And water vapor in the atmosphere acts as a kind of blanket at night retaining heat from the earth surface by absorption of infrared radiation. Main source of water in the stratosphere is the photochemical oxidation of methane:

$$CH_4 + 2O_2 + h\upsilon \rightarrow CO_2 + 2H_2O$$

The water is the source of stratospheric hydroxyl and as it is shown by the following reaction.

$$H_2O + h\upsilon \rightarrow HO + H$$

There is much other reaction but for want of space they have not been given here.

Now if these reactions are violated by human activities just because of polluted mind. To avoid these consequences we have a very powerful technique known as Vipassana meditation. Let us know what this means, Vipassana means seeing things as they are. Seeing things in a special way. This seeing in a special way purifies mind and purifies consciousness. It brings about awareness, the highest kind of awareness with stillness of mind. Equilibrium in mind and brain and yet dynamic and full of energy. It brings equilibrium between five elements such as earth, water, fire, air and space with that of consciousness and functions towards the achievement of enlightenment. Once the inner equilibrium is established the outer environment, which is full of pollution, can be purified. It is the mind basically that can evolve processes to prevent, protect and purify. Any kind of pollution, may it be air, water, sound, space, soil and social for that matter, can be purified by technique of meditation, the vipassana meditation. We know destruction of nature, natural beauty and serenity and the natural resources results from ignorance, greed and lack of respect for living things.

Once ignorance, greed and lack of respect are neutralized or annihilated, which can be very well done by this technique of meditation this earth will be free from all sorts of pollution.

Even at global levels a ethical law could be so evolved that it will take care of protecting and preserving the environments in such a manner that it will conserve both matter and mind. Hence conserve, earth and energy not only of planet earth but of space in cosmos.

This needs to be cultivated by meditation. Meditation *that cultivates act of* loving, kindness, stillness, simplicity and contentment, truthful communication and clear and radiant awareness. Therefore the phenomena of cause and effect should be kept in mind moment to moment through meditation.

References

Chitkara, M.G., 2000. *Buddhism and Environment*, Vol. IV. APH Publishing Corporation, New Delhi.

Manhan, S.E., 2000. *Environmental Chemistry*. Lewis Publishers, London and New York.

Sustainable Environmental Management *Pages 248–258*
Edited by: **Dr. L.V. Gangawane & Dr. V.C. Khilare**
Published by: **DAYA PUBLISHING HOUSE**

Chapter 28

Urbanization: A Worldwide Problem

Nilima Wahegaonkar

Department of Botany, Vasantrao Naik Mahavidyalya, Aurangabad – 431 003, MS, India

ABSTRACT

Since their origin, cities have been the centers of education, religion, business, communication and political power. Unfortunately, they have also been sources of pollution, crowding, diseases and misery. At present, nearly half of the world lives in urban areas. Demographers expect that by the end of 21st century, 80-90 percent of the population will live in the cities and that some giant interconnecting metropolitan areas with hundreds of millions of residents shall be developed.

Keywords: *Urbanization, Immigration, Population, Pollution.*

Introduction

Since their origin, cities have been the centers of education, religion, business, communication and political power. As cradles of civilization, cities have influenced culture and society to a large extent. Unfortunately, they have also been sources of pollution, crowding, diseases and misery. Until recently, only a small percentage of the world's people lived permanently in urban areas and the cities were small by modern standards. Majority of people lived in rural areas where farming, fishing, hunting, timber harvesting, animal breeding, mining or other natural resource based occupation were the support of their lives.

Since the beginning of Industrial Revolution, some 300 years ago, cities have grown rapidly both in size and power. In the developing country like India, the transition from an agriculture based society to an industrial one has been accompanied by urbanization. Industrialization and urbanization bring many benefits but they also cause many problems also.

At present, nearly half of the world lives in urban areas. Demographers expect that by the end of 21st century, 80-90 percent of the population will live in the cities and that some giant interconnecting metropolitan areas with hundreds of millions of residents shall be developed.

What is a City?

The US Census Bureau considers, any incorporated community to be a city, regardless of size, and defines any city with more than 2500 residents as urban. In a *rural area,* most residents depend on agriculture or other ways of harvesting natural resources for their livelihood. In an *urban area,* however, majorities of the people are not directly dependent on natural resources–based occupation.

A *village* is a collection of rural household linked by culture economics, family relationships and association with the land. A *city,* on the other hand, is a differentiated community that is specialized in art, craft, services or professions rather than natural resource- based occupations. The rural village has a sense of security, tradition and association while a city offers more freedom to experiment, to be upwardly mobile and to break from restrictive traditions that can be harsh and unfriendly.

An urban area with about 10 million inhabitants is a supercity or megacity. Megacities in many parts of the world have grown to enormous size.

Urbanization

The United States underwent a dramatic rural to urban shift in the 19th and early 20th century. Now many developing countries are going through a similar demographic movement. In 1850, only about 2 percent of the world's population lived in cities. By 2000, 47 percent of the people in the world were urban. Only Africa and South Africa are still predominantly rural but people have started moving to cities. Some urbanologists predict that by 2100, the whole world will be urbanized to the levels now seen in developed countries.

Over 90 percent of the population growth in the next 25 years is expected to occur in the less developed countries of the world. Most of that growth will be in the already overcrowded cities of the countries like India, China, Mexico and Brazil. The combined population of these cities is expected to jump from its present 6 billion to more than 8 billion by the year 2025. The rural populations of these countries are expected to remain constant or even decline as rural people migrate into the cities.

Recent urban growth has been particularly remarkable in the largest cities, especially of the developing world. In 1900, 13 cities had populations over 1 million; all except Tokyo and Peking (Beijing) were in developed world, Europe or N. America. By 1995, there were 235 metropolitan areas of more than 1 million people but only three of them were in developed countries. In 2005 there are 26 most populous cities in the world that have population above 8 million. Population estimates for these world's largest urban areas calculated in 2005 are given in Table 28.1:

A century ago, London was the only city with more than 5 million people; now 26 cities have populations larger than that. Some futurists predict that by 2050 at least 400 cities will have populations of 1 million or more and 93 supercities each with 6 million or more residents. Three-fourths of those cities will be in the developing world. In just next 25 years Mumbai, Delhi (India), Karachi (Pakistan), Manila (Philippines) and Jakarta (Indonesia) are expected to grow by at least 50 percent.

The actual population of many of the megacities is uncertain. Where to draw city boundaries and how to count every person is controversial. The characteristic feature of the city is that it gradually merges into the nearly villages. Many people live beyond the official city boundaries but depend on the urban economy for their livelihoods. In the megacities of the developing world, as much as half the urban population is of temporary workers that are residents of unplanned slums and shantytowns. The estimates show that cities like Mumbai and Kolkata already accommodate populations more than 15 and 12 million respectively and many more people are pouring in everyday.

Table 28.1

Sl.No.	City	Country	Population
1.	Tokyo–Yokoama	Japan	33,200,000
2.	New York	United States	17,800,000
3.	Sao Paulo	Brazil	17,700,000
4.	Seoul–Incheon	South Korea	17,500,000
5.	Mexico City	Mexico	17,400,000
6.	Osaka-Kobe-Kyoto	Japan	16,425,000
7.	Manila	Philippines	14,750,000
8.	Mumbai	India	14,350,000
9.	Jakarta	Indonesia	14,250,000
10.	Logos	Nigeria	13,400,000
11.	Kolkata	India	12,700,000
12.	Delhi	India	12,300,000
13.	Cairo	Egypt	12,200,000
14.	Los Angeles	United States	11,789,000
15.	Buenos Aires	Argentina	11,200,000
16.	Rio De Janeiro	Brazil	10,800'000
17.	Moscow	Russia	10,500,000
18.	Shanghai	China	10,000,000
19.	Karachi	Pakistan	9,800,000
20.	Paris	France	9,645,000
21.	Nagoya	Japan	9,000,000
22.	Istanbul	Turkey	9,000,000
23.	Beijing	China	8,614,000
24.	Chicago	United States	8,308,000
25.	London	United Kingdom	8,278,000
26.	Shenzhen	China	8,000,000

Causes of Urbanization

The growth in urban populations takes place by two ways; by natural increase *i.e.* due to more births than deaths and secondly by immigration. Natural increase is supported by improvement in food supplies, provision of better basic necessities and advanced medical assistance that improves the life span of the people. Natural increase is responsible for two thirds of urban population growth in Latin America and East Asia. In Africa and West Asia, immigration plays the major role in urban growth. Immigration to the cities can be caused by two factors; push factors that force people out of the rural areas and the pull factors that attract them into the city Government policies also play on important role in the shift of people from rural to urban areas.

1. Immigration: Pull factors
2. Immigration: Push factors
3. Government Policies

Immigration: Push Factors

Migration of the people to cities is for many reasons. Sometimes when the villages get overpopulated it just cannot support more people then the surplus populations are forced to migrate to the cities in search of jobs, food and shelter. In some places, economic forces or political, racial or religious conflicts drive the people out of their homes. Whatever is the reason but it depopulates the rural regions.

Land possession patterns and changes in agriculture also play a role in pushing the people out of villages. Poor farmers have a small piece of land or are hired by rich ones during the season and once the harvest is over they are left jobless. Natural calamities like drought, heavy rains, floods etc. as well contribute to the immigration of the poor farmers out towards the cities.

Pull Factors

The first and foremost factor that attracts the rural populations to cities in spite of its hectic life is the excitement, liveliness and opportunities. Cities offer jobs, housing, entertainment and freedom from the constraints of village traditions. The cities provide possibilities for upward social status, prestige and power that are not available in the villages. Cities provide wide choices in arts, crafts and professions and also markets for the same.

The modern communication systems, the televisions, that have reached the smallest villages, attract people to the cities by propagating the superficial and false images of luxury and opportunity. We generally believe that the beggars and homeless people on the streets of many cities of developing countries have accepted to live as they do out of helplessness, but the fact is that many of them fancy to live in the city. In spite of the miserable conditions, life in the city probably seems to be favored.

Government Policies

The biased policies of the government often favor urban over rural areas thus supporting both push and pull of the people into the cities. Developing countries commonly spend most of their budgets on betterment of urban life, especially around the capital city where the political leaders live, even though a very small population lives there and gets benefited directly from the investment. This results in a monopoly of the cities on good jobs, housing, education and attractive life-style. These opportunities drag the rural people in search of better job and better income for better life.

Governments often manipulate food prices for the benefit of politically more powerful urban population but at the expense of rural people. Buying foodstuff for lower prices pleases the city residents, but the local farmers find it very uneconomical. It comes difficult for them to continue to grow crops. As a result, more and more people leave the villages to join their other rural mates in the city. Here the find some odd job that pays, though low but regularly to keep their families alive.

Problems of Urbanization

Both developed and developing countries face similar problems of urbanization. The major challenges being providing the available needs to the large and fast increasing population and secondly face the adverse effect that results as by-product of dense population.

It has been predicted that 90 percent of the human population growth in the next century is expected to occur in the developing countries, mainly in Africa, Asia and South Africa. The growth will take place in those cities which already have troubles in supplying sufficient food, water, housing, jobs etc. for their residents. The unplanned and unmanageable growth of these cities causes tragic and severe urban environmental problems. Some these challenges are as follows:

1. Traffic and congestion
2. Degradation of water resources
3. Loss of agricultural land
4. Loss of landscape and wildlife
5. Depletion of energy resources
6. Air pollution
7. Sanitation and water pollution
8. Housing.

Traffic and Congestion

First visit to any megacity, especially in a developing country excites any visitor by its variety of vehicles and mess they create at every street. But within a few hours he realizes that the noise, congestion and confusion of traffic are suicidal. In the cities like Jakarta, Mumbai, Kolkota etc. the traffic is almost chaotic all the time. People spend three to four hours reaching the work place. An average person spends almost 44 days a year sitting in traffic jams. About 20 percent of the total fuel is lost during the vehicle is standing still and hours of work is lost each year is worth at least (40 to 50 billion rupees)

Degradation in Water Resources

All the buildings, highways, parking lots, driveways and other paved areas in the cities lead to increase in runoff water and decrease in infiltration of water in the soil. These result in flooding, depletion in ground water decrease in quality of water, drying of springs, and encroachment of seawater. The havoc created by heavy rains in Mumbai in 2005, is an example of floods and advancement of sea water onto the land that resulted due to violation of the land and encroachment over the sea to accommodate more and more people that are streaming in the megacity. The tragedy took many lives and the economic loss is in billions.

Loss of Agricultural Land

The most serious problem, caused by urbanization, in the long run is the loss of prime agricultural land. The extensive development of cities consumes thousands of acres of agricultural land every year. Area under cultivation is depleting day by day. This ultimately affects the economy of the country. Prices of the food grains get inflated. The population that suffers is mainly the poor, which cannot afford the good quality stuff and have to buy substandard grains. The farmers that loose their lands in this process turn to the cities in search of jobs and complete the vicious cycle.

Loss of Landscape and Wildlife

New developments in the city, buildings sprouting up continuously, lying down of better roads, highways and expressways are consuming land at an increasing pace. In 1992 to 1997 3.2 million acres of land per year were developed as against only 1.4 million acres a year from 1982 to 1992. New highways and expressways are routed through open areas that provide the least expensive path at the cost of the picturesque landscape and wild life. To build the highways the mountains are cut down and the vegetation is destroyed. The animal species are exposed and either killed or forced to migrate to nearby jungles. The nearby ecosystems are invariably already occupied that makes it extremely difficult for the displaced species to survive.

Another reason for the decline in the number of species from birds to amphibians is the increasing number of road kills. Today, much more wild life is killed by vehicles than by hunters.

Depletion of Energy Resources

Urbanization has increased the number of vehicles. The urban life style has become vehicle-dependant. The increase in fuel consumption has nearly tripled since last 50 years. The comfortable lifestyle of cities has increased the expenditure of water and electricity. Heavy usage of electrical appliances, mixers, grinders, ovens, refrigerators, coolers, air conditioners has increased the consumption of energy beyond limits, which has resulted in depletion of the energy.

Air Pollution

Heavy traffic of old and poorly maintained vehicle, smoky factories create air pollution in the cities. Lenient pollution laws, corrupt officers, inadequate testing equipment, ignorance about the sources and effects of pollution and lack of funds to correct hazardous situations usually exaggerates the problem. An estimated 60 percent of Kolkata's residents are considered to suffer from respiratory diseases linked to air pollution. Lung Cancer mortality in Shanghai city is reported to be four to seven times higher than in the rural areas. Mexico City, which is in a high mountain bowl with lot of sunshine, little rain, high traffic levels and frequent air stagnation, has one of the highest levels of photochemical smog in the world.

India has more than 20 cities with populations of at least 1 million, and some of them–including New Delhi, Mumbai, Chennai, and Kolkata- are among the world's most polluted. Urban air quality ranks among the worlds worst. Of the 3 million premature deaths in the world that occur each year due to outdoors and indoor pollution, the highest numbers are assessed to occur in India. Continued urbanization has exaggerated the problem of air pollution and the cities are unable to implement adequate pollution control mechanisms. One of the most affected cities is New Delhi. The airborne particulate matter (PM) has been registered at levels more than 10 times India's legal limit. Vehicles are the major source of this pollution with more than 3 million cars, trucks, buses, taxis and rickshaws already on the road.

Sanitation and Water Pollution

The cities in the developing countries are growing rapidly but very few cities can afford to build modern waste treatment systems that can cope with the population growth. The World Bank estimates say that only 35 percent of the urban populations in the developing countries have satisfactory sanitation services. In Latin America, the situation very desperate, only 20 percent of urban sewage is treated. In Egypt, sewage system was built about 50 years ago in Cairo and the same is being used with the present large population of 10 million people. Less than one-tenth of the t5otal 3000 cities and towns in India have only partial sewage systems and water treatment plants. About 150 million people of Indian urban population do not have access to proper sanitation facilities. In Colombia, the Bogot River, which flows from the urban area with 5 million people residing in it, still has an average fecal bacteria count of 7.3 million per liter. This count is 700,000 times more than the safe drinking water level and 3500 times higher than its limit for swimming.

According to the studies by the World Bank, about 400 million people, *i.e.* one third of the population in the cities of developing countries do not have safe water to drink. The city residents get comparatively clean water than the rural population. In cities, many a time, people have to buy water that costs 100 times more than that is supplied by the municipal corporation and secondly it may not be safe for drinking purpose. Many rivers and streams are a no better than sewers but still the poor people, though urban residents, have to use it for washing clothes, bathing, cooking and in worst cases for drinking. Water borne diseases like Diarrhea, Dysentery, Typhoid and Cholera are very common in these countries. Rate of mortality especially in infants is very high.

Housing

A survey by United Nations have estimated that at least 1 billion *i.e.* 20 percent of the world's population live in crowded, unhygienic slums of the cities, in vast shanty towns and squatter settlements that are present along outskirts of most of the cities. Around 100 million people have no home at all. In Mumbai, for example, as many as half a million people sleep on the streets, footpaths and traffic circles because they have no other place to live. Ten times of this number live in crowded, dangerous slums and shanty towns throughout Mumbai. In Sao Paulo, Brazil, about 1 million 'street kids' who have either run away from home or have been abandoned by their parents live however and wherever they can. The census study in India (2001) indicated that there were total 607 towns reporting slums. Out of the total population of 178,393,941 persons 18,808,326 people live in slums. Numbers in some of the states in India are given below. In 2001 there were 11 states having more than 1 million populations living in slums.

Table 28.2

Sl.No.	*State*	*Total Population*	*Slum Population*
1.	Maharashtra	33,624,960	10,644,605
2.	Andhra Pradesh	15,752,946	05,149,272
3.	Uttar Pradesh	18,791,750	04,156,020
4.	West Bengal	14,250,720	03,822,309
5.	Tamil Nadu	14,175,792	02,530,289
6.	Madhya Pradesh	09,823,309	01,388,517
7.	Delhi	10,979,341	01,421,839
8.	Gujarat	11,427,259	01,346,709
9.	Karnataka	11,021,192	01,267,759
10.	Rajasthan	07,453,084	01,206,123
11.	Punjab	05,652,211	01,151,864

(Census of India 2001, revised on 22nd January 2002)

Slums and shanty towns are the most tragic reality of the megacities. *Slums* are generally legal but inadequate multifamily housings, built to rent to poor people. The chawls of Mumbai, India are tall tenements built in the 1950s to accommodate the immigrants. These housing were never safe but extremely dirty and airless buildings with bare minimum facilities. These chawls have already started collapsing due to their faulty architecture and low quality building material. Large families stay in a single room sharing their bathroom facilities with 50 to 70 other members.

Shanty towns are settlements developed when people move on the undeveloped lands and build their own houses. The shacks are built of metal or aluminum sheets, discarded packing tins, plastic sheets or whatever building materials people can grab. Some shanty towns are illegal housings where people stay or lend so called houses on rent without the permission from Municipal Corporation. Bustees around every megacity of India are examples of shanty town.

Squatter towns are one more type of illegal housings observed in and around the cities. These are spontaneous settlements where people occupy any land without the owner's permission. Sometimes these settlements are occupied by thousands of people who move on unused land overnight, building huts, laying out streets, markets and schools before authorities can root them out.

About three quarters of residents of Aids Ababa, Ethiopia live in filthy refugee camps. Two thirds of the population of Kolkata lives in unplanned illegal housings and nearly half of the twenty million people in Mexico City live in uncontrolled, unauthorized shantytowns. Many governments try to clean out the illegal settlements by bulldozing the huts and using the police force to drive out the settlers, but people either move back in or locate another place to settle down.

These popular but unauthorized residential places usually lack sanitary facilities, clean water supplies, electricity and roads. Considering the desperate and inhumane conditions, in these slums and shanty towns, many people just survive here. They keep themselves clean, raise families, educate children, find jobs and save a little money to send home to their parents. They learn to live in a dangerous, confusing and rapidly changing world and have hope for the future. They have parties; they sing and dance, and laugh and cry. Their ability to adapt to the situation is amazing.

Efforts to Meet the Challenges

Proper planning for the future, participation of the public and intervention of the Government of the county play a role in meeting the challenges caused due to urbanization. Such efforts will help us provide a sustainable environment.

Following efforts can be taken to meet the challenges of urbanization:

1. Planning of cities
2. Redesigning of cities
3. Change in lifestyle
4. Government policies

Planning of Cities

Expansion of cities, especially onto agricultural land, is considered a waste of land and resources. Many people feel that growing cities decrease local agricultural resources and spoil the untouched countryside and quiet rural areas. Others think that expansion of cities is a good opportunity to decongest the central cities and build smaller, more pleasant, more livable communities that have the benefits of both technological and cities and rural villages.

An alternative to spreading the population across a wide area is to build upward. This model of the city is based on technology and has been called "*technopolis*", a vertical city. For many people the polished, climate controlled, artificial environment is an ultimate achievement of modern civilization. To others, it seems brutal and dehumanizing. The emerging supercities of the developing countries are also heading toward this style because of its association with wealth, power and progress.

Redesigning of Cities

Many ideas have been put forth in redesigning the cities and make them more diverse, flexible and energy efficient.

1. The size of the city should be around 30,000 to 50,000 people, large enough to be a complete city but small enough to be a community. The city should be surrounded by a green belt.
2. It should be determined in advance where development will take place so as to prevent disorganized development of the city.
3. Develop shopping malls that would not only be shopping centers but places for the people to stroll, meet friends and interact socially.
4. Every residential area should be associated with local shopping centers that can provide the daily needs of the people with convenience, less stress, less automobile dependence and less use of time and energy.
5. Walking, cycling or use of small low speed, energy-efficient vehicles must be encouraged for most of the nearby trips.
6. Housing must be planned in such a way that it should provide comfort and safety.
7. Cities should be self-sustainable by growing food locally, recycling wastes and water, using renewable sources of energy, reducing noise, air and water pollution and creating a cleaner and safer environment.
8. Development of green belts and gardens must be encouraged as it absorbs up to 70 percent of rain water and provide habitat for wild life.

Change in Lifestyle

Day by day the people are getting used to a more unnatural type of life style. They are becoming more dependent on luxurious and comfortable lifestyle. Such life style is good to cope with the fast growing world but it has made man depend more on mechanical utilities, electrical and electronic appliances and automobiles. They are going away form the nature. They prefer to sit inside the four walls and enjoy the unnatural controlled and environment. Though this is essential for long hours of work and maximum output without fatigue but one should also realize the need to be in natural surroundings.

A citizen must remember that he is more than just a resident of a city, he or she must be an active participant to enjoy the privileges and benefits but not to the extent that it would be disastrous to the community or environment he lives in.

Government Policies

Principles of environmental protection and sustainable development should be incorporated into local and national government.

A city is more than just a place to live; it is a community that survives on civic duties and civil behavior. Civil defense, civil law and civil rights all together are significant for mutual protection of man and environment. The government lays down the laws and the citizens are bound to follow them. They should be aware of their rights and at the same time should be conscious of their duties so that they help out the government in meeting up the challenges like pollution, industrialization, urbanization etc.

Indian Government has taken steps to meet the challenges of urbanization; industrialization as well as pollution. The Indian Government has laid down various Acts since the First UN Conference on Human Environment, held in Stockholm (Sweden) on 5th to 10th June 1972. The conference was specially held to express concern over depletion of resources and air and water pollution.

The Indian Government is meeting the challenges of urbanization in different ways as follows:

1. Organization of Human Settlement Schemes to deal with the problem of urban slums. The Housing and Urban Development Corporation (HUDCO) 1970 and National Commission of Urbanization, 1988, are taking steps.
2. Legislation of Water pollution Act, 1974, for prevention and control of water pollution.
3. Legislation Air Pollution Act, 1981, to prevent and control air pollution.
3. Legislation Environment Protection Act. 1986.
4. Establishment of non-conventional Energy Sources, for new renewable non-polluting sources of energy as solar, wind, tidal, biomass-based sources, biogas.
5. Recycling of wastes, and
6. Establishment of Policies on housing, occupational health, slum improvement etc.

Considering the volume of problems that are created by urbanization through out the world the time has come when people should be made aware of its intensity. They should actively participate and help the government to pull the country out of the disaster.

References

Asrari E., V.S. Ghole, P.N. Sen and Motessaddi, 2005. *Nature Environment and Pollution Technology*, Vol. 3(2): 195–198.

Cunningham, W.P. and Barbara Woodworth Saigo, 2001. *Environmental Science: A Global Concern.* McCraw Hill Companies, New York, pp. 646.

Fulekar, M.H., 2004. *Nature Environment and Pollution Technology*, 3(4): 447–450.

Government of India Report on Census of India, 2002. Published by Office of the Registrar General, India.

Gregory, Robin, 2000. *Environment*, 42: 34–44.

http://www.eia.doe.gov/eneu/cabs/indiaenv/html.

http://www.virtvalref.com/uncrd/_sub/s109.htmal.

http://www.adb.ord/documents/CAPs/IND/0302.

http://www.iied.org/HS/about.html.

http://popindex.princeton.edu/browse/v57/n2/c.html.

http://darwin.nap.edu/books/0309075548/html/175.html.

http://www.prenhall.com/wright.

http://www.mhhe.com/environmentascience.

Jain, Renu, Dev, Kumar Dwivedi and A.B. Gupta, 2004. *Nature Environment and Pollution Technology*, 3(4): 429–434.

Mahendra, S.P. and Krishnamurthy, 2004. *Nature Environment and Pollution Technology*, 3(4): 473–476.

Mohanraj, R., P.A. Azeez and S. Pattabhai, 2005. *Nature Environment and Pollution Technology*, 4(2): 621– 626.

Murali, M. and T. Indira, 2004. *Nature Environment and Pollution Technology*, 3(3): 249–253.

Pawar, C.T. and M.N. Joshi, 2002. *Nature Environment and Pollution Technology*, 1(4): 351–356.

Raina, A.K. and Bindu Aggarwal, 2003. *Nature Environment and Pollution Technology*, 2(4): 479–484.

Sharma, P.D–2003. *Environmental Biology*. Rastogi Publications, Meerut, pp. 416.

Tewari, D.N., 1987. *Victims of Environmental Crisis*. EBD Educational Pvt. Ltd. Dehardun, India, pp. 196.

Wright, Richard T. and Bernard, J. Nebel, 2002. *Environmental Science: Toward a Sustainable Future*, Prentice-Hall of India Pvt. Ltd., New Delhi, pp. 681.

Sustainable Environmental Management *Pages 259–266*
Edited by: **Dr. L.V. Gangawane & Dr. V.C. Khilare**
Published by: **DAYA PUBLISHING HOUSE**

Chapter 29

Transition from Hydrocarbon to Hydrogen: Global Initiatives to Harness Green Energy

A. Yogamoorrthi

Centre for Futures Studies, Pondicherry University, Pondicherry, (Union Territory), India

ABSTRACT

In view of growing concerns about climate change and depleting fossil fuels reserves, increased use of the international community in terms of its energy consumption and future energy demand, is forced to have a change in its options and alternatives. It is the need of the hour to opt or substitute, non-conventional sources of energy in all possible ways and means to minimize our dependency on non-renewables. This paper looks into the potential renewable sources of energy and high lights the avenues and opportunities in developing mass production of Biological Hydrogen using microorganisms.

Keywords: *Hydrocarbon, Hydrogen, Energy, Genetic engineering.*

Introduction

Energy is the driving force in the development of any country. Energy independence is a self-sustained economy, which will function well with total freedom from oil, gas or coal imports. In view of energy as a factor in global transformation, 19th century was based on ' Steam economy'; the 20th century on Oil Economy whereas 21st century will optimistically be based on the green energy economy *i.e.* renewable energy economy. There exists a strong relationship between economic growth and energy consumption. With regard to population, India is the second largest country in the world and has 17 per cent of the world population. The huge population, from 300 million in 1947 to over one billion people today, is putting strain on environment, infrastructure, employment and natural resources. This situation compels international community to attain ' energy independent economy.

Energy is drawn from conventional and non-conventional or natural sources. Among conventional sources, coal and fossil fuel are important while solar, wind energy and biomass is the main components of non-conventional energy source. Availability of the conventional source is not only finite but also very limited. India's mineral oils reserves are just 0.8 percent of the world's known reservoirs. The demand of fossil fuel or mineral oil is increasing day by day and is likely to reach 275 MMT by the year 2025. The production in the year 2003-04 was just 33.4 MMT *i.e.* 33 percent of the total consumption during this period. Accordingly, about Rs. 94520 crore were incurred on import of mineral oils, which account for 26.7 percent of total imports during the year 2003-04.Thus diversification of energy portfolio seems to be the answer to this gigantic problem.

Renewable energy sources are indigenous and can contribute towards reduction in dependency on fossil fuels. Renewable energy also provides national energy security at a time when decreasing global reserves of fossil fuels threaten long-term sustainability of the Indian economy. This article looks into the viability of clipping biological hydrogen as a potential renewable energy source for

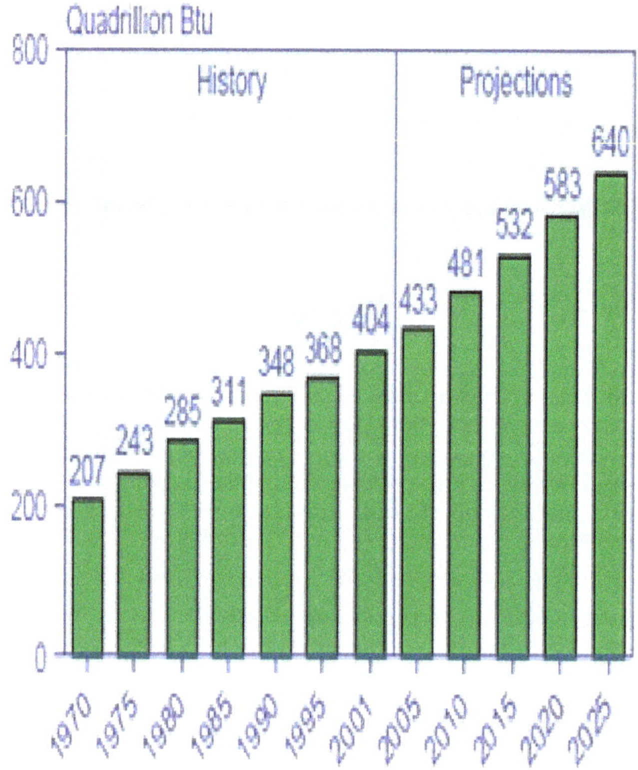

Figure 29.1: World Energy Consumption, 1970–2025

***Sources*: History: Energy Information Administration (EIA), International Energy Annual, 2001 (DOE/EIA-0219 (2001) (Washington, DC, February, 2003), Website: www.eia/doe/gov/iea. Projections: EIA, System for the Analysis of Global Energy Markets (2003).**

future world. It also looks into how best scientific capabilities could help the world to attain the status of sustainability in terms of energy by substituting renewal energy resources in general; biological hydrogen in particular.

Why Hydrogen?

Hydrogen gas (H_2) is a high caloric, clean energy carrier that may take a central place in our future society sustained but renewable energy source. Besides its large-scale application as a clean fuel, $H_{2 is}$ produced for industrial use as a chemical feed stock in the manufacture of ammonia, methanol, refined petroleum fuels, hydrogenate vegetable and animal oils and other chemical. It has its applications in the reduction of metals and production of synthetic natural gas. With the increasing burning of fossil fuel and consequent possible changes in global climate, the use of H_2 as a safe fuel. Hydrogen gas is seen as a future energy carrier by virtue of the fact that it is renewable, does not evolve the " greenhouse gas " CO_2 in combustion, liberates large amounts of energy per unit weight (122 kj/ g) on combustion *i.e.* its heat of combustion per unit weight is about 2.5 times that of hydrocarbon fuel, 4.5 times that of ethanol and 6.0 times that of methanol and its thermodynamic energy conversion efficiency of 30-35 per cent is grater than that of gasoline. And also easy to convert electricity using in fuel cells.

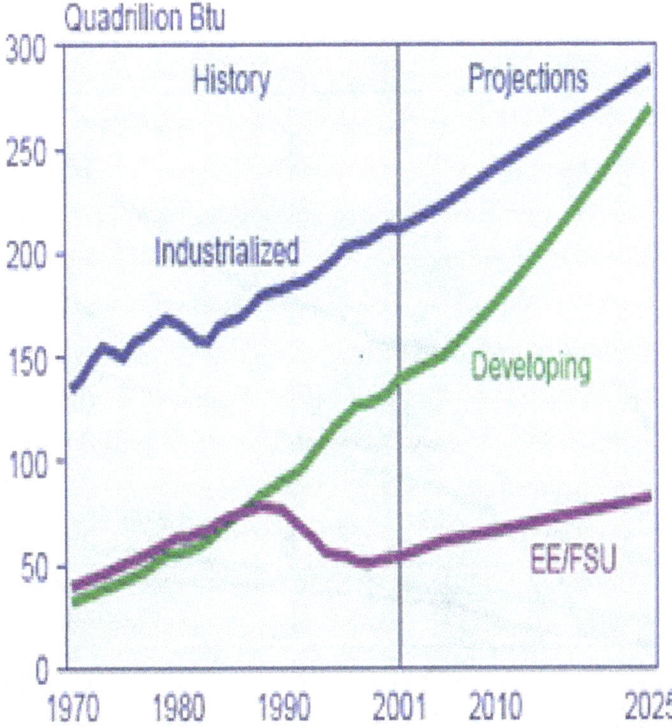

Figure 29.2: World Energy Consumption by Region, 1970–2025

***Sources*: History: Energy Information Administration (EIA), International Energy Annual, 2001 (DOE/EIA-0219 (2001) (Washington, DC, February, 2003), Website: www.eia/doe/gov/iea. Projections: EIA, System for the Analysis of Global Energy Markets (2003).**

At present, H_2 is seen as an expensive option, because it is costly to produce in large quantities at the industrial level, and because fuel cells rely on scarce metals such as platinum. H_2 and electricity could team to provide attractive options in transportation and power generation. Interconversion between these two forms of energy suggests on-site utilization of H_2 to generate electricity, with the electrical power grid serving in energy transportation, distribution utilization, and H_2 regeneration as needed. A challenging problem in establishing H_2 as source of energy for the future is the renewable and environmental friendly generation of large quantities of H_2 gas. Thus, processes that are presently conceptual in nature, or at a developmental stage in the laboratory, tested for feasibility. It has the potential to reshape the entire energy industry. If all vehicles were run on hydrogen fuel cells, it would be a huge step towards solving air pollution problem in cities. Thus it helps to reduce human health and visibility problem.

Global Concern Over Hydrogen Energy

A roundtable was organized by Winrock International India (WII) on April 1, 2005 to facilitate a US-India dialogue on hydrogen energy in New Delhi. The event was sponsored by the US Agency for International Development (USAID) as part of the ongoing 'Communications and Outreach Support' program under the USAID and National Energy Technology Laboratory (NETL) Greenhouse Gas Pollution Prevention Project.

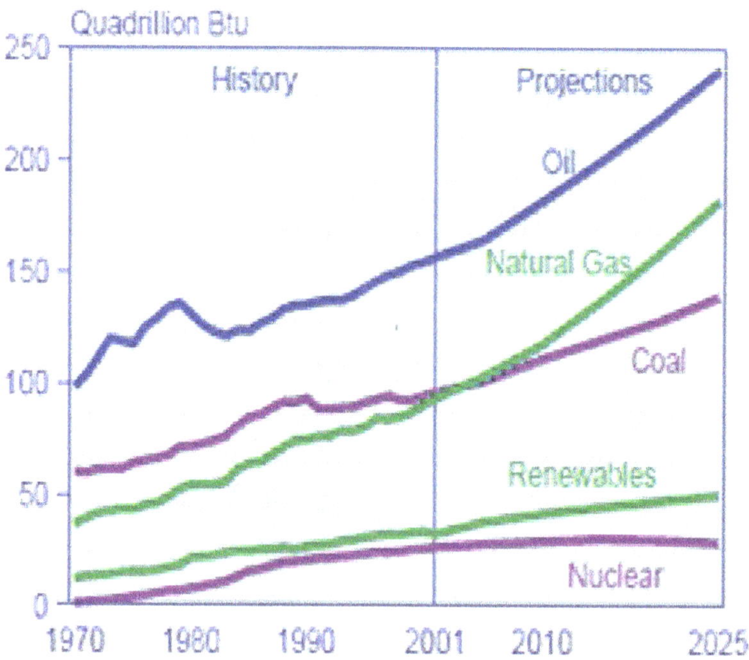

Figure 29.2: World Energy Consumption by Energy Source, 1970–2025

***Sources*: History: Energy Information Administration (EIA), International Energy Annual, 2001 (DOE/EIA-0219 (2001) (Washington, DC, February, 2003), Website: www.eia/doe/gov/iea. Projections: EIA, System for the Analysis of Global Energy Markets (2003).**

The objective of the roundtable was to develop an overall strategy and next steps on enhancing cooperation between the United States and India with a specific focus on the role of the private sector. Against this backdrop, a small group of 25 experts were invited to the roundtable to discuss, debate, and comment on key issues pertaining to the development of a hydrogen economy in India. These experts represented various stakeholder organizations such as academic (Indian Institute of Technology–Madras [IIT–Madras]), the private sector (Bajaj Auto), industry associations (Society of Indian Automobile Manufacturers [SIAM], Confederation of Indian Industry [CII]), energy companies (National Thermal Power Corporation [NTPC], Indian Oil Corporation [IOC], Indraprastha Gas limited [IGL]), US organizations (USAID, NETL, the US Embassy Science Office, and the United States Asia Environmental Partnership [USAEP]), as well as the Indo-US Science and Technology Forum and WII. The roundtable was structured into three sessions. In the first Plenary Session, experts from the United States (NETL) and India (IOC and NTPC) made presentations on the status of developments in hydrogen energy in their respective countries. The second Plenary focusing on "Discussion on the Potential of US-India Hydrogen Collaboration" began with a presentation by USAID on "Clean Energy and Hydrogen Initiative," followed by open house discussions on key issues concerning hydrogen energy development in India. The experts discussed these issues in detail and shared their concerns and views, and provided valuable suggestions to take forward the hydrogen work in India in cooperation with their US counterparts.

Hydrogen Energy Road Map

The urgency of making the transition from the present hydrocarbon energy economy (oil) to hydrogen energy economy has been recognized globally. Efforts are in progress not only in the developed countries but also in the developing countries like India, China, and Brazil. Receiving the road-map document, on 21 November 2005 at New Delhi, prepared by the Steering Group on Hydrogen 'the road map would put India on the forefront of the new global hydrogen energy economy and provide sustainable energy security to all citizens in the coming years'. provide a clean, reliable, sustainable, and alternative energy source for our growing energy needs. India could provide the Road map for other countries also in this area'.

The National Hydrogen Energy Road Map describes in depth the different aspects of hydrogen energy that include production, storage, transport, delivery, applications, safety, standards and codes, and capacity building and awareness. It highlights hydrogen production as a key area of focus in addition to existing methods of hydrogen production based on steam methane reformation, production of hydrogen from coal gasification, nuclear energy, biomass, etc. In the area of hydrogen storage that includes gaseous, liquid, and solid stage storage, various goals concerning efficiency of storage, useful cycle life, compactness, cost, and so on to be achieved by 2020 have been identified. The road map proposes two major initiatives: GIFT (Green Initiative for Future Transport) and GIP (Green Initiative for Power Generation). GIFT aims to develop and demonstrate hydrogen-powered IC (internal combustion) engine and fuel cell-based vehicles ranging from small two-/three-wheelers, cars/taxis, buses, and vans through different phases of development. It is envisaged that implementation of the National Hydrogen Energy Road Map, as proposed, would see one million hydrogen-fuelled vehicles on Indian roads by 2020. GIP, on the other hand, envisages developing and demonstrating hydrogen-powered IC engine/turbine and fuel cell-based decentralized power generating systems.

India Action Plan to Utilize Hydrogen

Shri Kariya Munda, (The Minister for Non-conventional Energy, Govt. of India) constituted a Steering Group under the Chairmanship of Ratan Tata, and the Steering Group report is expected in next few months to prepare an action plan for demonstration and commercialization of hydrogen and fuel cell powered vehicles and power generating systems in the country through public private partnership. The Ministry of Non-conventional Energy Sources setup the National Hydrogen Energy Board to guide the Indian Hydrogen Energy Programme. In it's first meeting various issues concerning coordinated development of hydrogen energy programme for India including production; storage, transport, distribution, safety, standards and applications were discussed.

Biological Hydrogen Production Using Microbes

Molecular hydrogen is widely used by microorganisms as a source of energy. A microbiological production of H_2 by fermentation of different substrates is a complicated process. The conversion of sugars and carbohydrates to H_2 is achieved by a multienzyme system. Two main classes of hydrogenases have been identified as: a) Iron (Fe) only and b) Nickle-Iron (Ni-Fe) containing. One of the best-studied arobic hydrogen oxidizers, the beta Proteobacterium Ralstoniaeutropha (formerly Alcaligeneseutrophus), harbors two distinct [NiFe]-hydrogenases which catalyst the heterolytic cleavage of H_2 into 2H+ and 2e-. The genes encoding the hydrogenase subunits are arranged in two large operons together with accessory and regulatory genes involved in hydrogenase biosynthesis. The distribution of (Fe)–hydrogenases was once thought to be limited to a small number of bacteria and a few peculiar H_2 production eukaryotes. However, it is now clear that [Fe]- hydrogenases are more widely distributed among eukaryotes than reports of H_2 production have suggested. Indeed genes bearing the hallmark signatures of [Fe]–hydrogenases are found both in our own genome and in the genomes of higher eukaryotes. At present, the functions of most of these new proteins remain unknown; it is not even known whether they can all make H_2, Radical new hypotheses have suggested that hydrogenases played a keyrole in the formation of eukaryotic cell. These unique enzymes have thus moved from the margins of eukaryotic biology to become the focus of intense speculation and interest.

The search for microbes conventionally involves isolation form different sources and characterization of the naturally occurring microbes. The process is tedious and time consuming. A rapid screening method for hydrogen–producers is not available and is one of the reasons why very few new hydrogen–producers are reported. Modern biological methods involving 16s–r RNA gene amplification/cloning and sequencing, micro arrays, proteomics, etc., can be exploited for searching novel microorganisms. On the other hand various metabolic and genomic databases can be restored to for searching novel and efficient hydrogen producers. Infect, with the availability of more than 300 sequenced genomes, bioinformatic tools have become handy for such analyses.

Enhancement of Hydrogen-producing Capabilities through Genetic Engineering

Although genetic studies on photosynthetic microorganisms have markedly increased in recent times, relatively few genetic engineering studies have focused on altering the characteristics of these microorganisms, particularly with respect to enhancing the hydrogen-producing capabilities of photosynthetic bacteria and cyanobacteria. Some nitrogen-fixing cyanobacteria are potential candidates for practical hydrogen production. Hydrogen production by nitrogenase is, however, an energy-consuming process due to hydrolysis of many ATP molecules. On theother hand, hydrogenase-dependent hydrogen production by cyanobacteria and green algae is "economic" in that there are no ATP requirements. The mechanism of hydrogen production is not however sustainable under light conditions. Water splitting by hydrogenase is potentially an ideal hydrogen-producing system. Asada and co-workers attempted to overexpress hydrogenase from Clostridium pasteurianum in a cyanobacterium, Synechococcus PCC7942, by developing a genetic engineering system for cyanobacteria. These workers also demonstrated that clostridial hydrogenase protein, when electro-induced into cyanobacterial cells is active in producing hydrogen by receiving aelectrons produced by photosystems.

Another strategy being examined is the enhancement of hydrogen-producing capabilities of photosynthetic bacteria. In nitrogenase-mediated hydrogen-producing reactions, a considerable amount of light energy, which is converted to biochemical energy by the photosystem, is lost through various biochemical processes. Control of the photosystem at an appropriate level for nitrogenase activity, would result in reduced energy losses, and thus improved light energy conversion. To this end, with the objective of utilizing genetic engineering techniques in controlling the photosystem level in the potent hydrogen-producing photosynthetic bacteria Rhodobacter sphaeroides RV, the puf operon encoding photoreaction center and light-harvesting proteins was isolated and characterized.

Researches are working on a bacterium that has potential for the mass scale hydrogen production for future energy use in large-scale applications. It is a non-phototrophic bacterium and does not require an expensive bioreactor for high-density cell culture and H_2 production in large scale. The bacterium known as Y19 grows fast (doubling time = I Hour) and has high H_2 production activity (27 mmol/hg cell). It can grow to a high density on organic carbon sources aerobically or anaerobically, and is easily shifted to CO dependent production state (anaerobic conditions in the presence of CO). Y19 can produce H_2 from various organic carbon sources including Glucose, Sucrose and Starch. More studies on long term stability, recovery from oxygen damage, metabolic engineering on carbon and energy flow are being carried out. Y19 is believed to have a great potential for mass-scale biohydrogen production for potential future energy applications.

Research and Development in Biological Hydrogen Production

Biological hydrogen production is now receiving much attention as an environmentally acceptable technology. Although a few research groups are active in the basic or applied fields of hydrogen production, recent world wide environmental problems have prompted the formation of national projects for biological hydrogen production. The German Federal Ministry for Research and Technology funded a biological hydrogen production project (1989–1994), in which universities undertook basic research. In Japan, the ministry of International Trade and Industry is promoting a project for biological hydrogen production by environmentally acceptable technology (1991–1998) through the Research Institute of innovative technology for the earth (RITE), with financial support of the New Energy and Industrial Development Organization (NEDO). The project includes development of total technologies for biological hydrogen production, the screening and breeding of microorganisms, and basic research and development of photo bioreactors and anaerobic bioreactors.

The Hydrogen Committee of the International Energy Agency (IEA, under the ouspices of the OECD) has rearranged Annex Committees for hydrogen technologies. The target of Annex 10 is the photoproduction of hydrogen. This consists of three subtasks: i) photoelectrochemical hydrogen production, ii) photobiological hydrogen production, and iii) standardization. The three-year plan (1995-1997) aimed to establish a closely collaborative worldwide research network promoting hydrogen production technologies.

Conclusion

NASA is the primary user of hydrogen as an energy carrier; it has used hydrogen for years in the space program. Hydrogen fuel lifts the space shuttle into orbit. Hydrogen batteries—called fuel cells—power the shuttle's electrical systems. The only by-product is pure water, which the crew uses as drinking water. Hydrogen fuel cells (batteries) make electricity. They are very efficient, but expensive to build. Small fuel cells can power electric cars. Large fuel cells can provide electricity in remote areas. Hydrogen may soon be added to natural gas, though, to reduce pollution from existing plants. Soon hydrogen will be added to gasoline to boost performance and reduce pollution. Adding just five percent hydrogen to gasoline can significantly lower emissions of nitrogen oxides (NOX), which contribute to ground-level ozone pollution. An engine that burns pure hydrogen produces almost no pollution. It will probably be 10-20 years, though, before you can walk into your local car dealer and drive away in a hydrogen-powered car.

References

Das, D. and Vezhiroglu,T., 2001. *International Journal of Hydrogen Energy*, 26 (1): 13–28.

Kalia *et al.*, 2003. *Trends in Biotechnology*, 21(4):152–156.

Kalia, V.C. and Sadhana Lal, 2005. *Bioenergy News*, 8(3): 12–18.

Lenz *et al.*, 2002. *J. Mol. Microbiol. Biotechnology*, 4(3): 255–262.

Sustainable Environmental Management *Pages* **267–273**
Edited by: **Dr. L.V. Gangawane & Dr. V.C. Khilare**
Published by: **DAYA PUBLISHING HOUSE**

Chapter 30

Biotechnology in Environment Management

S.R. Wate[1] and S. Bodkhe[2]

*[1]Deputy Director, Scientist and Head, [2]Scientist,
Environmental Impact and Risk Assessment Division, National Environmental Engineering Research Institute, NEERI PO, Nehru Marg, Nagpur – 440 020, India*

ABSTRACT

The use of bioremediation and allied biotechnologies for management of environmental pollution is at its zenith in recent years. Recent advancements in environmental engineering coupled with better understanding of molecular biology offer opportunities to make the treatment/ remediation process more efficient. Biotechnology thus represents a promising method of solving environmental pollution problems. In this paper, various ways of using the biotechnology in the environment are discussed. Although the biotechnological tool has gained excellent reputation towards solving environmental problems, the paper envisages the research needs in the field of biotechnology.

***Keywords**: Biotechnology, Environmental management.*

Introduction

Growing public awareness of health issues related to pollution and improved enforcement of legislation for industries by the Central and State Pollution Control Boards, strong business prospects are envisaged in Indian environmental management industry.

Bioremediation and allied technologies for environmental cleanup has achieved commendable success from a relatively insignificant international representation in the mid-1980s. Successful utilization of environmental biotechnology with the general growth of concurrent government, industrial, and public support has further perpetuated the accelerated use of biotechnological tool in the environmental management.

Environmental biotechnology is not a new field; composting and wastewater treatment technologies are familiar examples of "old" environmental biotechnologies. However, recent developments in molecular biology, ecology, and environmental engineering now offer opportunities to modify organisms so that their basic biological processes are more efficient and can degrade more complex chemicals and higher volumes of waste materials.

Although commendable success has been achieved, the potential benefits of the new environmental biotechnologies are far from fully realized and exploited. Advances in this arena are delayed not only by legal and social barriers but also by a dearth of basic scientific knowledge about organisms that are useful as a biotechnological tool.

Microorganisms (primarily bacteria and fungi) are nature's original recyclers. Their capability to transform natural and synthetic chemicals into sources of energy and raw materials for their own growth suggests that expensive chemical or physical remediation processes might be replaced or supplemented with biological processes that are lower in cost and more environmentally benign.

Biotechnology therefore represent a promising, albeit largely untapped resource for solving environmental pollution problems. Research is ongoing to verify the bioremediation potential of microorganisms. For example, a recent addition to the growing list of bacteria that can sequester or reduce metals is *Geobacter metallireducens*, which removes uranium, a radioactive waste, from drainage waters in mining operations and from contaminated groundwater's. Studies suggest that further exploration of microbial diversity is likely to lead to the discovery of many more organisms with unique properties useful in bioremediation.

A tiny fraction of the microbial diversity of the Earth has been identified and an even smaller fraction has been examined for its biodegradation potential. Research aimed at characterizing this diversity likely would lead to the discovery of novel mechanisms for biodegradation of pollutants. As the biochemical and ecological nature of microbial biotransformations of pollutants is getting evolved and understood, new bioremediation technologies based on these mechanisms are emerging continuously.

Role of Biotechnology in Environmental Management

Cleaner Production–Manufacturing with Less Pollution or Less Raw Material

Environmental biotechnology offers ways to make industrial processes work more efficiently and create less pollution in following ways:

1. Biotechnology may replace synthetic chemicals. For example, it is possible to make a variety of plastics from plant sugar rather than from petrochemicals by using specially tailored yeasts and microorganisms. The advantage of these products is that they are biodegradable
2. It is expected that bioleaching will progressively replace chemicals used in recycling or bleaching paper. Many laundry detergents use enzymes to replace phosphate detergents
3. Some heavy-duty stain removing bacteria have even been found in heavily-alkaline lakes where they have survived by break down of toxins in their environment
4. Scientists are looking for microorganisms in hostile natural environments (for example very hot, cold or oily places) to use the potential they have developed for industrial purposes.

Better Treatment of Wastewater and Solid Wastes

1. Most sewage treatment plants traditionally use a combination of chemical, physical and microbial treatment to break down organic matter in the sewage. It is now possible to use modern biotechnology techniques to analyse the conditions needed to optimize the performance of the biomass and the systems so as to tailor the technology to different uses
2. The treatment of organic waste produces valuable products such as biogas, and modern high-tech composting factories can turn tones of organic garbage into precious soil.

Bioremediation–Cleaning of Contaminated Sites

1. Naturally occurring micro-organisms, mainly bacteria and fungi, are being used to help clean up sites contaminated by heavy metals, acids, petroleum derivatives, chlorinated solvents and explosives
2. Certain plants have also been found to absorb toxic metals such as mercury, lead and arsenic from polluted soils and water, and scientists are hopeful that they can be used to treat industrial wastes
3. Oil spills; from small industrial units to massive ocean oil spills are often treated with oil-degrading bacteria.

Tracking the health of the environment through bio-monitoring

1. Two frequently harvested species of marine mollusk, the Sydney rock oyster and mussel, may be useful "bio-indicators" for the heavy metal toxin zinc and cadmium.

Biomass Energy

Biomass is plant and animal material that can be used as an energy source from the traditional wood to waste material such as bagasse from sugarcane, to specially grown energy crops that can be converted to ethanol using modern biotechnology techniques and used with petrol in vehicles.

1. Several companies are planning to build bio-diesel plants that convert vegetable oil or abattoir by-products into diesel fuel
2. Even human waste can be converted into a combustible fuel.

Improved Varieties

Genetic engineering is not the only tool for creating novel plants with special traits. Convention breeding is still used for developing useful plants.

1. By cross breeding, scientists are developing wheat plants to increase their salt tolerance so that they can grow in areas of greater salinity
2. Eucalypts are also being cross-bred to introduce natural genes from other species to create plants that tolerate salinity and dryland conditions.

Genetic Engineering for Environmental Solutions

Many of the biotechnology applications described above can be varied or refined using genetic engineering.

1. Several companies are trying to develop gene technologies to develop enzyme products that detoxify pesticide residues, which would be of particular value to the cotton, horticultural and rice industries
2. In a variation on bioindicators, bacteria have been genetically modified as 'bioluminescors' that give off light in response to several chemical pollutants. These are currently being used to measure the presence of some hazardous chemicals in the environment
3. Other genetic sensors that can be sued to detect various chemical contaminants are also being trialed. Some can be used to track how pollutants are naturally degrading in ground water.

India has a large number of identified polluted areas, including land, fresh water, and marine sites that, by law, must be cleaned up. The extent of contaminated agricultural acreage, industrial sites, and aquifers and other water bodies is unknown, but the magnitude of the problem is undoubtedly large and clean-up expenses could be considerably high. India is among several nations developing bioremediation technologies. Maintaining and enhancing the India position in this arena will require continued investment in the generation of new knowledge needed for the development of new technologies. Investment in bioremediation research has the dual benefits of solving important environmental problems while stimulating the growth of the Indian bioremediation industry. The complex environmental milieu in which bioremediation and other environmental technologies will be employed demands an holistic research strategy, the traditional, piecemeal approach will not be adequate. Recent developments in biology have provided new tools and approaches for monitoring the environment and engineering organisms with the capacity to degrade environmental pollutants. These developments have created unprecedented opportunities for significant advances.

Bioremediation research generally is conducted at one of three scales: laboratory, pilot scale, or field trial. To help ensure that results achieved at the first two scales can be translated to the field, the research program should be conceived as a continuum, with investigators working at each scale involved throughout the research conceptualization and planning process. The aim is to translate research findings from the laboratory into viable technologies for remediation in the field.

The biological and physical complexity of the field environment where research findings ultimately will be tested and applied demands a multidisciplinary research team composed of microbiologists, ecologists, engineers, hydrologists, and other specialists.

Field Studies

The successful development and application of bioremediation technologies depends on field-based research to verify the efficacy of planned approaches under natural conditions. Field research is a complex undertaking, in part because any number of problems can arise. For example,

1. An immunological probe may not be effective due to binding of the substrate to the soil or soil organic matter
2. An engineered microorganisms may be competitive with the indigenous population, or it may not survive at certain times of the year or under specific moisture regimes
3. Biocontrol and environmental monitoring technologies are unlikely to be effective in all regions, due to differences in plant communities, soils, climate, disease vectors, land management practices, and economic policies

4. Geological and soil conditions may preclude delivery of bioremediative microorganisms to polluted sites; or
5. Supposedly "clean" sites may continue to leach contaminants as chemical and physical conditions change over time.

A further challenge in field evaluations of bioremediation technologies is the need for data collection over long periods of time and at various spatial scales. It can be difficult, if not impossible, to find suitable field sites.

Research Needs in Environmental Biotechnology

Further research is needed on the basic ecology of microorganisms and interactions among microbial community members. In nature, microorganisms seldom exist or act as single species; instead, they act collectively as consortia. Additional research of this type will provide a framework for understanding how microbial communities respond to various environmental stresses; how to accelerate in situ bioremediation by native microbial communities; and whether the introduction of engineered microbes with enhanced bioremediation potential can survive and function within established communities and help remediate the site. Such studies are likely to provide corollary insights into aspects of microbial biochemistry important for bioremediation as well as the roles of microorganisms in biogeochemical cycling.

Physiology/Biochemistry Research

Much of the information relevant to bioremediation has come from studies of the genetics and physiology of aerobic bacteria. As a result, the best-known biochemical processes related to bioremediation are oxygen dependent. This characteristic limits their effectiveness in many polluted underground and underwater sits with minimal or no oxygen. Both biological and physical strategies for improving the supply of oxygen in such sites have been proposed, and research on aerobic organisms must be continued. However, long-term success in dealing with a wide array of polluted site with little or no oxygen will require information that can be obtained only through increased research on the genetics, physiology, and biochemistry of anaerobic organisms.

In nature, biodegradation in sites with little or no oxygen is mediated by anaerobic and microaerophilic microorganisms. Because of the difficulty of isolating and culturing such organisms in the laboratory, their metabolic diversity and their potential use in environmental biotechnology only recently have been appreciated. These technical obstacles are being overcome with improved cultivation methods, new technologies for identification of microorganisms, and new methods for studying their metabolism in situ.

New knowledge about anaerobic microorganisms has expanded opportunities for exploiting their metabolic diversity in bioremediation. Since the discovery in 1982 that some anaerobes can dehalogenate carbon compounds. However, to exploit anaerobes for bioremediation, more knowledge is needed about the biology of diverse anaerobic microbes, including how they respond to fluctuating oxygen levels.

Genetical Research

In the past, researchers have been unable to conduct genetic studies on these hard-to-culture organisms, but recent development in molecular biology made it possible to isolate and study genes of almost any organism. Scientist now can analyze genes that govern a wide variety of metabolic processes, including the degradation of environmental pollutants. Such genetic analyses provide information about mechanisms underlying the operation and evolution of degradation pathways.

While the isolation and characterization of novel microorganisms and genes are important in developing bioremediation strategies, knowledge about the general features of the microbial genome also is needed. Bacteria exhibit a high degree of genetic plasticity, or changeability. Analysis of complete microbial genomes will reveal the nature of this plasticity and suggest how genetic engineering can be used to modify organisms to impart the characteristics needed for bioremediation.

New knowledge about the diversity of microorganism and the organization of their genes also will help explain how environmental factors influence the expression of genes and the regulation of microbial metabolism. Research has revealed that genes expressed when one compound is present can play a role in the metabolism of a second compound. For example, some bacteria were found to degrade highly toxic trichloroethylene (TCE) when the less toxic compound toluene was present. Mutant organisms also have been isolated that can degrade TCE in the absence of toluene due to genetic changes that cause the TCE-degrading gene to be "turned on" continuously. However, bacteria capable of TE degradation may lack other traits–such as tolerance for heavy metals, salts, and acid soils–needed for their use in bioremediation. Genes for these traits could be added through genetic engineering, thereby increasing the degradation efficiency of native microorganisms.

Recombinant microorganisms with expanded degradation capabilities have been developed recently. Researchers in several countries studied a number of microorganisms, each of which degraded a restricted range of pollutants, and characterize the genes involved. Investigators then combined genes from different species into one strain of bacteria that can degrade multiple types of pollutants. Similar applications of modern genetic techniques should make it possible to tailor bacteria for bioremediation of sites contaminated with specific combination of toxic compounds.

Biosensors

Currently, in site degradation processes cannot be measured or validated directly; researchers must rely on tracers and gas generation to assess bioremediation processes. Increased investment in biosensor research could lead, over the long term, to improved tools for efficacy assessment. The development of biosensors promises to revolutionize the way pollutants are detected and monitored in the environment.

Unlike standard methods, which rely on analytical chemistry in measuring the total concentration of a pollutant, biosensors can detect the fraction available to microorganisms. Biosensors also have the advantage of being nondestructive and located on-line, meaning that samples do not have to be transported to a laboratory for analysis. Biosensors may utilize either whole bacterial cells or specific molecules (*e.g.*, enzymes of biomimetics) as a detection system. Combinations of biosensors in array can be exploited to deal with a diversity of toxicants and pollutants.

One type of biosensor involves linking a gene such as the mercury resistance gene (mer) or the toluene degradation (tol) gene to gene that code for bioluminescence within living bacterial cells. The biosensor cells can signal that extremely low levels of inorganic mercury or toluene in contaminated waters and soils by emitting visible light, which can be measured with fiber-optic fluorometers.

Conclusion

Due to its comparatively low cost and generally benign environmental impact, bioremediation offers an attractive alternative and/or supplement to more conventional clean-up technologies.

Bioremediation has been successful at many sites contaminated with petroleum products. However, it is not always the technology of choice because efficacy and the rate of degradation at any particular site cannot be predicated reliably. Improved predictive and process validation capabilities would help stimulate wider use of this technology. Research also could lead to development of biotechnologies to remediate areas contaminated by metals, pesticides, radioactive elements, other toxic materials and mixed wastes.

Research in environmental biotechnology has unique international status. International cooperation will be needed to help generate new scientific knowledge in this arena. In addition, environmental biotechnology has tremendous potential for use in developing nations seeking low-cost solutions to environmental problems, such as municipal waste disposal, conversion of agricultural wastes to energy sources, and cleanup of polluted areas. Thus biotechnology can be extremely useful tool towards environmental bioremediation and solving pollution problems.

Index

www.ingramcontent.com/pod-product-compliance
Lightning Source LLC
Chambersburg PA
CBHW061330190326
41458CB00011B/3952

* 9 7 8 9 3 5 1 2 4 3 7 9 3 *